CONVERGENCE
Illicit Networks and National Security in the Age of Globalization

Edited by
Michael Miklaucic and Jacqueline Brewer

With a Foreword by
Admiral James G. Stavridis, USN

Published for the
Center for Complex Operations
Institute for National Strategic Studies
By National Defense University Press
Washington, D.C.
2013

Contents

Foreword

Admiral James G. Stavridis, USN

Illicit networks affect everyone in our modern, globalized world. From human trafficking in Eastern Europe to drug smuggling in East Asia, to the illicit arms trade in Africa, to terrorist cells in East Asia and insurgents in the Caucasus, transnational illicit networks have tentacles that reach everywhere. The trade in illegal narcotics is perhaps most worrisome, but of growing concern is the illicit trafficking of counterfeit items, weapons, natural resources, money, cultural property, and even people by shrewd, well-resourced, and nefarious adversaries.

I have experience combating these threats personally at the tactical, operational, and strategic levels. As a young naval officer on a variety of ships, I spent a fair amount of time patrolling the global commons where transnational criminals in the guise of pirates and drug smugglers proliferate. I was in the Pentagon on 9/11 and personally experienced the global reach of modern terrorism. Later on, when I commanded U.S. Southern Command (USSOUTHCOM), one of my subordinate commands was Joint Interagency Task Force–South in Key West, Florida, a multinational and interagency/interministerial command that counters drug trafficking in the Western Hemisphere. I also experienced the pernicious effects that transnational crime has on our friends as it ranges throughout the entire Western Hemisphere. After leaving USSOUTHCOM to become commander at U.S. European Command (USEUCOM), I saw that Europe was also challenged by the same types of transnational crime. In response, I stood up the Joint/Interagency Counter Trafficking Center in Stuttgart, Germany, designed to counter transnational criminal networks in cooperation with our international partners.

When I took command of USEUCOM, I also became the Supreme Allied Commander, Europe (SACEUR). As SACEUR, I command Operation *Active Endeavor*, which counters trafficking in the Mediterranean, and Operation *Ocean Shield*, which is part of the international counterpiracy efforts off the Somali coast. In addition, we have pioneered responses to cyber threats. At the Supreme Headquarters Allied Powers Europe, we have the North Atlantic Treaty Organization Computer Incident Response Center, and at USEUCOM, we are taking steps to create a subunified cyber command that will have links to both U.S. Cyber Command in the United States and USEUCOM in Germany.

All of these organizations were designed to facilitate a whole-of-government approach, where all elements of national power work together in order to address emerging threats. Eventually, once societies understand the nature of the threats facing them, they will hopefully mobilize nongovernmental assets, adopting a "whole of society" approach. When all of these elements work together, governmental and nongovernmental groups can join international regional and global groups to form a "whole of international society" approach, allowing us to close the seams that exist between nations and regions. Only then will we be able to close these illicit transnational networks.

These networks have taken advantage of modern advances in communications and transportation to globalize. Narcotraffickers in the Andean Ridge, for instance, have expanded operations as far as their markets in the United States and Europe. Illegal arms merchants have expanded their operations around the world. Human smugglers have moved their slaves from underdeveloped countries to sex operations throughout the developed world. And, of course, we have all seen the global reach of modern transnational terrorism. No one is immune from this insidious threat.

It will take a combination of initiatives to defeat the threats created by illicit criminal networks. These transnational organizations are a large part of the hybrid threat that forms the nexus of illicit drug trafficking—including routes, profits, and corruptive influences—and terrorism, both home grown as well as imported Islamic terrorism. With the latest wave of globalization allowing for even more movement of people, goods, and information, these actors have spread their tentacles wider and deeper, breaking new ground. At the same time, they have demonstrated an ability to adapt, diversify, and converge. This has allowed them to obtain vast resources and to continuously reorganize themselves to stay ahead of efforts to combat them. They have achieved a degree of globalized outreach and collaboration via networks, as well as horizontal diversification.

Criminal networks have the advantage of three primary enablers. First are the huge profits realized by transnational criminal operations. Second is the ability of these organizations to recruit talent and reorganize along lines historically limited to corporations and militaries. The third is their newly developed ability to operate in milieus normally considered the preserve of the state, and often referred to as the diplomatic, informational, military, and economic elements of national power. In the past, although some corporations or cities wielded some of these elements of power, for the last 200 years or so, it was mainly states that did. Transnational criminal elements have recently been able to generate these state-like capabilities. Through resource development and reorganization, they now rival the capabilities of many states and overwhelm the capabilities of others.

Access to large-scale resources is made possible by the incredible amount of money that transnational crime generates. This in turn enables other criminal activities. Criminals suborn rule of law actors within countries and are able to operate across borders. The end result is corruption. They can often use these resources to train and equip themselves better than those who seek to halt them. For example, the narcoterrorists in Mexico are often better armed than the government forces that they face. Their communications networks are also world class because they can afford the best equipment.

These organizations have also adapted their recruiting and organizing. Drug cartels have been recruiting trained special operations forces and educating lawyers and accountants. Mara Salvatrucha (MS-13) has also adopted a technique of telling its more promising new members to avoid getting tattooed so that they can better blend into the societies they operate in. These organizations adapt in order to take advantage of the emerging situations that they find themselves in. The Revolutionary Armed Forces of Colombia, for instance, started out by guarding airfields for the cartels and quickly realized that they could take over the trade themselves. Over the space of a few years, they vertically integrated, morphing from terrorists to narcoterrorists. Other organizations such as al Qaeda have discovered franchising, enabling them to reach into areas that they previously could not organize in. For example, Hizballah has created connections with drug-trafficking organization in the Tri-Border Area of Argentina, Brazil, and Paraguay. This ability to change approach to both recruiting and organizing helps make these organizations extremely dangerous. And as these organizations spread out horizontally and meet fellow criminals, they form into modern networks, mimicking normal societies.

Although the fields of diplomacy, information, military, and economic power have generally belonged to states since the 1700s, modern illicit transnational networks have expanded their operations into these areas. Modern illicit criminal networks communicate regularly with states, using information to communicate diplomatic messages to those that may want to face them down. They also use information to recruit vulnerable persons, transforming them into sex slaves or suicide bombers. As mentioned above, some of these networks have developed a military capability, which allows them to go toe-to-toe with national security forces in places like Mexico and Afghanistan—and their economic strength allows them to operate with impunity.

The problems created by these networks are legion. Two of the most dangerous threats are corruption and cyber security. Corruption acts as an enabler for transnational criminals, spreading deep within societies, preventing the rule of law from ensuring the safety and well being of the citizens of countries where it is endemic. Although challenging in developed countries such as the United States and Germany, it is an existential threat in less developed countries that are struggling to defeat this menace. Where the government writ does not go, criminal networks fill the vacuum and provide alternative governance. Where the police and judiciary are affected, criminals prey on the population and do not pay for their misdeeds. Because no one will invest in these areas, the people will not have access to the jobs and educational opportunities that would empower them to improve their situations. Corruption is like an autoimmune disease, shutting down the immune system so the body can be overwhelmed by any challenge.

Cyber is a nascent threat, but one with tremendous future capability. There have been two proven cyber attacks, one against Estonia in 2007 and the other against Georgia in 2008. In both cases, distributed denial-of-service attacks ranged from single individuals using various low-tech methods such as ping floods[1] to expensive rentals of botnets[2] used for spam distribution that swamped Estonian and Georgian organizations including parliaments, banks, ministries, newspapers, and broadcasters—rendering many of the Web sites inoperable. Although the effects were localized to those countries, they do show what a cyber attack can produce. In the long run, cyber attacks will be able to shut down entire parts of the infrastructure of a state, which will directly threaten populations and significantly delegitimize states under attack. This danger will only grow as criminal networks hire their own hackers and eventually train them.

These illicit criminal networks threaten the United States both directly and indirectly. Directly, these criminals have attacked U.S. facilities and citizens throughout the globe. They also weaken the fabric of American society, which they touch through violence and corruption.

Indirectly, these organizations threaten the United States by attacking our allies and partners throughout the world. Imagine a world populated by narcoterrorist states. This situation would make it difficult for the United States to put forward its interests such as democracy and free trade globally. But at the root of the issue is the fact that narcotics and other illegal trafficking impact U.S. national security immensely by presenting the following dilemma: How best do we remain a free and open society—a land of opportunity and a partner of choice for our neighbor countries—especially in the post-9/11 era in which criminals and terrorists seeking to do harm continuously exploit our borders on land, at sea, and in the air?

All of these aspects of illicit networks and transnational crime are what make this book important. The Center for Complex Operations (CCO) has produced this edited volume, *Convergence: Illicit Networks and National Security in the Age of Globalization*, that delves deeply into everything mentioned above and more. In a time when the threat is growing, this is a timely effort. CCO has gathered an impressive cadre of authors to illuminate the important aspects of transnational crime and other illicit networks. They describe the clear and present danger and the magnitude of the challenge of converging and connecting illicit networks; the ways and means used by transnational criminal networks and how illicit networks actually operate and interact; how the proliferation, convergence, and horizontal diversification of illicit networks challenge state sovereignty; and how different national and international organizations are fighting back. A deeper understanding of the problem will allow us to then develop a more comprehensive, more effective, and more enduring solution.

Notes

[1] A *ping flood* is a simple denial-of-service attack where the attacker overwhelms the victim with Internet Control Message Protocol Echo Request (ping) packets.

[2] A *botnet* is a collection of Internet-connected computers whose security defenses have been breached and their control ceded to a malicious party.

Acknowledgments

Every published book depends on a network, and the network engaged in the development of this book is extensive. It includes many senior and very busy experts and leaders in the field, within and outside of the U.S. Government. The concept for the book was born of the impressive intellectual capital emerging from the Center for Complex Operations (CCO) conference, "Illicit Networks in an Age of Globalization," held at National Defense University, February 8–9, 2011. Our conference co-organizers, Andy Roberts, Justin Cutting, Jessica Larsen, and Bridget Yakley, without whose support the conference could not have been held, also provided valuable editorial assistance and constant support throughout the book development and editing process.

There are many more whose efforts and support were crucial to the translation of concepts introduced at a conference into an enduring product that can be used to inform and educate practitioners and policymakers, while stimulating ongoing research and discourse in the academic and expert communities. We are confident this book will enhance understanding in Washington and other capitals of a challenging and growing threat to national security, as well as in the schools that educate the national and international security leaders of today and tomorrow. We thank David Sobyra, Elizabeth Phu, Gary Barnabo, and the leadership of the Office of Counternarcotics and Global Threats at the Department of Defense; David Luna and Kristen Larson of the Office of Transnational Criminal Threats and Illicit Networks at the Department of State; and Alex Crowther of U.S. European Command for their support to this publication and related research projects. Please accept our gratitude.

A champion of the need to use a holistic approach to fight the illicit transnational criminal networks that plague the global commons, Admiral James Stavridis has been a staunch supporter of this research since the beginning. The editors want to express our sincere gratitude to the admiral for his contribution to the book, but more importantly for his leadership in this arena and his determination to continually take the fight to the adversary. Thank you, sir.

The Center for Complex Operations is lucky to have an in-house force multiplier—our volunteer intern corps—that is a critical enabler engaged in every aspect of the work of the CCO. Mickey Kupecz, who gave tireless hours reviewing many of the book's chapters, was particularly critical to the present effort; Miranda Taylor provided crucial production and graphics support; and Roxanne Bannon, Samuel Beirne, and Molly Jerome provided research and editorial assistance with the book. Other members of the CCO staff provided support and guidance throughout the process and we could not have done this project without them.

Acknowledgments

We would also like to acknowledge Dr. Hans Binnendijk and Ambassador John Herbst for empowering us to succeed in this and all our related endeavors.

Of course, we reserve special thanks for our authors: for their time, support, advice, and also their patience.

Introduction

Michael Miklaucic and Jacqueline Brewer

In the last 20 years, globalization has outpaced the growth of mechanisms for global governance. This has resulted in a lack of regulation—whether it be on the Internet, in banking systems, or free trade zones. The same conditions that have led to unprecedented openness in trade, travel, and communication have created massive opportunities for criminals. As a result, organized crime has diversified, gone global, and reached macro-economic proportions.

—Walter Kemp, United Nations Office on Drugs and Crime,
Organized Crime: A Growing Threat to Security

In some situations these new networks could act as forces for good by pressuring governments through non-violent means to address injustice, poverty, the impacts of climate change, and other social issues. Other groups, however, could use networks and global communications to recruit and train new members, proliferate radical ideologies, manage their finances, manipulate public opinion, and coordinate attacks.

—National Intelligence Council, *Global Trends 2025*

Acceleration. Magnification. Diffusion. Entropy. Empowerment. The global environment and the international system are evolving at hypervelocity. A consensus is emerging among policymakers, scholars, and practitioners that recent sweeping developments in information technology, communication, transportation, demographics, and conflict are making global governance more challenging. Some argue these developments have transformed our international system, making it more vulnerable than ever to the predations of terrorists and criminals. Others argue that despite this significant evolution, organized crime, transnational terrorism, and nonstate networks have been endemic if unpleasant features of human society throughout history, that they represent nothing new, and that our traditional means of countering them—primarily conventional law enforcement—are adequate. Even among those who perceive substantial differences in the contemporary manifestations of these persistent maladies, they are viewed as major nuisances not adding up to a significant national or international security threat, much less an existential threat.

This view continues to dominate mindsets within the U.S. national security community. It fails to fully appreciate the growing power of nonstate actors and adversaries, or the magnitude of the threat they pose and the harm they impose on the international system itself. It underappreciates the possibility—indeed likelihood—of the convergence, whether for convenience or growing ambition, of criminal, terrorist, and even insurgent networks. More important even than insufficient recognition of the efficacy of these adversaries is the absence of a plan to counter the threat they pose to both national security and the international state system. In short, this view does not see the big picture—the long view of the declining robustness and resilience of the global system of nation-states that has been dominant for centuries, and the unprecedented attacks on that system. The failure of that system—in which we and so many have prospered beyond belief or precedent in history—would be a catastrophic and existential loss.

Global trends and developments—including dramatically increased trade volumes and velocity, the growth of cyberspace, and population growth, among others—have facilitated the growth of violent nonstate actors, the strengthening of organized crime, and the emergence of a new set of transcontinental supply chains as well as the expansion of existing illicit markets. The resourcefulness, adaptability, innovativeness, and ability of illicit networks to circumvent countermeasures make them formidable foes for national governments and international organizations alike. Their increasing convergence gives them ever-improved ability to evade official countermeasures and overcome logistical challenges as well as ever better tools for exploiting weaknesses and opportunities within the state system, and attacking that system. Since illicit actors have expanded their activities throughout the global commons, in the land, sea, air, and cyber domains, nations must devise comprehensive and multidimensional strategies and policies to combat the complex transnational threats posed by these illicit networks.

This book is an attempt to map the terrain of this emerging battlespace; it describes the scope of the security threat confronting the United States and the international community from transnational criminal organizations and what is being done to combat that threat—from the strategic level with the release, in July 2011, of the U.S. Government's *Strategy to Combat Transnational Organized Crime: Addressing Converging Threats to National Security* and, in April 2011, of the Department of Defense (DOD) *Counternarcotics and Global Threats Strategy*—to the operational level with increased regional partnerships and dialogues.

A Clear and Present Danger

What are the nature, magnitude, and danger of this phenomenon? The first chapters of this book describe the gravity and the scale of the challenge posed by proliferating, diversifying, and globalizing illicit networks. Authors Nils Gilman, Jesse Goldhammer, and Steven Weber label this phenomenon "deviant globalization" and argue that it is first and foremost an economic phenomenon that cannot be detached or separated from the broader process of globalization. It is that portion of the global economy that meets the demand for goods and services that are illegal or considered repugnant in one place by using a supply from some other part of the world where morals are different or law enforcement is less effective. Illicit networks are globally integrated enterprises. Just as corporations no longer think of national borders as barriers, these networks will operate wherever a profit can be made. No corner of the world is untouched by transnational organized crime, and many illicit networks have multicontinental areas of operations. Take, for example, the Sinaloa Cartel, which operates on five continents, earns billions of dollars, and has successfully diversified beyond drugs to other criminal activities such as kidnapping and money laundering.

Beyond their global operations, deviant entrepreneurs worldwide have demonstrated that they can effectively shut down large urban spaces (such as the *favelas* of Brazil) and even whole subregions (such as the Tri-Border Area where Argentina, Brazil, and Paraguay meet), while usurping their host states' basic functional capacity. They can and do provide social services to local constituencies ranging from community protection to the provision of such public goods as disaster relief, dispute resolution, and even garbage collection. The examples are too numerous to mention in full but include Hizballah in Lebanon, criminal syndicates in the favelas of Brazil's megacities, the Movement for the Emancipation of the Niger Delta in Nigeria, and drug-trafficking organizations such as the now defunct Medellín Cartel in Colombia. Over time, as the illicit economy grows and nonstate actors provide an increasing range of social goods, a security and political vacuum emerges from the gradual erosion of state power, legitimacy, and capacity, and the state is displaced by the same illicit networks that are eating away at its legitimacy.

In addition to defying both borders and law enforcers, transnational illicit networks are unrestrained by legislation, unimpeded by morality, empowered by vast amounts of cash, and able to take advantage of all the latest technological developments. They use technology for communications, even encrypted, to identify and recruit individuals with needed skills and to exchange tradecraft secrets and lessons with other illicit groups and networks around the world, manage their global enterprises, and plan global collaboration and operations. Globalization involving the crossborder integration of value-added economic activity has dominated international economic activity in recent years. For the past several decades, this sort of economic activity has grown twice as fast as overall global economic growth. And while new technologies have drawn the world's communities into ever-closer physical and economic proximity, they have not eliminated local differences in economic opportunities, moral attitudes toward different sorts of economic activity, or the capacity of law enforcement agencies in different parts of the world, resulting in what might be referred to as a "lumpy" environment. Illicit networks depend on and benefit from the same technologies and innovations that help legal private industry navigate this lumpy environment. This presents a substantial challenge to government agencies that cannot adapt or adopt the technologies as rapidly as their networked tormentors.

The old paradigm of fighting terrorism and transnational crime separately, utilizing distinct sets of tools and methods, may not be sufficient to meet the challenges posed by the convergence of these networks into a crime-terror-insurgency nexus. Violent nonstate actors, including terrorist organizations and insurgent movements, seek to collaborate with criminal networks—and in some cases become criminal networks—in order to finance acts of terrorism and purchase the implements of destruction and killing. Terrorists and insurgents can tap into the global illicit marketplace to underwrite their activities and acquire weapons and other supplies vital to their operations. On the other hand, many transnational criminal organizations have adopted the techniques of terrorist violence—sometimes referred to as the "propaganda of the deed"—as a central element in their strategic communication plans and a way of intimidating their rivals. In the Liberation Tigers of Tamil Eelam we see the convergence of crime, terror, and insurgency into a potent set of capabilities—drawing from and contributing to the toolkits of each of the three species of illicit organization—that is particularly toxic if not unique. The conventional view that mutual disincentives discourage terrorists from associating with transnational criminal networks, and vice versa, appears to be an artifact of a bygone era.

The illicit economy operates in the shadows of the legal and official political economy. In order to understand the scope of the problem, both scholars and practitioners must develop credible estimates of the size of the illicit economy using accessible data sets, sound methodologies, and transparent assumptions. Otherwise, it is hardly possible to support any argument concerning the size, nature, and growth of illicit networks and markets. Estimates of the magnitude of the phenomena we call transnational organized crime range from 1 to 15 percent of global product. Justin Picard attempts to sharpen the analytic tools for developing meaningful metrics in this otherwise occluded space. It is also necessary to estimate the degree of harm caused by illicit markets. One must establish a relationship between illicit activities and the actual harm they cause to society in order to understand and counter the threat they pose. This cannot be done without collaboration among experts on illicit markets, economists, criminologists, environmentalists, diplomats, political scientists, and public health specialists, among others.

Illicit commerce affects numerous industries from banking, finance, information technology, real estate, telecommunications, transport, and import/export to industries many do not often think of including entertainment, music, film, textiles, and luxury goods. By some estimates, transnational criminal organizations are attaining the same high levels of revenue as Fortune 500 companies. These organizations invest in the licit economy—through real estate, businesses, and other entities—as both a means to launder money and a way to gain legitimate revenue, thereby creating a "gray economy" in which legality is sometimes difficult to distinguish from illegality. With skilled lawyers, financiers, and accountants, illicit networks are able to launder their profits and effectively generate further income, often from strictly legal activity, thus drawing many more players—sometimes unknowingly—into the domain of illicit activity. Phil Williams puts these developments in the context of a range of disturbing "mega-trends" that he argues will lead to a future of lawlessness and disorder.

Complex Illicit Operations

How do illicit networks operate? What are their interrelationships? What are their logistical challenges and responses? What is their tradecraft? As long as there is a market for illicit products, illicit supply chains will thrive and provide a supply to meet the demand. Duncan Deville describes the various agents manning these supply chains, their functional roles, and their extraordinary creativity in adapting to any attempt to curtail their efforts. When joint Colombian and international law enforcement efforts succeeded in reining in the activities of Colombian coca cartels in the 1990s, the cartels changed their roles within the supply chain, and distribution operations migrated to Mexico. The "balloon effect" is not new. What has changed is the skill of trafficking organizations in applying aspects of a commercial supply chain to support their illicit activities and get their products to the customer, and their payment—predominantly cash—back to the merchant.

Criminal and terrorist entities are mobile and increasingly migratory actors capable of shifting both their locations and their vocations in order to exploit geographic, political, enforcement, and regulatory vulnerabilities. They do not view state borders as impenetrable "castle walls." This process has led a range of bad actors to shift their operations to certain regions, countries, or cities that will allow them to function in a relatively unimpeded manner. All illicit networks need is one lax jurisdiction to create a center of operations. Certain specific locations such as Dubai, Turkey, Moldova, and the Tri-Border Area of South America have proven so appealing to various criminal or terrorist enterprises that they have emerged as "hubs" in the illicit global economy. Global crime networks collocate with members of radical political movements in ungoverned border regions in Central and South America; similar collocation has occurred in the Balkans.

Patrick Radden Keefe argues that terrorist and criminal groups have different needs for their operations centers. Organized crime is dependent on a baseline of infrastructure and services, while terrorist organizations can operate in unstable, chaotic environments or even where there is complete state failure. Many hubs in the illicit economy are found in major cities that have the advantages found in strong states—infrastructure, banking, and a baseline level of rule of law. Other key qualities of hubs include state degradation, borderlands where sovereign boundaries converge, corruption, poverty with large informal economies, and extensive tribal or kinship networks.

One of the least understood nodes in the illicit network is a new type of nonstate political actor, a "global guerrilla" or "super fixer."[1] Douglas Farah describes a small group of super fixers and enablers who allow the networks to function and facilitate global connectivity among illicit networks including linking criminal and terrorist groups. Not loyal to any single entity or network, they are valuable to multiple, even competing, networks. Among the fascinating attributes of these individuals is their ability to survive regime changes and political upheavals that on the surface would seem to presage their doom. For them there is no ideological hurdle to overcome, only business to be done. These individuals owe little allegiance to anyone or any cause but themselves. As quasi-independent operators, super fixers are the key links in the chains that move illicit commodities—cocaine, blood diamonds, "conflict" timber, human beings—because they have the relationships to reach out and connect to the next concentric circle of actors. These operators are vital links—both for rebels in the jungle or for Afghan warlords who have no understanding of the outside world—to the market for their illicit commodities. In the process they can gain immense wealth and influence. The geopolitical importance of super fixers is only likely to increase in the coming decade.

The ultimate enabler of both crime and terrorism is money. Any book on transnational organized crime must address the importance of money and of following the money. Both criminal and terrorist groups must raise, move, and hide or integrate their money. Louise Shelley explains how investment in real estate is used as a way both of laundering illicit proceeds and of generating new revenue through traditional real estate–related transactions, such as rental properties and property sales. Investment in real estate also provides the illicit network with safe house capacity from which to execute operations. The crime-terror nexus is thought to be growing because terrorist groups act to take advantage of the money-raising and movement strategies of criminal groups. Current strategies to map and combat threat finance—criminal money laundering and terrorist financing—use the authorities of law enforcement, intelligence operations, public designation, and international cooperation with partner nations. Danielle Camner Lindholm and Celina B. Realuyo show how all of these strategies are essential for fighting transnational organized crime.

The Attack on Sovereignty

Illicit networks, with their associated crime and violence, are undermining and co-opting whole states, capturing the instruments of statecraft and state power to use for their own benefit. Several authors in this book describe the causes and conditions that result in the loss of control by legitimate governments. John P. Sullivan discusses "the reconfiguration of power within states" and the impact of *narcocultura*. Fragile and consolidating states are at the greatest peril from the growing efficacy of illicit networks. With weak state institutions and incompetent or corrupt law enforcement organs, fragile and weak states often enjoy little support from their citizens. The public goods that states should provide such as personal security, dispute resolution, financial opportunity, and employment are as likely to be provided by cartels, gangs, militias, or insurgencies as by weak or fragile states. In countries where economies are stagnant and unable to provide gainful employment for burgeoning youth populations, illicit networks—that is, the agents of deviant globalization—have arguably as great a claim on public loyalty as the impotent state. As the state cannot ensure personal safety and public security, which instead are provided by gangs, cartels, or militias, the exclusive legitimacy of the state's use of coercive force is dubious. In such cases, state sovereignty is supplanted by a different kind of legitimacy heralding a new and likely less benign sociopolitical order.

States can also be destroyed from the inside. State corruption ranges from criminal penetration, infiltration, capture, and, in the direst cases, state criminalization. Michael Miklaucic and Moisés Naím seek to parse the varying degrees of criminalization of the state, while Vanda Felbab-Brown examines how illicit networks are pushing Afghanistan toward criminalization. In many cases, weak states may simply not have the ability to resist co-option by illicit organizations and networks. In other cases, violent nonstate actors can become a de facto government and provider of state services. In today's world, there are growing numbers of illicit actors who have money, economic influence, and the ability to deploy violence.

Global power has become multipolar. Indeed, power has diffused and devolved to a wide range of nonstate actors within the global system and alongside the traditional holders of power, the states. While some of those actors—for example, international nongovernmental organizations and commercial enterprises—are undoubtedly quite comfortable working within the international system of nation-states, others are less benign, and some are even dedicated to overthrowing that system. Fighting this challenge to the state system is ever more daunting as states come to terms with their limited and declining capabilities. In most states, government bureaucracies remain hierarchical and stove-piped while illicit networks are node-centric, agile, and opportunistic. The future competitive advantage may lie with the nonstate groups.

Fighting Back

Though national and international responses to the challenges described in this book may not have yet met the hopes of the more pessimistic contributors, key players are beginning to recognize the threats that converging illicit networks pose, and they are fighting back. To find new ways to tackle these threats, we must take a holistic look at what is currently being done by governments around the world. The U.S. Government and many multinational organizations are taking decisive steps to attack the networks. The White House's 2011 *Strategy to Combat Transnational Organized Crime* (SCTOC) concludes that "criminal networks are not only expanding their operations, but they are also diversifying their activities. The result is a convergence of threats that have evolved to become more complex, volatile, and destabilizing."[2] The U.S. strategy is an important step toward setting clear priorities for U.S. Government actions against transnational organized crime. The challenge is in the implementation. Subsequent strategies and policies created and implemented to combat the threats posed by criminal networks should be connected to other focal points for the U.S. Government—particularly cybersecurity, counterterrorism, counterproliferation, building partner capacity, and strengthening governance. Holistic approaches that recognize transnational organized crime not as a stove-piped "law enforcement problem," but as a nefarious feature of the global security environment that touches the whole world and has an impact on a multitude of vital U.S. interests, are the types of approaches most likely to succeed over the long term.

Implement an Interagency Response

Currently in the United States there is no long-term strategic planning system to synchronize all the elements of national power to meet the challenge of converging illicit networks. Furthermore, in an operational environment, where civilian agencies are the most appropriate lead with the military in support, there should be a campaign planning mechanism as well. In 2010, the U.S. Agency for International Development, the third "D" in the U.S. triad of Diplomacy, Defense, and Development, stood up a policy planning office, but this is only a small step toward inculcating a culture of planning within the civilian agencies. The State Department must improve this capability. In support of the SCTOC, the State Department is working to "build international consensus, multilateral cooperation, and public-private partnerships to defeat transnational organized crime." This involves a broad range of bilateral, regional, and global programs and initiatives that help to strengthen international cooperation with other committed partners including key allies such as Australia and the United Kingdom. David Luna describes how such enhanced coordination enables the international community to dismantle criminal networks and combat the threats they pose not only through law enforcement efforts, but also by building up governance capacity, supporting committed reformers, and strengthening the ability of citizens—including journalists—to monitor public functions and hold leaders accountable for providing safety, effective public services, and efficient use of public resources. The Department of Defense *Counternarcotics and Global Threats Strategy*, derived from national strategic and military guidance, according to William F. Wechsler and Gary Barnabo, lays the foundation for how DOD supports the fight against transnational illicit networks. To ensure that the resources expended against illicit networks have maximum strategic impact, DOD must also link its efforts against transnational organized crime to other national security priorities and agencies.

Many other departments and agencies are involved in fighting transnational organized crime. The Treasury Department and various law enforcement agencies lead various aspects of the nonkinetic effort to drain the illicit economy. The SCTOC proposed to create a legislative package targeted at improving the authorities available to investigate, interdict, and prosecute the activities of top transnational criminal networks across the Federal Government and in partnership with state, local, tribal, and territorial levels, as well as with foreign partners. Under existing authorities, law enforcement agencies have the lead for transnational organized crime, while DOD and the State Department have the lead on counterterrorism and counterinsurgency issues; however, it is crucial that all agencies work together. Recently, we have seen that an increase in two-way support—from DOD to law enforcement and from law enforcement to the military—is a strategically important evolution in how the U.S. Government organizes to combat transnational organized crime. Moreover, the increased use of interagency task forces such as the Joint Interagency Task Force–South, whose success combating drug trafficking in the Caribbean is widely acknowledged, is another way to use the strengths and tools of various agencies. Such efforts must continue. Collaboration is ever more critical as we look to best use existing congressional authorities for each agency.

Attack the Financial Resources

To stop international narcotics trafficking and money-laundering networks, we must limit their access to the global financial system. It is widely acknowledged that nearly 95 percent of money laundering goes undetected. Recent indictments of multiple banking institutions and convictions of a growing number of commercial banks are indicative of increased official initiative in this area, but are also likely reflective of the widespread presence of money-laundering activities. Stuart Levey, former Treasury Under Secretary for Terrorism and Financial Intelligence, stated that "any financial institution that collaborates in illicit conduct on this scale risks losing its access to the United States." We must make good on this threat.

Speaking at the National Defense University in 2011, Treasury Department Assistant Secretary Daniel Glaser noted that analysts often cite the low cost of conducting a terrorist attack as a reason it is difficult to stop these networks, but that explanation neglects the cost of running and maintaining an illicit network. Salaries, training, logistics, bribes, and other expenses all expose the organization to potential vulnerabilities. The international community must be more proactive and determined in pursuing the money that enables these groups to operate. Governments must aggressively engage the financial sector and forge partnerships that will reduce the vulnerability of financial institutions and increase the efficacy of anti-money-laundering procedures.

Just looking at Mexico, billions of dollars pass through the U.S. border into that country each year. U.S. law enforcement efforts have traditionally focused on stopping the northward flow of illicit commodities and have neglected the southward flow of weapons and drug money back to Mexico. With bulk cash as the primary method for moving funds, we must work harder to increase our interdiction percentage above the current 1 percent of estimated total funds. Over the past 5 years, the Treasury Department's Office of Foreign Asset Control has taken steps to institutionalize measures to counter threat finance and protect the financial system from illicit actors using financial intelligence. Several programs aggressively target "super-fixers" and other enablers, rogue states, proliferators of weapons of mass destruction, money launderers, and drug traffickers among other targets. These programs track money flows and assist with the policy and criminal justice work of other U.S. Government agencies.

Understand the Networks

To best combat the threat posed by converging illicit networks, more intelligence is needed on the networks—their cultures, motivations, incentives, operations, and structures. Going further, the SCTOC specifically calls out for the need to improve signals intelligence and human intelligence collection on transnational organized crime. So far, we have heard of a handful of efforts to do this. Mapping these networks and their interactions, logistics, operations, personnel, and finances must be a key policy focus area in the future. Patrick Radden Keefe's and Duncan Deville's chapters offer starting points for further research, both on the networks and on the hubs that corral their behavior.

From lawyers to scholars to practitioners, the contributors to this book represent a wide range of disciplines, agencies, and roles in the fight against transnational organized crime. By connecting these groups, we aim to continue to build the network. It is hoped that this book will continue the dialogue and stimulate new thinking on how both the U.S. interagency community and the broader international community can work together more effectively to attack the illicit networks and contain the threats posed by their globalization, horizontal diversification, and convergence.

Notes

[1] John Robb, *Brave New War: The Next Stage of Terrorism and the End of Globalization* (New York: Wiley, 2007); Australian Crime Commission, *Organised Crime in Australia* (Canberra: Australian Organised Crime Commission, 2011).

[2] *Strategy to Combat Transnational Organized Crime: Addressing Converging Threats to National Security* (Washington, DC: The White House, July 2011), cover letter by President Barack Obama.

Part I.
A Clear and Present Danger

Chapter 1

Deviant Globalization

Nils Gilman, Jesse Goldhammer, and Steven Weber

> *The black market was a way of getting around government controls. It was a way*
> *of enabling the free market to work. It was a way of opening up, enabling people.*[1]
>
> —Milton Friedman

This chapter introduces the concept of deviant globalization.[2] The unpleasant underside of transnational integration, *deviant globalization* describes crossborder economic networks that produce, move, and consume things as various as narcotics and rare wildlife, looted antiquities and counterfeit goods, dirty money and toxic waste, as well as human beings in search of undocumented work and unorthodox sexual activities. This wide and thriving range of illicit services and industries takes place in the shadows of the formal, licit global economy, and its rapid growth is challenging traditional notions of wealth, development, and power. In sum, deviant globalization describes the way that entrepreneurs use the technical infrastructure of globalization to exploit gaps and differences in regulation and law enforcement of markets for repugnant goods and services.

Repugnance is an emotion that sometimes informs moral judgments. We recognize the reality of repugnance, but we take the view in this chapter that an analysis of the problem and prescriptions for what to do about it are best made within a more austere analytic frame. Some readers may find the use of standard economic terminology and concepts off-putting as we apply this logic to repugnant activities. We do not eschew or criticize moral judgments; we just believe they belong to a different kind of conversation. Ultimately, we argue that to give in to a moralizing impulse gets in the way of designing policy interventions that can most effectively mitigate the worst aspects of deviant globalization.

The most critical point to understand about deviant globalization is that it is inextricably linked to and bound up with mainstream globalization: we cannot have one without the other. Both are market-driven activities. Both are enabled by the same globally integrated financial, communication, and transportation systems. Both break down boundaries—political, economic, cultural, social, and environmental—in a dynamic process of creative destruction. And both are present in virtually every globalizing platform: we may be disgusted by men who go online to prearrange their sexual liaisons before their Thai holiday, but their use of the Internet as an information service is not technically different from someone prearranging a trek in Chiang Mai. Moreover, a Chinese shipping company delivering a container of illegally harvested Burmese hardwood to a furniture manufacturer in Italy may be doing something illicit, but that makes it no less a part of the synchronized global intermodal shipping system, and the burgeoning cell phone networks of Pakistan are just as useful to Afghan heroin wholesalers as they are to members of the Pakistani diaspora sending remittances to their families back home. The infrastructure of the global economy is dual-use and value-neutral. As these systems become increasingly efficient, interconnected, and indispensable, they help not only the formal global economy to grow but also its conjoined, deviant twin.

Consider the situation in Mexico today, which illustrates well the dynamics of deviant globalization. Whereas mainstream accounts of globalization's economic impact on Mexico usually focus on topics such as *maquiladoras* (production facilities across the Mexican border that are tightly tied to export channels into the United States), wage arbitrage (the substitution of low-wage Mexican labor for American labor), and economic dependency on the United States, appreciating the significance of deviant globalization shifts the focus onto other flows, such as humans and drugs heading north, as well as guns and money heading south. Meeting Western appetites for illicit drugs has generated vast fortunes in Mexico. This drug money, in turn, funds a narcotics production, transshipment, and wholesaling industry that, by some estimates, directly employs upward of half a million people, larger than the entire Mexican oil and gas industry.[3] Whereas mainstream accounts of globalization's political impact on Mexico emphasize the loss of sovereign governance capacity to the North American Free Trade Agreement, International Money Fund, and multinational corporations, examining Mexico through the lens of deviant globalization highlights the way that some drug-trafficking organizations have emerged as cripplers of supposedly sovereign governments to a far greater extent than structural adjustment programs ever have.[4]

The dynamic in Mexico is not some outlier or exception, but rather is emblematic of deviant globalization's rapidly growing systemic threat to international stability and security. An incomplete understanding of the systemic nature of deviant globalization, however, continues to lead many governments to implement poorly designed policy interventions that in many cases are not only ineffective but also counterproductive.

We believe that policy can do better. Deviant globalization, like globalization more broadly, is a systems phenomenon that requires a systems-oriented solution. Powerful economic and political forces are driving (deviant) globalization forward. But many aspects of how those forces manifest in markets remain subject to shaping interventions. We need to understand these forces and their interdependencies before we pull policy levers. And, faced with complexity and hard choices, we must avoid adopting a kind of fatalistic attitude that cuts off debate and action. The fatalism we find most politically dysfunctional is to paint any argument that questions the status quo as calling for rampant, indiscriminate "legalization" of illicit businesses and using that objection to shut down alternatives. We argue later that particular aspects of deviant globalization would be better dealt with through simple legalization, but not all. Creative regulation and enforcement can serve as finely grained tools, with lots of choices that governments and other transnational actors can make use of to advance their objectives. Our goal here is to illuminate the real stakes in the tough choices that deviant globalization presents.

What Is Deviant Globalization?

Deviant globalization, in the first place, is an economic phenomenon: it is that portion of the global economy that meets demand for goods and services that are illegal or considered repugnant in one place by using a supply from some other part of the world where morals are different or law enforcement is less effective.

Seen through this economic lens, a reasonable first question to ask is what creates the market opportunities for deviant globalization. The answer is *we do*. When we codify and institutionalize our moral outrage at selling sex by making prostitution illegal, for example, we create a market opportunity for those who kidnap women and smuggle them into sexual slavery. When we decide that methamphetamine is a danger to public health and prohibit it, we create opportunity for drug dealers who delight in the high profit margins as they fill illicit orders. When we ban the sale of organs in our domestic market, we create incentives for entrepreneurs to act as brokers or facilitators between physically desperate patients and economically desperate donors in poor countries. Every time a community or a nation, acting on the basis of its good faith and clear moral values, decides to "just say no," it creates an opportunity for arbitrage.

Deviant globalization is thus an economic concept, but it is also a moral and legal one. Deviant globalization grows at the intersection of ethical difference and regulatory and law enforcement inefficiencies. Wherever there is a fundamental disagreement about what is right as well as a connection to the global market, deviant entrepreneurs pop up to meet the unfulfilled demand. In meeting our collective desires, they see the differences in notions of public good, morality, and health as bankable market opportunities. Neoliberalism's wide-open, market-oriented rules may govern globalization, but the game gets played on a morally lumpy field.

In contrast to some mainstream theories of globalization, which depict it as a process that annihilates differences across space,[5] the concept of deviant globalization highlights the continued importance of spatial differences in the structure of the global economy. The pathways of deviant flows are determined by not only border security and state authority but also the particular factor endowments (that is, the amount of land, labor, capital, and entrepreneurship that a country possesses) that generate comparative advantage and arbitrage opportunities for deviant entrepreneurs. Appreciating the geographic particularities of deviant flows is therefore crucial. Deviant flows move through cities—in a de facto archipelago that runs from the inner metropolitan cities of the United States to the *favelas* of Rio de Janeiro to the *banlieues* of Paris to the almost continuous urban slum belt that girds the Gulf of Guinea from Abidjan to Lagos. They move through towns and villages—along the cocaine supply route that links the mountains of Colombia to São Paulo and the waterways of West Africa to noses in the Netherlands. And they move through the "global nodes" that make up the world's financial infrastructure—from Wall Street to London to Tokyo's Nihombashi District. In sum, it is possible to find manifestations of deviant globalization in almost every city, every household, every shipping lane and port, as well as almost every IP address connected to the global economy.

Mainstream Globalization	Deviant Globalization
Companies such as Wal-Mart are using extraordinary supply-chain technologies to revolutionize logistics, generating jobs for workers in developing countries and bringing much cheaper mass-consumption goods to the global middle classes.	The same supply chain technologies are used to tune up the efficiency of the global supply chain for counterfeit goods. Many of the inputs for the factories that make counterfeits are competitively sourced on global markets at minimum price, and the products are transported with new efficiency to consumers.
The Internet facilitates the global distribution of information, enables the collective production of knowledge goods, and enhances freedom to speak and to listen.	The Internet has become the easiest entry point to global systems for hostile and exploitative technologies (malware); social exploits (scams and spam); the identification of remote targets for pederasts; and the dissemination of radical ideologies that oppose or negate freedoms.
Capital mobility across national borders improves the efficiency with which the global economy allocates investment and should thereby enhance productivity immediately and, in particular, over the long term.	Capital mobility makes all kinds of finance for illegal activities, including crime and terrorism, as well as the laundering of money from other illicit activities far easier and much more challenging for political authorities.
The spreading ideology of privatization and market allocation released a historic burst of entrepreneurial energy and raised on the order of a billion people out of abject poverty in less than a generation.	The same ideologies have lent legitimacy to the concept of "everything for sale" including human beings (both whole and in pieces). Privatization ideologies in particular have led to the collapse in public goods provision— for example, by dumping waste and garbage "elsewhere."

Why Deviant Globalization Matters

As the example of Mexico shows, it is a serious error to view deviant globalization as a mere sideshow to what "really matters" in the global political economy. On the contrary, we believe that deviant globalization is not only central to contemporary geopolitics, but is also actively changing the landscape and distribution of power in the world economy in ways nearly as profound as any openly visible politico-economic trend or event has since the end of the Cold War.[6]

Deviant globalization matters to the future of the global political economy for three interconnected reasons. First, it challenges cherished notions of what "development" and "entrepreneurship" are supposed to be all about. Nations, institutions, and nongovernmental organizations (NGOs) from the Global North dedicate huge amounts of time and immense sums of money trying to help nations of the Global South modernize, diversify, and grow their economies. Liberal proponents of mainstream globalization view these efforts as a set of market-building steps toward delivering on the promise of capitalism. Marxist opponents criticize development practices for fostering dependency and, paradoxically, a permanent state of (at least relative) poverty. What liberals and Marxists agree on, however, is that the sorts of illicit economies encompassed by deviant globalization represent a form of economic parasitism that diverts developmental energy, capital accumulation, human assets, and other valuable resources away from more productive uses; instead of providing a platform for self-sustained growth, such deviant markets appear merely to line the pockets of gangsters.

But the role of the deviant entrepreneur in the development of the Global South is more complicated than this view would suggest. Simply put, deviant entrepreneurs are some of the most audacious experimenters, risk-takers, and innovators in today's global economy. In their relentless search for competitive advantage, they engage in just about all of the activities that other entrepreneurs do—marketing, strategy, organizational design, product innovation, information management, financial analysis, and so on. In many cases, they create enormous profits that meaningfully contribute to local development while also extruding inefficiencies from huge markets.

In other words, what both liberals and Marxists fail to appreciate is that many people living in poor nations in the Global South are already engaged in radical experiments in actual development through deviant globalization. Moreover, participating in the production side of deviant globalization—one hopes as an entrepreneur, but at least as a worker—is a survival strategy for those without easy access to legitimate, sustainable market opportunities, which is to say the poor, the uneducated, and those in locations with ineffective or corrupt institutional support for mainstream business.[7] Behind the backs of, and often despite, all those corporations and development NGOs as well as the World Bank and International Monetary Fund, the poor are renting their bodies, selling their organs, stealing energy, stripping their natural environments of critical minerals and wildlife, manufacturing drugs, and accepting toxic waste not because they are evil people but because these jobs are often the fastest, best, easiest, and even in some cases the most sustainable way to make money. Even for the line workers in deviant industries, the money accumulated over a few years can often form a nest egg of capital to start more legitimate businesses.[8]

To make the claim that deviant globalization is a form of development is not to deny the awfulness and oppressiveness of many deviant industries, or the significant social and environmental externalities that deviant globalization often imposes, or the fact that many of the participants in deviant globalization are coerced into their roles. No doubt there are better and worse ways to improve one's lot in life, but it is rare that participating in deviant globalization is the worst available choice. Mining coal in China, for example, may be a more "legitimate" profession than pirating ships off of the coast of Somalia, but it is debatable whether the former is a better job than the latter. Most important, whatever our normative feelings, both professions contribute to a kind of development.

Indeed, in many cases, states that host deviant industries recognize and embrace their developmental benefits. Consider the importance of sex tourism to the economies of countries such as Thailand, the Philippines, or Cuba. Except for sporadic crackdowns when international scrutiny grows too intense, these governments knowingly wink at sex tourism—it has been the foundation of much of their draw on the international tourism scene, and it is the source of significant quantities of hard currency. In practice, sex tourism is a tacit part of their developmental strategy. Just as many countries are willing to tolerate physical pollution as a way to attract investment and jumpstart growth, so other countries are willing to tolerate what we might call social or moral pollution in order to achieve the same ends.[9]

Seen from this point of view, deviant industries are not just about crime; rather, they are wellsprings of innovation—"disruption" in Clayton Christensen's sense, "creative destruction" in Joseph Schumpeter's sense—for political economies that need investment and growth and that have a hard time producing them via licit channels. At the same time, of course, the sort of innovation that deviant entrepreneurs produce is not a direct substitute for the kind of development proposed by metropolitan NGOs. Deviant development is different along several crucial dimensions:

- It is less transparent and operates according to more fluid rules of the game because deviant innovators are, by definition, less constrained—but it also creates new degrees of freedom, allowing entrepreneurs to try things that they could hardly try elsewhere.

- It is less centered in formal organizations such as corporations because deviant entrepreneurs do not organize in that way until and unless they have to—but it also enhances flexibility and adaptability.

- It struggles to make, monitor, and enforce contracts because even if the normal instruments of enforcement are weak in developing country settings, deviant entrepreneurs have even less access to them. This can get in the way of bargaining, increase transaction costs, and the like—but it also spurs the development of "alternative" means of deal-making, monitoring, and enforcement. These alternatives often involve mass violence—though, to be clear, from the point of view of those participating in these markets, such incidents of violence are undesirable because they increase the cost of doing business.

- Its success does not result in the long-term production of public goods, both because the deviant entrepreneurs are (for reasons of selection bias if nothing else) not public-minded sorts and because the profits from deviant industries are rarely if ever taxed by the state, the classic provider of public goods—but what the state loses, local communities and organizations partly gain in the form of jobs and capital.

- It tends to foster mass antisocial behavior, such as corruption. While all entrepreneurs are risk-takers, deviant entrepreneurs, virtually by definition, have a heightened willingness to flout social norms and conventions. Their success in turn degrades respect for other social norms and conventions, which may be otherwise unrelated to the market in question, leading to generalized decay of the dominant social order.

Rather than representing a divergence from the liberal norm of licit growth in the formal economy, deviant globalization might better be conceived as a way for the globally excluded to find a space to be innovative, a space in which the rules of the game have not already been stacked against them. At the same time, this is no subversively heroic Robin Hood morality tale. Today's deviant globalizers are not proto-revolutionaries aiming to remake society in an inclusive and collectively progressive fashion. Rather, they are opportunists whose public personae and brands are built around unbridled capitalist spirits—living fast, dying hard, and letting the rest of the world go to hell. The form of development they are enacting, ironically, is in many respects an ultra-libertarian one—one that tacitly rejects what liberal political economy defines as "the public good," and denies the need for any sort of state.

The second major reason why deviant globalization matters follows directly from the insight that this phenomenon is really the truest manifestation of libertarianism: deviant globalization degrades state power, erodes state capacity, corrodes state legitimacy, and, ultimately, undermines the foundations of mainstream globalization in ways that are only now being fully recognized. More specifically, deviant globalization is creating a new type of nonstate political actor, what John Robb calls a "global guerrilla,"[10] the sort of super-empowered individuals whose geopolitical importance is only likely to grow in the coming decade.

Deviant entrepreneurs wield political power in three distinct ways. First, they have money. As we have seen, deviant entrepreneurs control huge, growing swathes of the global economy, operating most prominently in places where the state is hollowed or hollowing out. Again, corruption fueled by drug money on both sides of the U.S.-Mexico border exemplifies this point.[11] Second, many deviant entrepreneurs control and deploy a significant quota of violence—an occupational hazard for people working in extralegal industries, who cannot count on the state to adjudicate their contractual disputes. This use of violence brings deviant entrepreneurs into primal conflict with one of the state's central sources of legitimacy, namely its monopoly (in principle) over the socially sanctioned use of force. Third, and most controversially, deviant entrepreneurs in some cases are emerging as private providers of security, health care, and infrastructure—that is, precisely the kind of goods that functional states are supposed to provide to their citizens. (However, since they are provided privately, to the deviant entrepreneurs' personal constituents, they are not "public goods" in the sense of goods equally accessible to all citizens.) Hizballah in Lebanon, criminal syndicates in the favelas of Brazil, the Movement for the Emancipation of the Niger Delta in Nigeria, and narcotraffickers such as the Zetas in Mexico are all deviant entrepreneurs who not only have demonstrated they can shut down areas of their host states' basic functional capacity, thereby upsetting global markets half a world away, but are also increasingly providing social services to local constituencies.

What makes these political actors unusual is that, rather that seeking to build or capture institutionalized state power, they thrive in (and indeed prefer) weak-state environments, and their activities reinforce the conditions of this weakness. Deviant entrepreneurs generally do not start out as political actors in the sense of actors who wish to control or usurp the state. In the first iteration, deviant globalization represents an entrepreneurial response to the failure of mainstream development: an effort by bootstrapping individuals to get ahead in a world where the state is no longer leading the way. Once these deviant industries take off, however, they begin to take on a political life of their own. The state weakness that was a permissive condition of deviant globalization's initial local emergence becomes something that the now empowered deviant entrepreneurs seek to perpetuate and even exacerbate. They siphon off money, loyalty, and sometimes territory; they increase corruption; and they undermine the rule of law. They also force well-functioning states in the global system to spend an inordinate amount of time, energy, and attention trying to control what comes in and out of their borders. Although deviant globalization may initially have flowered as a result of state hollowing, as it develops, it becomes a positive feedback loop in much the same way that many successful animal and plant species, as they invade a natural ecosystem, reshape their ecosystem in ways that improve their ability to exclude competitors.[12]

What is distinctive in this dynamic is that very few of these political actors have any interest in actually taking control of the formal institutions of the state. Deviant entrepreneurs have developed market niches in which extractable returns are more profitable, and frankly easier, than anything they could get by "owning" enough of the state functions to extract rents from those instead. Organizations such as the First Command of the Capital in Brazil, the 'Ndrangheta in Italy, or the drug cartels in Mexico have no interest in taking over the states in which they operate. Why would they want that? This would only mean that they would be expected to provide a much broader and less selective menu of services to everyone, including ungrateful and low-profit clients, those so-called citizens. None of these organizations plan to declare sovereign independence and file for membership in the United Nations. Corrupt the state? Of course. Own the state? No, thanks. What they want, simply, is to carve out autonomous spaces where they can do their business without state intervention.

This underscores a crucial point about deviant globalization: it does not thrive in truly "failed" states—that is, in places where the state has completely disappeared—but rather in weak but well-connected states, in which the deviant entrepreneur can establish a zone of autonomy while continuing to rely on the state for some of the vestigial services it continues to furnish.[13] Alas, states and deviant entrepreneurs are unlikely to find a sustainable equilibrium. On the one hand, the more deviant industries grow, the more damage they do to the political legitimacy of the states within which the deviant entrepreneurs operate, thus undermining the capacity of the state to provide the infrastructure and services that the deviant entrepreneurs want to catch a free ride on. On the other hand, the people living in the semi-autonomous zones controlled by deviant entrepreneurs increasingly recognize those entrepreneurs rather than the hollowed out state as the real source of local power and authority—if for no other reason than the recognition that if you cannot beat them, you should join them. Of course, just because these deviant providers of alternative governance functions end up seeming "legitimate" in the eyes of local stakeholders (if only because of the economic "development" benefits they provide), this type of governance is usually poorly institutionalized and nontransparent about both ends and means. Nonetheless, as these groups take over functions that would have been expected of the state, their stakeholders increasingly lose interest in the hollowed-out formal state institutions.[14] Thus, even though deviant entrepreneurs have no desire to kill their host state, they may end up precipitating a process whereby the state implodes catastrophically. Something like this took place in Colombia in the 1980s, in Zaire/Congo since the 1990s, and may be taking place in Mexico today.

What Is to Be Done? Think Regulatory Harmonization, Not Eradication

So where does the concept of deviant globalization leave us from a policy perspective? Let us begin by naming what has not worked. Most contemporary policy efforts aimed at addressing deviant globalization are dysfunctional—along a spectrum from simply ineffective to stunningly self-defeating. America's "war on drugs" is now recognized as overall an expensive and perhaps tragic failure. The U.S. Government has been fighting this (declared) war since the early 1970s at an estimated cost so far of $2.5 trillion.[15] Americans have rewritten criminal law, filled jails with convicts, deployed vast military capabilities abroad, revised the rules of city policing, and ruined millions of individual lives—and that is just the start. With some achievements (Colombia, Panama) notwithstanding, no one doubts who is losing this war. The winners, of course, have almost always been the narcotics supply chain entrepreneurs and their financiers.

Why have policymakers persisted so long with policies that clearly are not working? First, policymakers represent voting constituents, and these constituents, in general, put policymakers under intense pressure to make morality a primary lens through which to make policy choices. This involves labeling deviant entrepreneurs, their employees and associates, as well as their customers, as "bad actors" or simply as "criminals." Unfortunately, this moralizing, criminological approach creates an analytic deficit: by focusing on quashing individuals, it loses sight of the complex dynamics of the system in which the actors participate. And ignoring the system dynamic creates a policy situation that at best devolves into the legal equivalent of whack-a-mole, or pushes the problem over the horizon to some other neighboring jurisdiction.

More commonly, however, the moralizing approach actually makes the problem worse because it strengthens the system in which the deviant entrepreneurs are operating, much like pouring partially effective antibiotics into a rich microbial ecology. Governments' taking an aggressively prohibitionist approach are most likely to weed out weaker players, clearing the field for more clever, more ruthless deviant entrepreneurs. On the other hand, if governments are lucky, persistent, or powerful enough to take down some of the strongest deviant entrepreneurs, then they leave the second-tier players highly incentivized to evolve and innovate so as to avoid the mistakes of their predecessors. Adaptability is, after all, the hallmark of deviant globalization. Putting up roadblocks, such as a long, technically sophisticated fence at the U.S.-Mexico border, invites experimentation and risk-taking that leads to new and more effective smuggling routes for illegal immigrants, more powerful methamphetamines, and, ultimately, surprisingly (but not unpredictably) destructive outcomes.[16] Policymakers who put selective pressure on deviant industries without considering how the system will compensate and adapt are almost sure to be surprised at just how much progress the deviant entrepreneurs will show in response.

Despite this seemingly pessimistic perspective, it is our view that policymaking around deviant globalization can improve. But achieving that goal requires making some self-conscious (and politically painful) decisions to stop doing the things that appease our moral outrage but are actually making the problem worse. The prerequisite of improved policy performance is for policymakers to stop treating deviant globalization as if it is a removable cancer in an otherwise healthy global economy. So long as we have market prohibitions in a morally lumpy, globally connected world, we will have deviant globalization. The next step is to play for gains as we would in a strategic interdependence situation, an iterated game where the two sides can neither defeat nor control each other, but where each and every action taken affects the options that the other will have in the next round. In that context, there are real policy opportunities to curb the most pernicious of the deviant industries, to channel some of the deviant energy into less harmful channels, and to turn some of what deviant entrepreneurs know and are good at into goods that benefit a broader swathe of humanity and at lower cost. These are less ambitious but more realistic goals than the eradication of deviant globalization.

To be specific, policymakers should avoid indulging locally specific moral codes since that simply creates arbitrage opportunities for bad actors. Policymakers need to look critically at what their real options are: if they cannot universalize/globalize both the underlying moral principle and the regulatory/enforcement capacity, then they either have to put moral principles partially to the side or accept that the uneven efforts to impose them are likely to end up empowering bad actors who will profit off of moral outrage.[17] In some cases, various degrees of legalization may prove helpful, as has been illustrated by American experimentation with liquor laws over the past century. In other cases, effective management of deviant globalization may require artful regulation and enforcement that increase transparency and reduce arbitrage opportunities. In other words, by attempting to create a globally uniform regulatory framework, the Convention on International Trade in Endangered Species of Wild Fauna and Flora (referred to as CITES) for illicit wildlife trafficking and the Basel Convention for toxic waste (to cite two examples) are on the right track—though the latter has huge loopholes and the former does not address capacity issues.[18] What is certain is that unilateral approaches are almost certainly doomed to failure because that is the source of the arbitrage opportunities that incent deviant entrepreneurs.

Analysts who treat deviant globalization as a systems problem are able to consider a wider range of options for managing its impacts including, practically speaking, making deviant globalization less harmful. Encouraging deviant entrepreneurs to exit a given line of business, for example, requires understanding the local economic, legal, and political conditions—and, just as important, how these local conditions form part of a global system—that create arbitrage opportunities for deviant entrepreneurs to exploit. Understanding the local opportunity structure facing these entrepreneurs creates policy levers for changing those structures, whether through strategic investment in improved local law enforcement (to focus on reducing the scope for arbitrage), improved scrutiny of the supply chain (to increase the cost of transportation of the good or service from its low cost point of origin to its destination), or enhancement of legitimate local economic opportunities (to reduce the relative attractions of the illicit business). Such analysis will also improve the ability to anticipate economic diversification strategies for deviant entrepreneurs, even absent interventions into the opportunity space.

Notes

[1] Milton Friedman, PBS interview, October 1, 2000, available at <www.pbs.org/wgbh/commandingheights/shared/minitext/int_miltonfriedman.html>.

[2] This chapter is drawn in part from material in Nils Gilman, Jesse Goldhammer, and Steven Weber, eds., *Deviant Globalization: Black Market Economy in the 21st Century* (New York: Continuum Books, 2011).

[3] Philip Caputo, "The Fall of Mexico," *The Atlantic Monthly*, December 2009.

[4] William Finnegan, "Silver or Lead: A Drug Cartel's Reign of Terror," *The New Yorker*, May 31, 2010.

[5] Cultural analyses of mainstream globalization tend to highlight this "deterritorializing" aspect of globalization; see, for example, John Tomlinson, *Globalization and Culture* (Chicago: University of Chicago Press, 1999); and Nikos Papastergiadis, *The Turbulence of Migration: Globalization, Deterritorialization and Hybridity* (New York: Wiley, 2000).

[6] Phil Williams, *From the New Middle Ages to a New Dark Age: The Decline of the State and U.S. Strategy* (Carlisle Barracks, PA: Strategic Studies Institute, June 2008).

[7] Sudhir Alladi Venkatesh, *Off the Books: The Underground Economy of the Urban Poor* (Cambridge: Harvard University Press, 2006).

[8] Jeremy Seabrook, *Travels in the Skin Trade* (London: Pluto Press, 2001), and Rachel G. Sacks, "Commercial Sex and the Single Girl: Women's Empowerment through Economic Development in Thailand," *Development in Practice* 7, no. 4 (1997), describe the way that Thai "bar girls" use savings from their days in prostitution to start sewing or hairdressing businesses back in their villages of origin. Another example is the way that rappers (legendarily) parlay profits from drug dealing into starting a recording company; see the obituary by Jon Pareles, "Eazy-E, 31, Performer Who Put Gangster Rap on the Charts," *The New York Times*, March 28, 1995.

[9] Ryan Bishop and Lillian S. Robinson, *Night Market: Sexual Cultures and the Thai Economic Miracle* (New York: Routledge, 1999); and Vidyamali Samarasinghe, "Female Labor in Sex Trafficking: A Darker Side of Globalization," in *A Companion to Feminist Geography*, ed. Lise Nelson and Joni Seager (Malden, MA: Blackwell Publications, 2005).

[10] John Robb, *Brave New War: The Next Stage of Terrorism and the End of Globalization* (Hoboken, NJ: John Wiley & Sons, 2007).

[11] See Judith Miller, "The Mexicanization of American Law Enforcement," *City Journal* 19, no. 4 (2009), available at <www.city-journal.org/2009/19_4_corruption.html>, quoting Robert Killebrew, a retired U.S. Army colonel and senior fellow at the Washington-based Center for a New American Security: "'This is a national security problem that does not yet have a name' . . . The drug lords . . . are seeking to 'hollow out our institutions, just as they have in Mexico.'"

[12] John P. Sullivan and Robert J. Bunker, "Drug Cartels, Street Gangs, and Warlords," *Small Wars & Insurgencies* 13, no. 2 (2002); Max G. Manwaring, *Street Gangs: The New Urban Insurgency* (Carlisle Barracks, PA: Strategic Studies Institute, 2005); and Enrique Desmond Arias, "The Dynamics of Criminal Governance: Networks and Social Order in Rio de Janeiro," *Journal of Latin American Studies* 38, no. 2 (2006).

[13] One critical misconception promoted by the liberal enthusiasts of globalization, most prominently by Thomas P.M. Barnett in *The Pentagon's New Map: War and Peace in the Twenty-First Century* (Putnam, 2005) is that the cause of poverty and insecurity and ultimately state fragility is "disconnectedness" from the world economy. In fact, all of the most seriously "failed" states—Congo, Somalia, Afghanistan—are deeply connected to the global economy, albeit in ways that are hard to see clearly from London, New York, or Washington. While it is true that they remain weakly connected to the formal and legal parts of the global economy, such places are in fact highly deviantly connected—via the illicit trade in minerals, via piracy, or via the global drug trade, and so on. The crucial issue, in other words, is not connectedness or disconnectedness, but rather *what kind* of connectedness.

[14] Diane E. Davis, "Irregular armed forces, shifting patterns of commitment, and fragmented sovereignty in the developing world," *Theory and Society* 39, nos. 3–4 (2010). Globalization, by undermining national political institutions, also undermines the political centrality of national identity. But what appears to be replacing the national is not a "global" political identity (as "cosmopolitical" dreamers have hoped), but rather a return to more localized identities rooted in clan, sect, ethnicity, corporation, and gang.

[15] Claire Suddath, "The War on Drugs," *Time*, March 25, 2009.

[16] Jerome H. Skolnick, "Rethinking the Drug Problem," *Daedalus* 121, no. 3 (1992).

[17] See Bruce Yandle, "Bootleggers & Baptists: The Education of a Regulatory Economist," *Regulation* (May–June 1983).

[18] Jennifer Clapp, "Toxic Exports: Despite Global Treaties, Hazardous Waste Trade Continues," in *Deviant Globalization*.

Chapter 2

Lawlessness and Disorder: An Emerging Paradigm for the 21ˢᵗ Century

Phil Williams

The 20ᵗʰ century will probably go down in history as the exemplar of geopolitical interstate conflict with two World Wars centered in Europe followed by over four decades of the Cold War between the United States and Soviet Union. The 21ˢᵗ century, in contrast, could well become a period of lawlessness and disorder—a century in which states are in long-term decline; new violent actors challenge states and one another; resources such as food, water, and energy become a central focus of violent competition and of large illicit markets; demographic and environmental trends pose challenges to sustainability, security, and stability; and the severity of problems is significantly increased by the interconnections and often perverse interactions among them, or what Thomas Homer-Dixon termed "negative synergies."[1]

This agenda of challenges is highly diverse and complex, and these qualities alone could overload the management capacity of both individual states and the international community. Individual states, for the most part, are entering an era of retrenchment characterized by reduced budgets on everything from welfare to national security. While this might result in enhanced efficiencies and greater creativity, it could also reduce capacity and resilience at the national level. And even though there is broad agreement that many threats and challenges cannot be dealt with unilaterally, multilateral responses might prove equally problematic. Indeed, multilateral governance structures and strategies, while the darlings of liberal institutionalists, have inherent and fundamental weaknesses such as buck-passing, lowest-common denominator responses, and free riding. The search for international consensus often becomes no more than a cloak or a rationale for procrastination and inaction. Many states engage in cosmetic conformity with international norms and standards but tacitly defect from these regimes, violating agreements and ignoring obligations while continuing to pay lip service to the collective. In other words, both unilateral and multilateral approaches to emerging security challenges are likely to prove fundamentally inadequate.

Some commentators wholeheartedly reject such arguments. Indeed, many observers are sanguine about the current and emerging threat environment. It has been argued, in particular, that:

+ the threat to states from networked actors, whether terrorists or criminals, has been greatly exaggerated and networks can be defeated by states that are largely composed of hierarchical bureaucratic organizations[2]

+ the "core" of advanced economies is spreading both economic prosperity and stability to countries in the "gap" and gradually integrating them into a stable political and economic order[3]

+ the United States currently is not confronting a clear and present danger but enjoying "clear and present safety."[4]

Regarding the first of these arguments, it is hard to disagree with the assertion that networks can be defeated. Of course they can, but only after enormous and asymmetrical investment of time and resources, not to mention blood and treasure. U.S. military forces in Iraq, for example, ultimately adapted, and learned and internalized lessons more effectively than the fragmented insurgency opposing them.[5] Indeed, one of the key lessons was that a networked insurgency provides opportunities for wedge-driving, something that was done to good effect in the Anbar Awakening.[6] Yet even in Iraq, it is not clear that the defeat is absolute; the resilience and capacity for resurgence of both criminal and terrorist networks should not be ignored. Even when a network is completely dismantled, long-term consolidation of victory is sometimes confounded by what might be termed network succession, with one set of networked violent actors replacing another. For example, the Medellín and Cali cartels were defeated, but what was arguably their Colombian successor, the Revolutionary Armed Forces of Colombia (FARC) insurgency, as well as their Mexican successors, the Sinaloa Federation and the Zetas Organization, are if anything even more formidable. Moreover, when networks are a key component of illicit markets, destroying some of the networked organizations does not readily translate into anything more than a temporary disruption of the market. In some cases, such as Mexico, state success in weakening, decapitating, or destroying major drug-trafficking networks simply creates "vacancy chains" that encourage even greater violence among those that are left.[7]

The second thesis is little more than a modernized variant of Wilsonianism with an emphasis on the spread of free market economies as a major accompaniment to the powerful attraction of liberal democracies—both of which are deeply enshrined in the core states. Like traditional Wilsonianism, it suffers from a high degree of wishful thinking. Unfortunately, it is not clear that the core is integrating into the gap. In fact, the relationship between the core and gap seems to be going in the opposite direction from that envisaged by Thomas Barnett, with the gap enlarging at the expense of the core. Problems are spilling over from disorderly parts of the world into what are widely considered the zones of peace, stability, and order. What we are witnessing is not the success of the U.S. civilizing mission, but rather what Lillian Bobea termed "the revenge of the periphery," a trend that is evident in illegal migration, transnational crime, terrorism, piracy, and even insurgency.[8]

The authors of the third argument regarding a clear and present safety do not actually deny the existence of complex security threats and challenges; rather, they argue that current threats are inflated and the responses are overly militarized while emerging threats are largely ignored, especially in terms of resource allocation.[9] Their argument is far more compelling than the title of their article suggests, not least because the authors see clear and present safety as transitory. In their view, "the main global challenges facing the United States today are poorly resourced and given far less attention than 'sexier' problems, such as war and terrorism. These include climate change, pandemic diseases, global economic instability, and transnational criminal networks—all of which could serve as catalysts to severe and direct challenges to U.S. security interests."[10] This assessment is, in many respects, very compelling. The difficulty is that the more traditional security threats have not simply gone away to be replaced by a new set: new and old challenges coexist as part of a more crowded agenda in which traditional, resurgent, nascent, and truly novel and unfamiliar security threats demand attention in a climate dominated by economic austerity. Great power conflict, for example, has not disappeared and, as discussed below, against a backdrop of climate change and intensifying competition for diminishing natural resources could well come to the forefront again. Adversaries with the desire or capacity to do harm to the United States cannot be ignored now or in the future. The danger is that they will present new challenges more difficult to delineate, define, deter, and defend against. The inherent opportunity cost of managing one set of challenges inevitably results in attention and resource deficits in relation to other problems. In this connection, one of the most serious problems is the bias toward the familiar, which accentuates the inherent difficulties of "managing the unexpected."[11] Most of the emerging and nontraditional security challenges are novel or unfamiliar, extending well beyond the comfort zones of policymakers and officials. As such, they will likely prove impervious to the usual range of policy options embedded in standard operational procedures. Bureaucracy and budget managers are good at responding to routine and familiar challenges, but when the challenges are new and different, the level of competency tends to decline and resources are constrained at the very time they are most needed.

Adversaries cannot be ignored. Nevertheless, one of the key assumptions of this chapter is that security challenges emanate not only from hostile actors and pernicious interactions such as security dilemmas and arms races, but also from certain conditions that promote violence and instability—and in turn are worsened by them. The United States, in particular, tends to be overtly focused on adversaries or enemies who are then typically characterized as embodiments of evil. A paradigm shift is needed to create at least an accompanying focus on the kinds of conditions and the confluence of trends that create instability, disorder, and chaos. Some of these conditions and trends might not be susceptible to preventive measures; even so, acknowledgment of their debilitating consequences might make it possible, under some circumstances, to develop mitigation strategies. Yet successful mitigation might be the exception rather than the rule, especially as more security challenges will likely fall into the category of wicked problems that are not amenable to easy or readily available solutions.

In this connection, not only is governance inadequate to meet the challenges, as discussed above, but also governance, or more accurately poor or insufficient governance, is one of the challenges. In fact, we are seeing a crisis of governance that was effectively delineated by James Rosenau in his seminal *Turbulence in World Politics*.[12] Rosenau's analysis identified a crisis in authority structures that was immediately manifest in the collapse of the Soviet Union and the disintegration of the Warsaw Pact. During the two decades since then, more and more states have been characterized by weakness or fragility, and a few have actually fallen into the category of failed states. In other words, poor governance is itself a major source of instability, insecurity, and disorder. Indeed, many of the security challenges of the 21st century are likely to have more to do with inadequate governance than traditional great power rivalries.

Another important dimension of the emerging security environment is that unpredictable interactions among security challenges increase the potential for instability and disorder. Analysts have become quite skillful at identifying individual trends or silos of change but are far less adept at looking at the space between the silos and the ways in which different trends and challenges interact—often in surprising and highly damaging ways. What are too often treated as separate security threats and challenges are, in fact, interconnected in ways that increase the prospect for unexpected crises and unfavorable outcomes. Nowhere is this more obvious than in the *favelas* of Brazil.

Poverty has always been a problem, but in returning to the favelas after 30 years, Janice Perlman found that the pervasive impact of drugs and violence had transformed marginalization of the population into a reality and made it far more difficult for people to get out.[13] Other observers have discussed the toxic mix of kids, drugs, and guns that exist in parts of Latin America and that have made some favelas more dangerous than war zones.[14] Similarly, West Africa has long suffered from poverty, rapid urbanization and large slums, youth unemployment, corrupt governance, and weak institutions. In recent years, however, the influx of Latin American drug-trafficking organizations using the region as a transshipment point for cocaine en route to Europe has added an incendiary component to this already volatile mixture.[15] This has been most evident in Guinea-Bissau. Since its independence in 1974, Guinea-Bissau has suffered from chronic instability and a series of military coups. In recent years, however, the drug trade has exacerbated the situation, with allegations that both political and military elites have assisted the drug-trafficking organizations. In April 2012, the civilian leadership was ousted in yet another military coup that some commentators believe was designed to protect military profits obtained through facilitating the drug trade.[16]

Against this background, this analysis identifies a set of global trends or megatrends, some of which have and will continue to shape the security environment, while others will determine the nature and the severity of the security challenges faced by the United States and the international community during the next several decades. The notion of megatrends typically refers to "great forces in societal development that will affect all areas—state, market and civil society—for many years to come. . . . In other words, megatrends are our knowledge about the probable future. Megatrends are the forces that define our present and future worlds, and the interaction between them is as important as each individual megatrend."[17] Sometimes megatrends will have an offsetting, softening, mitigating, or neutralizing impact on each other, yet they can also be mutually reinforcing in their impact, creating tipping points to disorder, insecurity, and illegality. The discussion of each of these megatrends is inevitably brief and superficial but perhaps sufficient to tease out the implications for security and the development of illicit activities and markets.

Few trends or developments, however, are wholly negative in their impact. Urbanization, for example, has some positive consequences. Cities are engines of growth, drivers of innovation, concentrations of cultural vitality, and, as Edward Glaese has argued, provide "the clearest path from poverty to prosperity."[18] Indeed, the urban poor are generally better off than the rural poor, which is why cities attract people from rural areas. Nevertheless, cities—as suggested below—also pose enormous challenges to security and stability. The issue is not an either/ or; it is about the mix of positives and negatives. And the argument here is that sensitivity to the negatives needs to increase significantly.

The Drivers of Insecurity and Illegality

Globalization

Much has already been written about globalization, which is readily understood in terms of an enormous increase in the speed, ease, and density of transactions, reduction of transaction costs for the movement of commodities, development of various kinds of global flows, and increased interactions of individuals, societies, and economies. Although many benefits accrue from globalization, it is clear that it has a serious downside. Globalization has increased insecurity in a variety of ways—challenging cultural identity, disrupting social and political norms, and increasing transnational global flows that no single government can control and that are increasingly difficult to monitor. Globalization has facilitated various forms of trafficking and empowered criminal enterprises, has had disruptive consequences that have created incentives for criminality, and has enabled criminals and other violent armed groups to share operational knowledge and, in some cases, to collaborate.

Indeed, globalization has become an important facilitator for transnational organized crime, drug traffickers, terrorists, and the like. Those involved in trafficking illegal commodities (prohibited, regulated, counterfeit, or stolen goods) including drugs, endangered species, small arms, nuclear material, cultural property, and counterfeit pharmaceuticals are able to hide them in the vast amount of licit trade, creating both needle-in-a-haystack and needle-in-a-needle-stack challenges to state entities seeking to combat trafficking. Although Stephen Krasner is correct in his observation that states "have never been able to perfectly regulate transborder flows," it is arguable that they have never before had to contend with the sheer volume, speed, and diversity of the people and commodities that cross their borders both legally and illegally.[19] In many respects, the prosaic and unheralded symbol of globalization is the intermodal container, a development that has transformed the scale of global trade by reducing transaction costs and—in spite of such measures as the Container Security Initiative rolled out by U.S. Customs and Border Protection—largely denied states the ability to control what comes across their borders unless they are willing to place global trade on hold. As the author of the definitive history of the container noted:

> *With a single ship able to disgorge 3,000 40-foot-long containers in a matter of hours, and with a port such as Long Beach or Tokyo handling perhaps 10,000 loaded containers on the average workday, and with each container itself holding row after row of boxes stacked floor to ceiling, not even the most careful examiners have a remote prospect of inspecting it all. Containers can be just as efficient for smuggling undeclared merchandise, illegal drugs, undocumented immigrants, and terrorist bombs as for moving legitimate cargo.*[20]

Consequently, what Carolyn Nordstrom termed "the illusion of inspection" is likely to remain the dominant approach with "taxes and tariffs" acting as "obstacles not obligations" and borders as boundaries not barriers.[21]

Globalization has also been as empowering for organized crime as it has for legitimate business. This is reflected in what might be termed an expanding and more diversified pool of criminal enterprises with a transnational reach and in some cases a global presence. Russian criminals, for example, have followed the sun, establishing themselves in Israel, Spain, Mexico, the Caribbean, and Thailand among other places.[22] Chinese criminal organizations have a presence in North America and much of Europe as well as in their more traditional haunts in Asia. Perhaps no criminal organizations, however, are as ubiquitous as Nigerians, who operate in countries as diverse as China, Thailand, Paraguay, Holland, and Germany, and are involved in a mix of criminal activities ranging from drug trafficking and women trafficking to various forms of financial fraud. Mexican drug-trafficking organizations import precursor chemicals from China, work with a resurgent Sendero Luminoso in Peru, have branches, franchises, or affiliates in many U.S. cities, preserve links with the 'Ndrangheta in Italy, and maintain a presence in West Africa for moving cocaine to Europe.[23]

In addition to empowering organized crime and encouraging it to go transnational, globalization has had highly disruptive consequences for many societies, thereby ensuring a steady flow of recruits to criminal organizations. Globalization has winners and losers—and the pain for the losers can be enormous. In an odd way, organized crime can help to mitigate this pain. In zones of social and economic exclusion, informal and illegal markets provide critical safety nets while organized crime not only offers one of the few forms of employment but also provides a degree of governance and even public goods. In the absence of opportunities in the legal economy, people simply migrate to the illegal economy. As James Mittelman notes, "transnational organized crime groups operate below as well as beside the state by offering incentives to the marginalized segments of the population trying to cope with the adjustment costs of globalization. These groups reach down and out to the lower rungs of social structures—the impoverished—a substratum that does not lend itself to the easy strategies prescribed by the state."[24] While it is sometimes argued that globalization has slowed down as a result of economic recession, this merely reinforces the argument about organized crime as an alternative career path: where the licit economy is not providing jobs or career opportunities, people migrate to the illicit economy and opt for criminal careers.

Population Growth and Demographic Trends

Population growth will continue to be a driver for the next few decades. After passing 7 billion in late 2011, world population is expected to surpass 9 billion people by 2050 and—although the rate of increase will decline in the latter half of the century—to exceed 10 billion by 2100.[25] Disentangling the impact of such increases on security, order, and stability is not easy. At the most obvious and simplistic level, increased population places new demands on the environment, but also on food, water, and energy supplies. Unless there are commensurate advances in food production, water availability, and the development of alternative energy sources, these areas will be subject to increased competition, supply deficits, and disruptions. Accordingly, they will look increasingly attractive to organized crime. This is discussed more fully below.

The issue, however, is not simply one of increasing numbers. As Jack Goldstone has noted, "international security will depend less on how many people inhabit the world than on how the global population is composed and distributed: where populations are declining and where they are growing, which countries are relatively older and which are more youthful, and how demographics will influence population movements across regions."[26] In this connection, as the United Nations Department of Economic and Social Affairs has noted, "most of the additional 3 billion people from now to 2100 will enlarge the population of developing countries, which is projected to rise from 5.7 billion in 2011 to 8.0 billion in 2050."[27] In many of these countries, state capacity is limited, employment opportunities are constrained, and there is a significant gap between the "youth bulge" and available employment opportunities. Indeed, many developing countries are "faced with the necessity of providing education or employment to large cohorts of children and youth even as the current economic and financial crisis unfolds. The situation in the least developed countries is even more pressing because children under 15 constitute 40 percent of their population and young people account for a further 20 percent."[28]

The implications for security and stability are sobering. Young men, in particular, are a major source of violence and unrest in societies. In Latin America, and increasingly in parts of Africa, they have gravitated toward gangs, drugs, and guns, and there is nothing to suggest that such a trend will change as the youth population continues to increase. Urban youth with few prospects are not a force for stability in society. Gangs provide a sense of identity and belonging as well as a degree of empowerment and rationale for violence. For those who do not enter gangs, migration—legal or illegal—might appear an attractive option, and it seems likely that both human smuggling and human trafficking will experience major increases over the next few decades. As more and more young people look for opportunities simply not available in their countries of origin, unless developed countries liberalize immigration, the opportunities for criminal entrepreneurship in the people-moving business will increase significantly. This conclusion is reinforced by general migration trends. As a study by the United Kingdom Ministry of Defence has noted, "The number of international migrants has increased from a total of 75 million a year in 1965, to 191 million a year in 2005 of whom around 10 million are refugees, and up to 40 million are illegal migrants. That number may grow to 230 million by 2050."[29]

If young populations feed into criminal activities on the supply side, older populations might do so on the demand side. According to Jack Goldstone, "The European countries, Canada, the United States, Japan, South Korea, and even China are aging at unprecedented rates.... In 2050, approximately 30 percent of Americans, Canadians, Chinese, and Europeans will be over 60, as will more than 40 percent of Japanese and South Koreans."[30] While there are concerns that this will lead to declines in productivity, one sector that will flourish is health care: "As populations age, they will demand more health care for longer periods of time."[31] When this is combined with the advances in medicine, it seems likely that the demand for replacement organs will increase. Although organ trafficking was long dismissed as an urban myth, it is clear that it has occurred in several countries. In one case in Kosovo, for instance, organized crime was clearly involved. According to one commentary, the organ-trafficking network "included criminals from countries such as ... Moldova, Ukraine, Turkey, Russia and Israel."[32] Such cases are all too likely to expand in scope and frequency. After all, there are increasing numbers of young, healthy potential donors in the developing world, increasing numbers of potential recipients in the developed world, and more than enough criminal middlemen and unscrupulous healthcare workers willing to match them up regardless of laws and regulations.

Urbanization

Cities have always been engines of economic growth, repositories of wealth, power, and entrepreneurship, and centers of culture, scholarship, and innovation. But they have also been breeding grounds of disease, concentrations of poverty and crime, and drivers of instability and revolution. For a variety of reasons, the future development of cities is more likely to create negative rather than positive effects and precipitate instability rather than create order. Some urban areas will even degenerate into what Richard Norton has termed "feral cities," which are effectively ungovernable and out of control.[33] What makes this all the more important is that in the next few decades, there will be more cities, more larger cities, and more globally connected cities. Big cities mean big problems, and globally connected cities mean global problems. Before examining these problems in more detail, however, it is necessary to offer some brief observations about the growing importance of cities. The analysis then considers ways in which continued urbanization might impinge on stability and security.

One of the most dramatic facets of urbanization is the growth of megacities. Inclusion in this category requires 10 million or more inhabitants. Yet, as one scholar has noted:

> The number increased dramatically from three megacities (Mexico City, New York, Tokyo) in 1975 to around 20 in 2007 and is projected to reach almost 30 worldwide by 2025. The emerging megacities whose populations range from five to ten million have likewise experienced a notable increase. The number is expected to increase from currently about 30 to almost 50 in 2025. Together, "established" and emerging megacities now account for roughly 15% of the world's total urban population.[34]

The sheer size of such cities creates major environmental hazards, generates serious law and order problems especially in poorer areas of poorer cities, and strains infrastructures that are already overstretched. Yet the problem is not simply one of size; even smaller cities of a few million can become violent and dystopian as has happened in Ciudad Juarez, Mexico, and Caracas, Venezuela, both of which have seen homicide rates surpass 200 per 100,000.

The most important facet of the link between urbanization and security is probably the growth of slums. The UN-Habitat report on *The State of the World's Cities 2006/7* described slums as the "emerging human settlements of the 21ˢᵗ century."[35] The report also noted that "urbanization has become virtually synonymous with slum growth, especially in sub-Saharan Africa, Western Asia, and Southern Asia."[36] Characterized by lack of durable housing, sufficient living area, access to water and sanitation, and security against eviction, slums can be understood best in terms of Castell's notion of zones of social and economic exclusion.[37] They are generally areas where the state, in terms of the provision of services, is either minimalist or absent. Many arriving from rural areas in search of economic opportunity find that that they have merely traded rural destitution for urban destitution. In these circumstances, resentment and alienation are likely to be widespread. This dynamic was captured in 1996 by Wally N'Dow, head of the Habitat II Conference, who noted that there were already more than 600 million people officially homeless or living in life-threatening urban conditions and warned that "a low-grade civil war is fought every day in the world's urban centers. Many cities are collapsing. We risk a complete breakdown in cities. People feel alienated."[38] These comments, although remarkably prescient, were if anything an underestimation of the problem. If present trends continue, the world slum population will be 2 billion in 2030.[39] In other words, by then the number of people living in the conditions highlighted by Dow will have more than tripled since he made his comments.

Another key dimension is the enormous potential for pernicious interactions between urban growth on the one side and economic crises, high levels of unemployment, and weak and inadequate governance on the other.[40] Although cities provide new opportunities for some, the zones of social and economic exclusion will expand enormously. Few cities in the developing world will have the capacity to generate sufficient jobs to meet the demand of their growing populations. And one group that tends to suffer disproportionately from unemployment consists of youths and young men. The problem here, as Thomas Homer-Dixon has observed, is that "underemployed, urbanized young men are an especially volatile group that can easily be drawn into organized crime."[41] They also provide the main recruitment pool for street gangs, terrorists, rioters, predatory criminals, and the like. In other words, the youth bulge has increasingly become an urban youth bulge with dangerous consequences.

In many parts of the world, rapid urbanization has created such overwhelming problems that the state has simply stepped back and abandoned any pretense of providing governance in many of the massive slums. Although the Brazilian government is currently trying to re-introduce state power into selected favelas in Rio ahead of the 2014 World Cup and the 2016 Olympic Games, the long-term success of this initiative remains uncertain. Elsewhere, large parts of many cities have become extremely hazardous for the instruments of state power and representatives of state authority, especially law enforcement and military forces. Cities in the 21ˢᵗ century will continue to provide enormous opportunities for "hit and run" tactics, disruption of efforts to establish the authority and legitimacy that are essential to state-building, and kidnappings and other criminal activities that help to fund insurgency and sectarian violence. The concomitant of this is the growth of fear and insecurity. In some cities, fear has become so pervasive that it has led to the creation of fortified communities and "protected enclaves."[42] Cities have become segmented with the fortified enclaves representing one type of no-go zone and slums and high violence areas representing another. This segmentation or "spatial transformation" into "safe" and "dangerous" localities dramatically manifests the stark juxtaposition of wealth and poverty in cities.[43] As such, it could well increase the long-term potential for violence.

Unfortunately, the challenges of urbanization have not evoked a commensurate response. City governance, like state governance, is often corrupt, ineffective, and patchy at best. It is not surprising, therefore, that bottom-up or organic but informal governance mechanisms emerge as a substitute for governance from above. This is both positive and negative. It is positive in that it provides some degree of order, limited but real economic opportunities, and rudimentary services; it is negative in that the providers are often criminal organizations simply using paternalism to enhance their own security. This phenomenon is visible in the Cape Flats in South Africa where, as Andre Standing has noted, "the criminal economy delivers employment and goods to thousands of individuals who are socially excluded," while the "criminal elite provides ... 'governance from below' ... by performing functions traditionally associated with the state."[44] These functions include dispute settlement, a degree of social protection, and even private philanthropy that is at least a partial substitute for state provision of welfare.[45]

If anything, the phenomenon is more striking in the slums of Kingston, Jamaica, where the "dons" provide employment, services, and protection and in some cases "enjoy respect and greater legitimacy than formal state actors and institutions."[46] In the favelas of Rio de Janeiro, the emergence of both the parallel economy and alternative forms of governance was well described by Elizabeth Leeds in the mid-1990s. Leeds emphasized that the organizers of the drug business "are first and foremost businessmen who are using the physical space of the favela ... as the locus of operation for a highly lucrative informal-sector activity. For that space to be available and protected, they must offer something in return."[47] In other words, the provision of employment, services, and protection is based on self-interest rather than altruism. Nevertheless, it has provided benefits to communities that have been ignored by the state apart from their victimization and exploitation by corrupt police forces. The long-term consequence of alternative governance, however, is that the legitimacy of the state is further eroded while the power of organized crime and other violent armed groups that are also service providers is greatly enhanced.

Natural Resources and Global Climate Change

Another factor that seems likely to be a source of disorder, criminality, and the expansion of illicit markets is the rapid depletion of global resources. Michael Klare, in a series of pioneering studies, has shown that natural resources such as oil, gas, and certain minerals and metals are becoming less readily available and, as such, are increasingly a source of contention both in interstate relations and within states as governments and corporations compete for what is left. In Klare's words, this competition "will be ruthless, unrelenting, and severe. Every key player in the race for what's left will do whatever it can to advance its own position, while striving without mercy to eliminate or subdue all the others."[48]

Criminal organizations are likely to participate in this competition even if they are on the margins rather than at the center. In this connection, examples of criminal exploitation of natural resources abound. The diversion, theft, and bunkering of oil (loading it from small vessels into tankers) in and off the coast of Nigeria's Niger Delta is a major problem and is carried out by a mix of criminalized militias, corrupt military and civilian officials, and transnational criminal enterprises.[49] Although the militias claim to pursue political agendas, the profit motive seems to have become increasingly important. Something similar appears to have happened in the Democratic Republic of the Congo where rebel groups have become deeply involved in the exploitation and smuggling of minerals such as cassiterite (tin), gold, coltan, and wolframite (tungsten), and at least some of the violence has been over the control of mining sites and smuggling routes.[50] Another example is Iraq and the lack of copper resources in the Middle East. Because copper is not produced in the region, it can fetch high black market prices. After major U.S. combat operations in Iraq in 2003, the electric pylons that had been downed were systematically stripped of their copper, which was then smuggled elsewhere in the region, a process that reportedly reached "industrial scale" proportions.[51] In the event, this complicated the task of restoring Iraq's power grid. Another example of exploitation of minerals is the FARC, which in recent years has become heavily involved in both the extortion of gold mining companies and the illegal mining of gold. It is also possible that the FARC will become involved in illegal mining of coltan in Venezuela. Many of the coltan deposits are close to the Venezuelan border with Colombia in areas where the FARC has a significant presence.[52] Consequently, it would be natural for the FARC to extend its activities into coltan mining. In other instances, criminal and armed groups are implicated in the timber trade, with some estimates suggesting that half of all tropical timber in the international market is either illegally logged or illegally exported.[53] The point of all this is to emphasize that violent armed groups and criminal organizations of one kind or another are already deeply involved in exploiting natural resources. The end of easily accessed resources and the increase in prices that will result will encourage even more extensive involvement of organized crime either in the illegal exploitation of these resources or in the theft and diversion of products after they have been obtained legally.

Closely linked to this are the possible consequences of global climate change. Although the scientific debate over global climate change is ongoing, it is clear that even if only some of the more moderate predictions are fulfilled, the consequences could be dire. According to one study, even the most conservative and cautious expected climate change scenario would bring in its wake "heightened internal and cross-border tensions caused by large-scale migration conflict sparked by resource scarcity, particularly in the weak and failing states of Africa; increased disease proliferation, which will have economic consequences; and some political reordering."[54] It is also very likely that even this level of global climate change would create disruptions and shortages of food and water. Gwynne Dyer has persuasively argued that "the first and most important impact of climate change on human civilization will be an acute and permanent crisis of food supply."[55] It is probable that climate change will create significant disruptions in agricultural production leading to a sporadic rather than steady supply of foodstuffs, significant increases in food prices, and a commensurate increase in political instability. Increased food prices are believed to have been one of the drivers of the Arab Spring and are typically associated with revolution and upheaval.

Water would also be at risk. Indeed, the coordinated if not collective judgment of the U.S. Intelligence Community suggests that "Between now and 2040, fresh water availability will not keep up with demand absent more effective management of water resources. Water problems will hinder the ability of key countries to produce food and generate energy, posing a risk to global food markets and hobbling economic growth."[56] One of the problems inhibiting more effective management of the water sector is that in many countries corruption in this sector is already pervasive and enduring. Even more seriously, water problems and the increasing scope and severity of droughts will be one of the more immediate consequences of global climate change. This in turn could create new conflicts or exacerbate existing conflicts. Cattle-raiding in East Africa, for example, has already reached new levels of violence as a result of an extended drought.[57] Furthermore, "by 2020, between 75 million and 250 million people" in Africa will likely "be exposed to increased water stress due to climate change."[58] In some countries disputes over water frequently become violent. In Yemen, according to one report, water disputes account for as many as 4,000 deaths a year.[59] And this pattern is likely to become prevalent elsewhere. As the Organisation for Economic Co-operation and Development noted:

> Within states, at the local level, competition over water use, its availability and allocation can lead to low-scale violence, which can escalate into instability within states and across sub-regions. Tensions between citizens and authorities over water issues may initially manifest themselves in the form of civil disobedience. They may, however, also escalate into acts of sabotage and violent protest if adequate political participation is not possible.[60]

In other words, the linkages between water stress, violence, and political instability could all too easily become uncomfortably close.

Disruption of food supply and shortages of water would also provide attractive new opportunities for criminal organizations. Indeed, the creation of large and lucrative illicit markets in licit products would be a natural response to temporary interruptions or permanent shortages. It is no accident that black markets emerge rapidly in the aftermath of war and disaster as they did in Germany and Japan in World War II. They also become features of wartime, as Peter Andreas illustrated so compellingly in his study of the siege of Sarajevo.[61] In a world where demand for staple products significantly surpasses supply and shortages become the norm, black markets are likely to be large and enduring. The energy sector in some parts of the world (Russia, Nigeria, Iraq, and Mexico in particular) has already been infiltrated by organized crime. Water in refugee camps has long been a precious commodity and has sometimes been used for extortion. It is also worth noting that organized crime has already penetrated the food sector in some countries. According to Italy's largest farmers' association, "organized crime has spread its involvement through the entire food chain from acquisition of farmland to production, from transport to supermarkets," with criminal groups often "undercutting prices paid to farmers for their products and inflating prices paid by consumers in food stores."[62] The result of sporadic food and water supplies would be to offer transnational criminal organizations unprecedented opportunities to embed themselves very deeply in sectors that, the Italian example notwithstanding, have not traditionally been among their priorities including major areas of activity or markets of influence.

The other development of particular relevance is the impact of rising sea levels on population movements. Rising sea levels are projected to have an impact on countries as diverse as Bangladesh and the United States as well as many African coastal states. It is the countries of South and Southeast Asia, however, that are most vulnerable, especially in the coastal lowlands of Bangladesh, China, Japan, Vietnam and Thailand. Many of these areas have a high population density. The implication is that widespread population displacement will occur and environmental refugees will become an increasing segment of the global refugee population. Climate change in general could expand the displaced population to unprecedented levels with short-term surges resulting from extreme weather events such as floods, droughts, hurricanes, and the like. Much, of course, will still depend on state responses. A broadly liberal policy toward the immigration of environmental refugees would allow the problem to be managed through legal means. If states adopt strict controls on immigration, which is likely given the political pressure on governments to look after their own populations, organized crime will receive an enormous boost for its human-smuggling activities.

On top of all this, climate change could alter strategic relationships among great powers, lead to interstate water conflicts, and create new and unpredictable vectors for disease transmission. Unless states are able to manage the resulting crises with a higher degree of competence than existing governance mechanisms and institutions, which are already under considerable stress, they could rapidly become scapegoats and targets. And it is to the future of the state that attention must now be given.

The Decline of the State and the Rise of Alternative Governance

States clearly retain the formalities of sovereignty and the capacity to impose laws and regulations that determine the incentives and opportunities as well as the obstacles for organized crime and traffickers of various commodities. Yet the Westphalian state as such is past its peak—a peak symbolized most dramatically in the total wars of the twentieth century. Indeed, states face two fundamental and interconnected challenges: they are often unable to meet the economic needs and expectations of their citizens, and they are unable to elicit the loyalty and allegiance of significant portions of these same citizens.

The inability of most states to meet the needs of their citizens reflects the rise of complex or wicked problems that are resistant to short-term or ready solutions as well as what might be called the long-term demography of unemployment. The job creation capacity of most countries of the developing world is already modest at best and will become even less adequate to meet the needs of growing populations, while even in developed countries large segments of immigrant populations remain unemployed, underemployed, or employed only for the most menial tasks. For countries such as Nigeria, even if they succeed in overcoming the mix of corruption and incompetence that pervades governance structures, it is unlikely they will create sufficient opportunities for a rapidly growing population. The result is that the disenfranchised and alienated segments of society will grow, as will disputes over resources. This is also likely in other African societies where the state, rather than being above politics, is simply the prize of politics. In these circumstances, politics becomes a zero-sum game and the distribution of the spoils is heavily skewed in the favor of the group, tribe, clan, ethnic, or sectarian faction in power.

Even where "governance" is not a zero-sum game, however, the weaknesses of the state are often debilitating. These weaknesses can be understood in terms of capacity gaps and functional holes. Gaps in state capacity lead to an inability to carry out the "normal" and "expected" functions of the modern Westphalian state and to make adequate levels of public goods or collective provision for large parts of the citizenry. In Latin America, this has resulted in what Gabriel Marcella described as "inadequate public security forces, dysfunctional judicial systems, inadequate jails which become training schools for criminals, and deficiencies in other dimensions of state structure such as maintenance of infrastructure."[63] Indeed, Marcella goes on to argue that "at the turn of the 21st century, Latin American countries have essentially two states within their boundaries: the formal and the informal. They are separate entities often walled off from each other, though they interact with the informal state supporting the other."[64] Similar observations have been made by Rapley, who has argued not only that the state "lacks the largesse needed to buy the loyalty of an ever-increasing number of players," but also that other informal forms and structures of governance move in to replace the state.[65] "Where the State can no longer provide employment, build houses, pave roads or police the streets, or where the police are so woefully underpaid that they supplement their incomes from corruption, sometimes turning on the very citizens they are meant to protect, in such cases, private armies and mini-states might fill the vacuum left behind by a retreating state."[66] In this connection, it is worth emphasizing that one reason for the resurgence of Sendero Luminoso in Peru has been that in most respects, the state does not exist outside Lima. Other states too have leaders who hold the office of president or prime minister but are really little more than the mayor of the capital city. Indeed, in many countries in the developing world, the state has failed to expand its remit far beyond the capital in terms of either extraction or provision for the citizens. Such limitations put both the authority and legitimacy of the state into question.

Even where governments have long been viable and effective and have enjoyed a high level of legitimacy, the state often appears to be in retreat. As reducing public debt becomes the first priority of governments, provision for social welfare and social safety nets is contracting. The welfare state was really limited to a small subset of countries and was fully embraced only in Western (and especially Northern) Europe, Canada, Australia, and New Zealand, and even some of these states are backing away from the levels of support they provided in the last several decades. In some instances such as Greece and Great Britain, the retrenchment process has provoked mob violence. For its part, the United States, having gradually created a partial and symbolic set of safety nets, is now in the process of restricting and reducing them. In an era characterized by austerity and concerns over public debt, the reality of the welfare state is increasingly restricted. Ironically, however, the idea of the welfare state and the expectation that the state is the provider of public goods remains prevalent, setting the scene for continuous disappointment and ultimately serious disaffection.

Since the provision of collective goods, including public security, has in many instances been patchy and incomplete, with the current phase characterized by the entrenchment and withdrawal of the state, there are numerous opportunities for alternative forms of governance, often local in scope and criminal in character, to emerge as challengers to the state. Sometimes the inadequacies and inequities of the state provoke insurgent movements; in other cases they encourage the rise of warlords, militias, and gangs; in yet other cases they offer incubating spaces for criminal organizations that extend patronage and even some forms of philanthropy out of self-interest rather than altruism. Whatever, the precise nature of these groups or the circumstances of their emergence, they typically fill at least some of the gaps left by the state. This trend is likely to continue so long as governments focus on debt reduction rather than the effective provision of a wide range of public goods.

Another major problem for states stems from alternative loyalties of significant portions of the population. During much of the 20th century this problem seemed a thing of the past as nation and state became increasingly synonymous and wartime cohesion solidified relationships between states and their societies. Since the end of the Cold War, however, divisions and fissures within states have become increasingly apparent. Indeed, in some countries the lack of primary affiliation with the state and the resurgence of primordial loyalties—to family, clan, tribe, ethnic group, religion, or sect—has created a crisis of loyalty among significant and often growing segments of "national" populations. As David Ronfeldt has argued, "Even for modern societies that have advanced far beyond a tribal stage, the tribe remains not only the founding form but also the forever form and the ultimate fallback form." [67] Indeed, "many of the world's current trouble spots—in the Middle East, South Asia, the Balkans, the Caucasus, and Africa—are in societies so riven by embedded tribal and clan dynamics that the outlook remains terribly uncertain for them to build professional states and competitive businesses that are unencumbered by tribal and clan dynamics."[68] In Somalia, Iraq, Yemen, and Afghanistan, clans and tribes, to say the least, have complicated efforts to engage in state-making. Similarly, in some African countries tribal loyalties have a significant impact on both patronage networks and distribution of state resources. Tribes and clans and the warlords who sometimes lead them typically define both their interests and their identities in ways that implicitly or explicitly challenge notions of public interest and collective identity as symbolized in state structures and institutions.[69] In turn, these definitions significantly skew patterns of resource allocation. The gap between favored beneficiaries and the rest of the population further undermines the legitimacy of the state while strengthening the legitimacy of those who have succeeded in tilting the playing field in favor of their tribe or clan. None of this should be surprising. As Zonabend has observed, "The lineage or clan is more than a group of relatives united by privileged ties; it is also a corporate group, whose members support each other, act together in all circumstances, whether ritual or everyday, jointly own and exploit assets and carry out, from generation to generation, the same political, religious or military functions."[70] Few states have this kind of unity except under conditions of total warfare. And the age of total warfare appears to have passed.

Significantly, criminal organizations also exhibit some of these features of tribes and clans. Many criminal organizations, although certainly not all, have an ethnic, family, tribal, or even geographical basis. Even when that is not the case, bonding mechanisms—which can include time spent together in prison or simply working together in risky conditions—play an important role. This sense of affiliation to nonstate and often violent groups, while it sometimes coexists easily with loyalty to the state, can also subvert or weaken the state. Moreover, in an era of state budgetary retrenchment, these alternative loyalties and the organizational forms that accompany them will become increasingly important. From the state perspective, this is a highly negative synergy.

A closely related problem for the state is the rise of subnational and transnational challengers. In part this reflects the fact that many states have inadequate social control mechanisms and weak law enforcement and criminal justice systems. Yet there are other considerations that have also fed in to the rise of transnational criminal organizations. When states are failing or inadequate in terms of economic management and the provision of social welfare, the resulting functional holes create pressures and incentives for citizens to engage in criminal activities. Amid conditions of economic hardship, extralegal means of obtaining basic needs often become critical to survival. For countries in which there is no social safety net, resort to the informal economy and to illicit activities is a natural response of many citizens. From this perspective, the growth of organized crime and drug trafficking, along with the expansion of prostitution, can be understood as rational responses to dire economic conditions. Such activities are, in part, coping mechanisms in countries characterized by poverty, poor governance, and ineffective markets. Organized crime is a highly effective form of entrepreneurship, providing economic opportunities and multiplier benefits that would otherwise be absent in economies characterized by inadequate economic management. Illicit means of advancement offer opportunities that are simply not available in the licit economy. The difficulty, of course, is that the filling of functional spaces by organized crime exacerbates the weakness of the state.

In contrast, the power of criminal organizations (along with that of clans, warlords, and ethnic factions) is increased by connections outside the state. According to Shultz and Dew, "One of the more disturbing trends of non-state armed groups is the extent to which such groups, including these clan-based groups, are cooperating and collaborating with each other in networks that span national borders and include fellow tribal groups, criminal groups, and corrupt political elements."[71] Even where such collaboration does not occur, many criminal networks operate in a transnational manner, engaging in jurisdictional arbitrage both to maximize profits (selling illicit and trafficked commodities where the price is highest) and to minimize risk, often by operating de facto safe havens. Ironically, when necessary these groups use national sovereignty for defensive purpose even though their transnational (illicit) business activities trample all over sovereignty and respect for national borders.

Consequences and Conclusions

The preceding discussion makes it hard to avoid the conclusion that the four sets of drivers examined and elucidated are providing a "perfect storm" that could have profoundly unsettling implications for future security and stability. First, it is very akin to the "coming anarchy" that Robert Kaplan warned about in his seminal article in 1994.[72] Although most intelligence forecasts for the next several decades and beyond are deliberately nonalarmist and sometimes even bland in tone, there is an increasing recognition of the scope and severity of the challenges and the way in which they might interact. The interactions make each of the challenges less malleable. Wicked challenges become even more wicked when they intersect and overlap. As John Sullivan pointed out, the urban poor, for example, are likely to be the most significantly harmed by global climate change, especially as many large urban centers are coastal and particularly vulnerable to rising sea levels and extreme weather events.[73] Similarly, Global Business Network has captured the perverse and potentially disastrous interaction of these megatrends in a brilliant analysis focusing on the possible

> interplay of the collapse of civil order and the dynamics of urbanization. If a climate change-induced system disruption reduces the ability of the government to deliver political goods (Katrina being an obvious example), it also reduces political legitimacy and halts economic activity, thus driving local populations to rely upon primary loyalties (families, neighborhoods, religious organizations, gangs, etc.) for daily survival. This dynamic in the political system is often (and will increasingly be) played out in urban settings—physical spaces that require intensive external flows of goods and services to survive, and that are also highly (and increasingly) interconnected and networked via transport and telecommunications infrastructure. Collapsing civil order within urban settings will offer extreme economic rewards in the form of smuggling and black markets; indeed, these may be the only functioning markets, making virtually everyone in these spaces a "bad actor." Those unwilling or unable to profit from the chaos will radiate outward through refugee flows, exporting social conflicts to adjacent locales.[74]

Second, even if the worst-case scenarios are avoided, in the continuing dialectic between the forces of order and the forces of disorder, it is clear that the latter are gaining strength while the limits of state power are increasingly apparent. The megatrends discussed above, for example, can hardly fail to be advantageous to violent nonstate armed groups. Not only will transnational criminal organizations be empowered by the availability of whole new sets of black market opportunities, but also groups engaged in political violence will almost invariably find criminal activities an increasingly attractive funding mechanism. At one level, this is old news. Several decades ago Paul Clare examining the IRA's use of extortion described it as the organization's "Capone discovery."[75] This has only become more pronounced since then as terrorist organizations and insurgencies have appropriated the methodologies of organized crime. Ironically, the more powerful and prominent these organizations become, and the more that is expected of them as an alternative form of governance, the more deeply implicated they become in criminal activities. For example "Hezbollah's financial needs have grown alongside its increasing legitimacy . . . as it seeks to rebuild after its 2006 war with Israel and expand its portfolio of political and social service activities."[76] At the same time, Iranian direct support seems to have declined as a result of the sanctions imposed by the international community. "The result, analysts believe, has been a deeper reliance on criminal enterprises—especially the South American cocaine trade—and on a mechanism to move its ill-gotten cash around the world." [77] Recent revelations about Hizballah's connections to cocaine trafficking from Latin America to Europe through West Africa and in money laundering through the Lebanese Canadian Bank SAL certainly seem to bear this out. Although the extent of Hizballah's direct involvement in trafficking—as opposed to taxing traffickers and benefitting from the activities of supporters such as convicted trafficker Chekry Harb—has not yet been determined, one investigator described the group as operating "like the Gambinos on steroids."[78]

The corollary is that although some strong legitimate states will endure, even those will face a complex and potentially overwhelming set of challenges, while the number of what might be termed qualified, restricted, notional, or hollow and collapsed states will only increase. Moreover, many of the weaker states will be neutralized, penetrated, or in some cases even captured by organized crime or other violent armed groups. In effect, we will continue to see a world of formal state structures but at least some of these will be little more than fronts for criminal structures. The emphasis on state sovereignty will do little to obscure the dispersal of real authority and power among what Rapley described as "autonomous political agents, equipped with their own resource bases, which make them resistant to a reimposition of centralized control."[79] One of the corollaries of this is the increasing spread of disorder from the zone of weak states and feral cities in the developing world to the countries of the developed world. In effect, "the revenge of the periphery" will become even more pronounced.

The third and closely related development is that good governance and the rule of law will become even more infrequent and restricted than they are at present. Most modern societies are at least nominally based on the rule of law and on the notion of the state providing security for its citizens not only against external military threats but also against internal criminal threats. Yet in more and more instances, not only is the state unable to provide for the security of its citizens at the domestic levels, but increasing portions of the population are seemingly becoming unwilling to obey laws or accept certain norms of behavior. This is a form of anomie, which is generally understood as a kind of behavioral sink, a degeneration of rules and norms and the emergence of forms of behavior unconstrained by standard notions of what is or is not acceptable. In some instances this collapse of social norms—which typically reinforce the rule of law—results from external shocks. In such cases, not only are the restraints removed but norms and inhibitions that were once in place no longer apply or guide behavior. This is partly because the penalties for noncompliance with the norms have suddenly been removed. Yet it also represents something much more fundamental: a willingness to put morality and decency to one side, a marked absence of respect for fellow citizens who become simply targets to be exploited for financial gain, and a readiness to engage in forms of behavior that are normally regarded as reprehensible. As defined by Passas, anomie is a withdrawal of allegiance from conventional norms and a weakening of these norms' guiding power on behavior.[80] In some cases, the descent into anomie is rapid; in others it is a long-term trend. Passas, for example, has argued that not all anomie should be linked to strain theory or seen as in terms of sudden collapse. Rather, he argues, it is often a result of structural contradictions in society, which create a gap between expectations and the opportunity to fulfill them—a gap that typically results in social deviance or criminality.[81] The result, however, is the same: a decline of behavioral norms and standards, the spread of both organized and disorganized crime, and the growing pervasiveness of violence in society. An anomic world is one in which social capital is replaced by criminal capital, violence is no longer under the monopoly control of the state, and inhibitions on barbaric behavior (such as decapitations, mutilation of bodies, etc.) have largely disappeared. Some of the drug violence in northern Mexico has this quality.

In many ways, Malcolm Gladwell's notion of contagion accords fully with the notion of anomie and spreading criminality. Gladwell spends much time discussing the "stickiness" or attraction of certain ideas and forms of behavior.[82] From this perspective, criminal careers as a source of employment appear very attractive in countries where opportunities in the legal economy are limited. Moreover, imitation and emulation are the order of the day. This is not surprising: those who behave within the law are disadvantaged compared to those who do not. Moreover, operating outside the law is likely to appear increasingly attractive in the event that urban expectations remain unfulfilled, states reduce the provision of collective goods, or energy interruptions and food and water shortages become the norm.

It is possible to go even further: global anomie is becoming a problem at the state level as well as the individual level. This is apparent in the spread of corruption in government, in the development in many countries of a "political-criminal nexus" between the elites and organized crime, and the willingness of some governments to protect criminals and drug traffickers.[83] Although some governments are willing to confront criminal organizations and seek to constrain criminal activities, others acquiesce because of weakness, accept corrupt payments as forms of rent, or even develop collusive and collaborative relationships with criminals. Sometimes organized crime takes over the state; in a few instances—Serbia under Milosevic, North Korea, Russia under Putin and a revitalized intelligence apparatus, and possibly Bulgaria—the state takes over organized crime. One of the best examples of how a state can succumb to the attractions of lawlessness is Guinea-Bissau, which, as discussed above, has become a transshipment hub for Latin American drug-trafficking organizations moving cocaine to the European Union. From one perspective, embracing this as an opportunity is a rational decision for a government and people with limited resources and a high level of de-pendence on a single commodity (cashew nut) for export. For a country that can attract little foreign direct investment, the influx of money associated with drugs can appear enormously attractive. But the result has been to increase tensions between political and military leaders, distort economic sectors, and make an already fragile country even less stable. Furthermore, the way in which Guinea-Bissau and other West African states have embraced the drug business suggests that the illicit global economy will become a much larger component of the overall economy. Indeed, in a complex system, positive feedback loops strengthen tendencies toward behavior that works and weaken those that do not. Criminal behavior works to a degree that makes it very attractive. Even states that emphasize the rule of law are not immune to this trend, as is evident simply by looking at the increase in corruption cases on the southwest border of the United States. All this could lead to a growing divide between states in which the rule of law remains intact and those in which it has broken down. Indeed, it is conceivable that the major global divide of the 21st century will be caused not by competing ideologies, the struggle for power, or the "clash of civilizations," but by clashes between states that uphold law and order and those that are dominated by criminal interests and criminal authorities. This, in turn, would make it difficult to achieve the collective approach to common problems such as global climate change that is the only way in which strategies of mitigation (prevention is no longer an option if it ever was) might be successfully implemented.

There is perhaps one ray of light in what is otherwise a very gloomy picture. Technology has the potential to inhibit or reverse some of the trends that have been identified in this analysis. Technologies that help to increase food production, facilitate better water management, reduce difficulties and costs of desalination, and offer cheaper and more plentiful sources of energy, for example, could confound some of the more dire predictions outlined above. Nevertheless, it would not be prudent to reply on magic bullets of this kind. Instead, the first step in avoiding a lawless century is to acknowledge the portents of a gathering storm. Yet there is little evidence of this in the United States, as the obsession with Iran, North Korea, and China is inexorably nudging airpower and seapower back to the center of U.S. strategy. Ironically, intelligence agencies are already highlighting the ways in which both current and emerging challenges suggest an alternative threat paradigm that goes well beyond the parameters of conventional strategic thinking. Unfortunately, these agencies appear to suffer from the same curse as Cassandra: in spite of being compelling, the warnings are largely being ignored.

Notes

[1] Thomas Homer-Dixon, *The Upside of Down* (Washington, DC: Island Press, 2006), 106.

[2] Mette Eilstrup-Sangiovanni and Calvert Jones, "Assessing the Dangers of Illicit Networks: Why al-Qaida May Be Less Threatening Than Many Think," *International Security* 33, no. 2 (2008), 7–44.

[3] Thomas P.M. Barnett, *The Pentagon's New Map* (New York: Putnam, 2004).

[4] Micah Zenko and Michael A. Cohen, "Clear and Present Safety: The United States Is More Secure Than Washington Thinks," *Foreign Affairs* (March/April 2012).

[5] Chad Serena, *A Revolution in Military Adaptation: The U.S. Army in the Iraq War* (Washington, DC: Georgetown University Press, 2011).

[6] Austin Long, "The Anbar Awakening," *Survival* 50, no. 2 (April 2008), 67–94.

[7] The concept of vacancy chains is developed in H. Richard Friman, "Forging the vacancy chain: Law enforcement efforts and mobility in criminal economies," *Crime, Law and Social Change* 41, no. 1 (February 2004), 53–77, and applied to Mexico in Phil Williams, "The Terrorism Debate Over Mexican Drug Trafficking Violence," *Terrorism and Political Violence* 24, no. 2 (Special Issue: Intersections of Crime and Terror, 2012), 259–278.

[8] Private correspondence with author.

[9] Zenko and Cohen.

[10] Ibid.

[11] Karl Weick and Kathleen Sutcliffe, *Managing the Unexpected: Assuring High Performance in an Age of Uncertainty* (San Francisco: Jossey-Bass, 2001).

[12] James N. Rosenau, *Turbulence in World Politics* (Princeton: Princeton University Press, 1990).

[13] Janice Perlman, *Favela: Four Decades of Living on the Edge in Rio de Janeiro* (New York: Oxford University Press, 2010).

[14] "Freedom from Fear in Urban Spaces: Discussion Paper," Human Security Outreach Program, available at <geo.international.gc.ca/cip-pic/library/Freedom%20From%20Fear%20in%20Urban%20Spaces.pdf>.

[15] James Cockayne and Phil Williams, *The Invisible Tide: Towards an International Strategy to Deal with Drug Trafficking Through West Africa* (New York: International Peace Institute, 2009).

[16] Peter Fabricius, "West Africa hit by coups," *Daily News*, April 24, 2012, available at <www.iol.co.za/dailynews/opinion/west-africa-hit-by-coups-1.1282707>.

[17] Gitte Larsen, "Why Megatrends Matter," available at <www.cifs.dk/scripts/artikel.asp?id=1469>.

[18] Edward Glaese, *The Triumph of the City* (New York: Penguin, 2011), 1.

[19] Stephen D. Krasner, "Abiding Sovereignty," *International Political Science Review* 22, no. 3 (2001), 229–252.

[20] Marc Levinson, *The Box* (Princeton: Princeton University Press, 2006), 7.

[21] Carolyn Nordstrom, *Global Outlaws* (Berkeley: University of California Press, 2007), 116, 159.

[22] See Jonathan M. Winer and Phil Williams, "Russian Crime and Corruption in an Era of Globalization: Implications for the United States," in *Russia's Uncertain Economic Future*, ed. John E. Hardt, 97–124 (Washington, DC: M.E. Sharpe, 2003).

[23] Grace Wyler, "Mexico's Drug Cartels Are Spanning The Globe," *Business Insider*, June 21, 2011, available at <http://articles.businessinsider.com/2011-06-21/politics/30053839_1_mexican-cartels-sinaloa-european-markets>.

[24] James Mittelman, *The Globalization Syndrome* (Princeton: Princeton University Press, 2000), 210.

[25] Department of Economic and Social Affairs: Population Division, *World Population Prospects The 2010 Revision—Highlights and Advance Tables* (New York: United Nations [UN], 2011), xiii, hereafter cited as Population Division.

[26] Jack A. Goldstone, "The New Population Bomb: The Four Megatrends That Will Change the World," *Foreign Affairs* (January/February 2010), 31–43.

[27] Population Division, xiii.

[28] Ibid.

[29] DDDC, UK Ministry of Defense, *Global Strategic Trends—Out to 2040* (January 1910), 95.

[30] Goldstone, 34.

[31] Ibid., 35–36.

[32] "Canadian man testifies at Kosovo organ trafficking trial," CTV News, March 23, 2012, available at <www.ctv.ca/CTVNews/World/20120323/Canadian-man-testifies-at-Kosovo-organ-trafficking-trial-120323/#ixzz1qdKZPeeI>.

[33] Richard Norton, "Feral Cities," *Naval War College Review* 56, no. 4 (Autumn 2003), 97–106, available at <www.nwc. navy.mil/press/Review/2003/Autumn/pdfs/art6-a03.pdf>.

[34] Dirk Heinrichs et al., *Risk Habitat Megacity* (Berlin: Springer, 2012), 4.

[35] UN-Habitat, *State of the World's Cities 2006/7* (London: Earthscan, 2006), 19.

[36] Ibid., 11.

[37] The notion of social exclusion is discussed in Manuel Castells, *End of Millennium* (Oxford: Blackwell, 1998), 71–72.

[38] Wally N'Dow, quoted in "The Urbanizing World: Megacities; Bane or Boon," *World Press Review* (August 1996), 8.

[39] "Slum Dwellers to double by 2030: Millennium Development Goal Could Fall Short," UN-Habitat, *21st Session of the Governing Council*, Nairobi, Kenya, April 16–20, 2007.

[40] This point is emphasized by Thomas Homer-Dixon, "Standing Room Only," *Toronto Globe and Mail*, March 6, 2002, available at <www.homerdixon.com/download/why_population_growth.pdf>.

[41] Ibid.

[42] Karina Landman, "Alley-gating and neighbourhood gating: Are they two sides of the same face?" Paper delivered at the conference *Gated Communities: Building Social Division or Safer Communities?* Glasgow, September 18–19, 2003, available at <www.csir.co.za/Built_environment/Planning_support_systems/gatedcomsa/docs/Glasgow_paper_v5.pdf>.

[43] Ibid.

[44] Andre Standing, *The social contradictions of organized crime on the Cape Flats*, ISS Paper 74, June 2003, 7.

[45] Ibid., 9.

[46] Rivke Jaffe, "Between the street and the state: Crime and citizenship in urban Jamaica," available at <www.irmgard-coninx-stiftung.de/fileadmin/user_upload/pdf/urbanplanet/collective_identities/Jaffe_Essay.pdf>.

[47] Elizabeth Leeds, "Cocaine and Parallel Polities in the Brazilian Urban Periphery: Constraints on Local-Level Democratization," *Latin American Research Review* 31, no. 3 (1996), 75.

[48] Michael T. Klare, *The Race for What's Left: The Global Scramble for the World's Last Resources* (New York: Macmillan, 2012).

[49] The author has benefited on this issue from work done by two graduate students at the University of Pittsburgh, Tomas Malina and Brian Gray.

[50] The author is grateful to Giacomo O'Neill, a graduate student at the University of Pittsburgh, for his research and analysis of this issue.

[51] "Addressing Organized Crime and Drug Trafficking in Iraq: Report of the UNODC Fact Finding Mission," United Nations Office on Drugs and Crime, August 5–18, 2003 (Vienna, August 25, 2003).

[52] Edward Fox, "FARC Set to Exploit Venezuela 'Conflict Mineral,'" March 12, 2012, available at <http://insightcrime.org/insight-latest-news/item/2336-farc-set-to-exploit-venezuela-conflict-mineral>.

[53] Tim Boekhout van Solinge, "Eco-Crime: The Tropical Timber Trade," available at <www.rodolfocordero.com/silvia/Lab-fall09/NESSF1011/Silvia/Boekhout%20%20timber%20crime.pdf>.

[54] Kurt M. Campbell and Christine Parthemore, *National Security and Climate Change in Perspective*, in *Climatic Cataclysm*, ed. Kurt M. Campbell, 19 (Washington, DC: The Brookings Institution, 2008).

[55] Gwynne Dyer, *Climate Wars* (Brunswick, Australia: Scribe, 2010), xi.

[56] "Global Water Security," *Intelligence Community Assessment, ICA 2012-08,* February 2, 2012, available at <www.dni.gov/nic/ICA_Global%20Water%20Security.pdf>.

[57] Christian Parenti, *Tropic of Chaos: Climate Change and the New Geography of Violence* (New York: Nation Books, 2011).

[58] "Climate Change—Health and Environmental Effects: International Impacts," U.S. Environmental Protection Agency, available at <www.epa.gov/climatechange/effects/international.html>.

[59] "Under Pressure: Social Violence over Land and Water in Yemen," *Small Army Survey,* Issue Brief Number 2, October 2010, available at <www.yemen-ava.org/pdfs/Yemen-Armed-Violence-IB2-Social-violence-over-land-and-water-in-Yemen.pdf>.

[60] "Water and Violent Conflict," *OECD Development Assistance Committee,* available at <resource_en_92767.pdf>.

[61] Peter Andreas, *Blue Helmets, Black Markets* (Ithaca: Cornell University Press, 2008).

[62] "Organized crime controls Italy's food industry," Reuters, January 19, 2012, available at <www.havocscope.com/italy/>.

[63] Gabriel Marcella, *American Grand Strategy for Latin America in the Age of Resentment,* SSI Monograph, No. 10, available at <www.strategicstudiesinstitute.army.mil/pubs/display.cfm?pubID=811>.

[64] Ibid., 24.

[65] John Rapley, "Keynote Address—From Neo-Liberalism to the New Medievalism," Australian National University Conference, *Globalization and Governance in the Pacific Islands,* October 2005.

[66] Ibid.

[67] David Ronfeldt, *In Search of How Societies Work: Tribes—the First and Forever Form,* RAND Working Papers (Santa Monica, CA: RAND, 2006), 53, available at <www.rand.org/pubs/working_papers/2007/RAND_WR433.pdf>.

[68] Ibid., 5.

[69] William Reno, *Warlord Politics and African States* (Boulder: Lynne Rienner, 1998).

[70] Quoted in Ronfeldt, 32.

[71] Richard Shultz and Andrea Dew, *Insurgents, Terrorists and Militias: The Warriors of Contemporary Combat* (New York: Columbia University Press, 2006), 53.

[72] Robert D. Kaplan, "The Coming Anarchy," *The Atlantic,* February 1994, available at <www.theatlantic.com/magazine/archive/1994/02/the-coming-anarchy/4670/>.

[73] John P. Sullivan, "A Catastrophic Climate," in *Global Biosecurity: Threats and Responses,* ed. Peter Katona, John P. Sullivan, and Michael D. Intriligator (New York: Routledge, 2010).

[74] Nils Gilman, Doug Randall, Peter Schwartz, *Impacts of Climate Change* (San Francisco: Global Business Network, January 2007).

[75] Paul K. Clare, *Racketeering in Northern Ireland: A New Version of the Patriot Game* (Chicago: Office of International Criminal Justice at the University of Illinois at Chicago, 1989), 16.

[76] Jo Becker, "Beirut Bank Seen as Hub of Hezbollah's Financing," *The New York Times,* December 14, 2011, available at <www.nytimes.com/2011/12/14/world/middleeast/beirut-bank-seen-as-a-hub-of-hezbollahs-financing.html?pagewanted=all&_r=0>.

[77] Ibid.

[78] Ibid.

[79] Rapley.

[80] In this section, I have drawn heavily on Nikos Passas, "Global Anomie, Dysnomie, and Economic Crime: Hidden Consequences of Neoliberalism and Globalization in Russia and Around the World," *Social Justice* 27, no. 2 (2000), 16–44.

[81] Ibid., 19.

[82] Malcolm Gladwell, *The Tipping Point* (New York: Back Bay, 2000).

[83] R. Godson et al., "Political-Criminal Nexus," *Trends in Organized Crime* 3, no. 1 (Autumn 1997). For a more recent analysis, see Moisés Naím, "Mafia States," *Foreign Affairs* (May/June 2012), 100–111.

Chapter 3
Can We Estimate the Global Scale and Impact of Illicit Trade?

Justin Picard

The Need for a Better Measurement of Illicit Trade

"Fighting a war" is the longstanding idea of how the problem of illicit markets should be addressed. President Richard Nixon initially coined the term *war on drugs* during a press conference in 1971, and it has since been adopted for other forms of trafficking such as counterfeiting and so-called modern slavery. But illicit trade is now increasingly characterized as a "whack-a-mole" problem:[1] while the trade may decrease following successful law enforcement efforts, it usually resumes after a time or switches to a different illegal commodity. One might note, however, that Darwinian whack-a-mole is the order of the day, as enforcement efforts may have the perverse effect of facilitating the "selection" of the most apt or powerful criminals. According to Nils Gilman's fourth rule on deviant globalization, "Once a deviant industry professionalizes, crackdown merely promotes innovation."[2]

Two important themes of this book are that illicit markets are connected problems primarily driven by supply and demand, and that their impacts are more important than the criminal actors themselves. This chapter addresses the challenge of measuring the collective size and impact of illicit markets and networks. According to the 2011 White House *Strategy to Combat Transnational Organized Crime*, the issue appears to demand urgent policy attention: "The expanding size, scope, and influence of transnational organized crime and its impact on U.S. and international security and governance represent one of the most significant of those [21ˢᵗ-century] challenges."[3] Such statements on the size and growth of the problem and its impacts are often heard, but rarely supported with reliable statistics and sound measurement. One goal of this chapter is to set the foundations for which the qualitative claims that have been made on the size and harm caused by illicit activities can be translated into a quantitative form.

Certainly, some data are already available that suggest the impressive magnitude of illegal markets. For example, the Web site Havocscope.com aggregates publicly available sources of information on black markets. It provides multiple rankings of black market size by country, region, product, and market. By summing up various estimates of different illicit activities, it calculates that the size of the illicit global economy was between $1.69 trillion and $1.92 trillion at the time of measurement.[4] However, in its "Black Market Data Documentation and Standards," Havocscope acknowledges that "the very nature of the black market creates information that is limited and difficult to verify. To be a credible platform, it is essential that all data presented by Havocscope is properly sourced,"[5] and that "all data listed within Havocscope is collected from credible public sources, such as newspapers, government reports and academic journals."[6] We do not question the fact that Havocscope is collecting and citing its sources very consistently, but more often than not the cited sources themselves do not explain how it came up with the data and figures, or how the statistics were produced.

It is well known that measuring illicit markets involves developing statistics on activities that, by their nature, are intentionally difficult to observe. Unfortunately, the hidden nature of the phenomenon has led to the propagation of estimates based on poor methodologies, undisclosed or questionable assumptions, and biased or inaccessible data sets. The lack of reliability of current estimates, and their use and abuse for influencing policies, have led to a backlash from scholars, who challenge the dominant school of thought on the frightening development of criminal networks in a globalized world. With justification, they question the use of dubious statistics to influence the policy debate and perpetuate wrong perceptions of the issues. This politicization of numbers has become a topic of academic research in its own right, as a recent collection of essays demonstrates.[7]

For Peter Andreas, a professor of political science at Brown University, "Popular claims of loss of state control are overly alarmist and misleading and suffer from historical amnesia."[8] He continues, "At base, much of what makes organized crime transnational involves some form of profit-driven smuggling across borders. Transnational organized crime is therefore simply a new term for an old economic practice."[9] Thomas Naylor, a professor at McGill University specializing in the history of black markets, criticizes the assumption that "something called 'globalization' and the accompanying spread of modern communications and transportation technology have been a godsend to criminal cartels. . . . The beauty of globalization is that, much as with organized crime, everyone can agree about its enormous impact because no one knows what it really is."[10] Naylor believes that "although it is certainly true that there is more economic crime across borders today, there is also much more legal business, and absolutely no proof that the ratio of illegal to legal business is increasing."[11]

These contrarian statements cannot simply be brushed aside. Indeed, there are also valid arguments that illicit trade may be decreasing. As trade is progressively harmonized, import tariffs have globally decreased, reducing some of the incentives for smuggling. While technological innovations may be helping producers of illicit goods and traffickers, they also make it harder to hide illicit activities, as trails of evidence are not erased easily in the digital world. Technology also makes it easier, in principle at least, for law enforcement to cooperate, exchange information, and detect suspicious activities or illicit shipments. Therefore, one can argue that it is unproven, and indeed very difficult to prove, that illegal markets are increasing, at least in proportion with global economic activity.

Yet if one considers the growth of illicit financial flows that have followed progressive financial deregulation in the last 50 years, which has facilitated the legitimization of criminal gains and corruption and tax evasion on a massive scale, it can be argued that the scope of the illicit economy is much larger than before, and its impacts extend far beyond illegal markets. In commenting on President Barack Obama's plan to focus on tax havens to recover tax money, Raymond Baker and Eva Joly regretted that "the [Obama] administration is largely missing a far more devastating problem related to offshore finance: money gained from criminal and other illicit sources. With the use of tax havens and other elements of an increasingly complex 'shadow' financial network, vast sums of illegal money are being shifted throughout the global economy virtually undetected."[12]

This discussion indicates that a clear definition of the problem and better metrics on the scale and the impacts of illicit activities are badly needed. In fact, several ingenious techniques have been developed to measure illicit activities, leading to estimates that may not claim perfect accuracy but are at least scientifically substantiated and useful to a better understanding of the issue. The first contribution of this chapter, then, is to bring more transparency to some of the existing measurements of the following five illicit markets: illegal drugs, counterfeits, human trafficking and forced labor, excise goods, and environmental crime.

A prerequisite to measurement is to determine which elements should be included in the calculation. For that, it is essential to have a reasonable definition of the problem that facilitates classification. An implicit part of the definition is the name given to the problem: *transnational organized crime* refers essentially to groups or associations of criminals; *global black markets* are about illegal markets that circumvent prohibition, taxation, or regulations; *illicit* or *deviant globalization* can be understood as the set of illegal or unethical activities that can be related to the internationalization of economic activity—a vast category indeed. Yet terms like *globalization* and *illicit* are vague, with the latter allowing for many interpretations by referring to activities that are either illegal or merely scorned by society. And social norms are not uniform; therefore, what may qualify as illicit trade will vary according to the perspective taken. This is confirmed by a historical perspective on illicit trade: "Much of what we define today as illicit traffic was not criminalized a century ago and was not even considered a problem."[13] Therefore, *illicit trade* can be understood in a narrow way as the trade in certain illegal commodities, or far more broadly as including any illicit economic activity. For example, the Global Agenda Council on Illicit Trade proposed the following broad definition: "illicit trade involves money, goods or value gained from illegal and also from generally unethical activity."[14] Considering the different ways to name the problem and interpret it, we will not attempt to propose a one-size-fits-all definition. Rather, the second contribution of this chapter is to propose a framework in which different definitions can coexist.

This chapter proposes to replace the usual measurement based on a summation of the revenues generated by the different illegal markets—a method that does not highlight the multiple dimensions of illicit trade and its connections to other major global issues. Thus the third contribution of this chapter is to suggest an impact-based measurement that draws on explicit and transparent assumptions to estimate how much a given issue is affected by illicit activities and networks. The new measurement focuses on the direct and indirect effects of illicit activities, such as the crime-related costs, and the social, political, and environmental harms. The economic value of impacts can also be estimated and aggregated in various ways. This measurement is applied to the five illegal markets mentioned above, yielding a total estimated economic cost of close to $1.5 trillion.

Statistics on Illicit Markets

This section reviews several of the most significant estimates of illicit activities, ordered by declining (claimed) market size, and assesses how much we should trust them. According to sociologist Joel Best, good statistics on social issues should be based on more than guessing. They should draw on clear and bounded definitions, reasonable measurement, and representative samples. As we shall see, fulfilling those requirements in the case of illicit markets is a fairly challenging task, as illicit activities are often hard to define or measure and intrinsically difficult to detect.

Illegal Drugs

The leading international authority on illegal drugs, the United Nations Office on Drugs and Crime (UNODC), publishes a World Drug Report annually. The 2011 report estimates that in 2009, the retail value for the global cocaine market was $85 billion and the heroin market $68 billion.[15] UNODC no longer publishes an estimate of the cannabis market, the last one in 2005 being $142 billion.[16] The same 2005 report provided the last available estimate on the value of the global illicit drug market: a retail value of $322 billion. Unfortunately, the margin of error is not indicated even though a factor of 5 to 10 would not be uncommon between the low-end and the high-end range of such an estimate. For example, a former UNODC employee, Francisco Thoumi, relates that an unpublished study by Peter Reuter for the Financial Action Task Force "is probably the most serious attempt to ascertain the size of the world illegal drug market and resulted in an estimated range between $45 billion and $280 billion."[17] He notes that "the wide range of the estimates reflects the diversity in possible assumptions required at several stages in the production, smuggling, and marketing chain."[18]

The illicit drug market can be measured by estimating either supply or demand. The possibility of using seizures is discussed in the next section on counterfeiting. Estimates of the drug market, whether based on supply or demand, are inherently imprecise for two reasons. First, they are based on multiplying different uncertain quantities, leading to an amplification effect on the overall uncertainty. Second, they require a series of assumptions that can be adjusted to greatly influence the end result. According to Thoumi:

> Any estimate of the size of the illegal drug industry requires that a series of steps be taken and that assumptions be drawn at each one of them. These include estimates about coca and poppy acreage, the frequency of coca leaf harvests, the drug content in coca and opium, the quality of the chemists employed, the amounts of drugs seized, the amount consumed in different markets, and drug prices at each stage of the production and marketing chain.

A 2010 RAND report on demand-side estimations of global drug markets[19] calculates that global retail expenditures on cannabis range from €40 to €120 billion, with their best estimate being close to half of the previous global estimate of approximately €125 billion (USD142 billion). The report explains: "Total consumption, therefore, is constructed as the sum of user-specific amounts consumed in a given year. The amount consumed, in turn, is the product of the number of days in which the drug was reportedly consumed, the typical amount consumed on those days, and the number of users who fall into a specific user-group category."[20] Finally, it obtains total expenditure by multiplying the amount by the typical price.

The RAND report also notes:

> that the UNODC figures imply that retail cannabis expenditures in the U.S. are close to €40 billion—more than three times the figure we generate as our best estimate. This is not entirely surprising since the UNODC assumes that every past year user consumes on average 165 grams whereas we assume an average of 96 grams. Further, the UNODC applies an average retail price that is more than twice as high as the figure we use (€4.8 and €12.5, respectively).[21]

The point is that small differences between two methods on, say, the measurement of drug price and drug consumption can lead to strikingly different estimates. And in its conclusion, the same report notes that "surprisingly little is known about typical quantities consumed of illicit drugs, which makes generating demand-side estimates difficult."

Calculating demand-side estimates is even harder for other drugs. According to the RAND report, "Given the popularity of cannabis across the globe, there is relatively more information available about cannabis prevalence and consumption patterns. Thus, more confidence can be placed in these estimates than those for the other drug markets." However, for the two other major drugs, supply-side estimates can be more precise since, according to UNODC, "unlike opium and coca, for which relatively reliable production data can be obtained, estimates on cannabis production [are] often based on perception and scientifically valid monitoring systems are the exception."[22]

Given the overall lack of reliability of estimates, or their large variations when indicated, one may wonder if knowing more precisely the size of the illicit drug markets is of such great importance. In its critique of existing drug estimates and their influence on drug policymaking, Reuter asks:

> Does this mismeasurement [of the global drug market] matter? For those interested in the size of the underground economy, the answer is clearly yes; estimates of the largest illegal market are potentially of considerable significance. However, the estimates were not developed for those purposes but to help in the development of drug policy. If policymaking regarding drugs were rational, or at least as analytically driven as say monetary policy, then the exaggeration would be a serious problem.[23]

Regarding the contribution that more accurate figures on the drug market would have on our understanding of the impacts, Thoumi argues that "what has been important is not merely the size of the trade, but its ability to alter social behaviors, increase corruption and crime, and fund insurgent and counterinsurgent guerrillas. The size of the illegal drug industry is not particularly relevant as a cause of these social developments."[24] In short, the effects of illicit markets matter more than their size. In the section below, where our proposed methodology to measure the impacts of illicit markets is presented, results of studies that measured the economic cost of illegal drugs will be used instead of the market size estimates.

Counterfeiting

In the early part of the last decade, 5–7 percent of international trade was the commonly used figure for counterfeiting. However, this estimate from the International Chamber of Commerce turned out to be, at best, an educated guess. At worst, it was seen as a conveniently large number created to draw attention to the problem. As this and other estimates came under severe criticism, the Organisation for Economic Co-operation and Development (OECD) was mandated to launch a study on counterfeiting in order to produce a more solid statistical estimate. Its 2008 report stated that "analysis carried out in this report indicates that international trade in counterfeit and pirated products could have been up to USD 200 billion in 2005." This figure was later updated to $250 billion based on 2005–2007 world trade data. This estimate appears lower than the previously cited figure, corresponding to a bit less than 2 percent of international trade. Also, strictly speaking, the basis of measurement is not the international trade value but the value of counterfeit goods reported by customs, which can be significantly higher than the shipment declared value.

Also there are very significant problems with the employed methodology, of which even the authors of the report seem to have been aware. They pessimistically noted, "The overall degree to which products are being counterfeited and pirated is unknown and there do not appear to be any methodologies that could be employed to develop an acceptable overall estimate." As the team of economists and statisticians discovered when they started their study, little data were available and the lack of harmonization between data sets made it very difficult to combine them. With limited funding and considerable time pressure, the quickest way forward was to send questionnaires to custom agencies through the World Customs Organization to collect data on the estimated retail value of seizures.

The definition of counterfeiting used was based on the Trade-Related Intellectual Property Standard agreement, and it is the only definition compatible with custom seizure. However, this definition is far from being universally accepted because it disproportionately represents the interest of brand owners. The definition of counterfeiting is a subject of controversy that has resulted in intense debates and the freezing of anticounterfeiting efforts of the World Health Organization (WHO). It will suffice to say that from the standpoint of consumers, there are many situations where intellectual property (IP) infringement does not necessarily involve counterfeiting (for example, patent infringement). On the other hand, goods may be deliberately and fraudulently mislabeled to deceive consumers without necessarily infringing any intellectual property right.

There is also lack of clarity in the report on the basis for recording interceptions. Customs authorities have indeed used the declared customs value, the reported market value, or the legitimate item value. This is problematic because with counterfeit goods, there can be an order of magnitude difference between these estimates.

Regardless, the study's worst problem was with sampling. Because seizures are a very small part of the market, it makes no sense that seizures should be used to estimate overall market size. Certainly, interception data are useful in determining tendencies over time, the relative likelihood that counterfeits originate in certain countries, or whether they affect certain categories of products. However, it is not a reliable indicator for estimating the global extent of counterfeiting: extrapolating the volume of counterfeit trade from seizure data, even roughly, is impossible without knowing the interception rate. To produce an estimate, the OECD used the responses to its questionnaire, and calculated that "the value of interceptions/seizures from 35 economies totaled USD 769 million in 2005." Basic mathematics shows that it is simply not possible to infer the total traded value in counterfeit goods without knowing the interception rate of customs. Nevertheless, by making some unsubstantiated assumptions for the upper limit on the counterfeit export rate of certain categories of goods and for certain economies, OECD inferred a ceiling on counterfeit and pirated goods in world trade of $100 billion. The upper ceiling of $200 billion was reached by doubling this number to account for statistical uncertainties. The authors noted that this would mean customs only intercepted 0.5 percent of counterfeit goods on average. The purely hypothetical and speculative nature of the reasoning that led to the $200 billion upper ceiling is, however, not obvious in the first sentence of the executive summary: "Analysis carried out in this report indicates that international trade in counterfeit and pirated products could have been up to USD 200 billion in 2005."

Other studies have approximated the counterfeit market size by making assumptions on the customs interception rate. In its report *The Globalization of Crime*, UNODC estimates an interception rate of 7 percent for footwear entering Europe and, by extrapolating this rate to other goods, deduces that $8.2 billion in consumer products entered Europe in 2008. If this interception rate had been used for the OECD study, the value of international trade in counterfeit goods would drop from $200 billion to less than $15 billion. To be sure, if this number were circulated instead of the OECD number, counterfeiting would appear to be a much smaller problem to policymakers.

However, not a single customs organization in the world reports its interception rate, and it is likely that most have little idea what this interception rate might be. In addition, according to a Government Accountability Office (GAO) report, which is the only indication that we have found in publicly accessible documents, there is evidence that this interception rate varies significantly by port of entry and type of product. Therefore, statistically speaking, it is not clear how an average interception rate of counterfeit goods could be computed. The same report also points out the relative infrequency of IP infringement. According to Customs and Border Protection's (CBP's) Compliance Measurement Program, a statistical sampling program, IP violations have been found in less than one-tenth of 1 percent of exams. Of 287,000 random exams, 0.06 percent had IP violations. It is unlikely that CBP detected all IP violations in those 287,000 shipments. Nevertheless, combining this detection rate with the approximately $2 trillion in imports for the United States, and assuming that counterfeit shipments have roughly the same international trade value as other shipments, the total international trade value of counterfeit goods going through U.S. customs each year would be a mere $1.2 billion. The point is not that IP violations are grossly overstated, but rather that if custom seizure data are useful to observe trends, they do not bring much clarity on the magnitude of counterfeiting.

The Government Accountability Office reviewed the current knowledge on the economic effects of counterfeiting and piracy. It concluded that three widely cited U.S. Government estimates of economic losses resulting from counterfeiting could not be substantiated due to the absence of underlying studies and highlighted the difficulties of coming up with reliable estimates on the global extent and impact of counterfeiting. The GAO review concluded that no single approach for quantifying impacts of counterfeiting and piracy can be used.

As early as 2002, a report for the European Commission had already established that "although many existing estimates of the size of the counterfeiting and piracy problems are based on extrapolating from the number of seizures, arrests or convictions made by enforcement agencies, we do not recommend this approach except in rare circumstances." The only exceptions are when the detection rate is known to be high (over 75 percent) or can be known with confidence. In practice, those conditions are virtually never met. This same report develops a detailed methodology to estimate counterfeiting for each of 19 different industries, including an estimate of the costs. Unfortunately, there was no followup to the report.

The fundamental difficulty with estimating counterfeiting is that the brand owners are best positioned to know the significance and prevalence of the issue, or how to differentiate genuine articles from counterfeits. Generally speaking, however, the industry is unwilling to share the data it has collected, even though this is harming public health, consumers, and its own long-term interests. Industry associations, which have an interest in boasting large numbers, publicize estimates but often without any supporting methodology or access to underlying data.

The Business Software Alliance (BSA) is one of the exceptions, as it publishes a detailed report annually on software piracy and has developed a methodology to measure the amount of counterfeit software. The BSA reaches its figure on the amount of unlicensed software by estimating how much personal computer software was deployed in a given year, estimating how much was paid for or otherwise legally acquired during the year, and then subtracting one from the other. The difference between these two quantities corresponds to the commercial value of installed illicit software, which was valued at $59 billion for 2010.[25] The BSA does not attempt, however, to relate the official retail value of pirated software to the actual losses of the software industry.

Interestingly, this technique of differentials is somewhat analogous to such methods as the World Bank Residual Model, which estimate illicit financial flows. The difficulty of performing direct observation is partly circumvented by measuring two quantities such as supply and demand that should in principle be equal in the absence of illicit activity. The difference between these two quantities corresponds to the amount of illicit activity. We could imagine using this principle to measure illicit activities for which it is hard or costly to make direct observation, but which can be measured as a difference between an input and an output. For example, counterfeiting could in some cases be measured as the difference between the legitimate supply and the total demand (which includes counterfeits) if the two quantities can be known or approximated, as is the case for the amount of deployed software.

Excised Goods

While the illicit trade in cigarettes is the most studied black market for the purpose of excise tax avoidance, it is likely that other illicit markets in excisable goods amount to tens of billions more. Illicit trade in tobacco creates a black market that is or can be relatively accurately measured. Defining the problem is relatively straightforward: illicit tobacco is traded for the primary purpose of avoiding excise taxes, either by smuggling tobacco from low to high taxation states or countries, by counterfeiting, or by local tax evasion. Identifying illicit tobacco is generally easier than other industrial goods because there is usually a tax stamp, which in the case of fraud is either absent, counterfeit, or does not match its intended destination. Furthermore, many cigarette companies have implemented forensic or digital means to identify fraudulent packs. Unlike many illicit markets, there is little ambiguity in measurement as well: the direct cost is the amount of unpaid taxes to countries. For sampling, the tobacco industry has developed detailed methodologies for avoiding biases when sampling the market.

After having been extensively involved and complicit in the illicit trade because "smuggled cigarettes have helped the companies expand sales, make inroads in markets they cannot enter legally, increase their market share in competition with rivals, keep cigarette prices down generally or win legal import status or production in another country," tobacco companies now recognize illicit traders as an important competitor. Therefore, it is vital for them to understand the size, nature, and dynamics of the illicit segment. Having reliable metrics on illicit tobacco is seen as important for resource allocation purposes and for dialogue with the government. In many ways, this industry, despite being still perceived as backward-minded and uncooperative, is leading the way for other industries that are still somewhat reluctant to address their problems with illicit trade.

By combining individual country estimates, a study prepared for the WHO conference on the Framework Convention on Tobacco Control estimated illicit trade at 10.7 percent of total cigarette sales, which could represent a loss to government revenue of $40 to 50 billion annually. However, accounting for the fact that a number of smokers of illicit cigarettes would quit if they had to pay the full price of excise taxes, governments could recover up to $31 billion.

Human Trafficking

Various estimates have been produced on the global scale of human trafficking. A GAO report is an excellent starting point to gain insights on the way these estimates have been produced and their assumptions, strengths, and limitations. The report describes the methodologies, key assumptions, and limitations used by the U.S. Government, the International Labor Organization (ILO), UNODC, and the International Organization for Migration to produce their estimates.[26] In particular, the report raises doubts about U.S. Government estimates that 600,000 to 800,000 persons are trafficked across international borders annually. On the other hand, the methodology developed by the ILO appears fairly rigorous, considering the complexity of producing a global value on such a complex phenomenon.[27] In practice, there is often a very fine line between considering whether a person is in a forced labor or an over-exploitation situation, or between choosing whether a person is being smuggled or trafficked (with concrete and sometimes cruel consequences for the concerned person in the latter case). To reduce the ambiguity as much as possible in determining whether a person is in a forced labor or trafficking situation, the ILO uses definitions adopted in international conventions and developed a typology of forced labor that reduces subjectivity in the classification of cases.

The ILO estimates that 12.3 million people were in forced labor situations between 1995 and 2004. Of that number, 2.5 million were trafficked either internally or internationally.

The study was based on a variety of sources that include reported cases, which were validated by ILO experts. Forced labor is peculiar among illicit activities in that it extends over a long period, which allows for models that can be used to estimate the probability that events are reported over time. Therefore, it is theoretically possible to exhaustively review available sources, although this is not practical given the millions of publications that would need review. Therefore, an astute variation of the "capture-recapture" method was used. Capture-recapture methods are commonly employed in ecology to estimate population size when it is not possible to count all individuals present in the population, in which case an estimate of the total population can be inferred from the degree of overlap between different counts (the lower the overlap, the larger the estimate). Adapting this method, two separate teams searched for cases in a variety of databases for a period of 6 months and the estimate was extrapolated from the degree of overlap.

The GAO report does point out a few methodological weaknesses, most notably on the extrapolations needed to overcome the gap between reported and unreported victims. Nevertheless, the transparency of assumptions, the effort to use a clear definition, and the overall availability of data sources provide a certain degree of confidence in this figure. The ILO has also used standard methods for estimating the annual revenue generated by the 2.5 million people in trafficked forced labor. The estimated annual profits are broken down by region of the world, and for sexual versus other commercial exploitation. The estimated total stands at $32 billion and represents an average revenue of $12,800 per person.[28]

Environmental Crime

According to TRAFFIC, the wildlife trade monitoring network, the value of legal, international wildlife trade was nearly $292.7 billion in 2005, based on declared import values. This includes timber ($190 billion), wild fisheries ($81.5 billion), and other wildlife ($21.2 billion), which includes plants, animal products, and live animals. Below we review the estimates on the global extent of illicit trade for each category.

Illegal logging can be defined broadly to include violations of any number of international, national, or local laws and regulations. However, as pointed out in a report prepared for the American Forest & Paper Association,[29] if illegal logging clearly signifies legal abuses, there is considerable debate about what might be considered illegal. Some illegal activities "appropriately rise to a level of international concern," such as harvesting without authority in designated national parks, without or in excess of concession permit limits, or violating international agreements such as CITES (Convention on International Trade in Endangered Species of Wild Fauna and Flora). A 2004 report from the World Bank on forest sustainability placed the annual revenue loss to governments from failure to collect taxes from forest concessions at more than $5 billion, and the annual market value of losses from illegal cutting of forests at over $10 billion.[30] While both elements are illegal, only the latter is directly contributing to deforestation, which places the illegal market at slightly more than 5 percent of the global market. However, the American Forest & Paper Association reports that the World Bank estimate was published without reference to a supporting methodology, adding that "No matter how broad or narrow illegal forest activity might be interpreted, its extent is impossible to know with any degree of certainty. Reported estimates are generally supported only through anecdotal information and supposition." Yet after making a review of the literature on illegal logging and checking whether the estimates are credible in light of trade and other relevant data, the report concludes that illegal logging represents on the order of 8–10 percent of global wood products trade.

Illegal Fisheries. A 2009 study was apparently the first attempt to undertake a worldwide analysis of illegal and unreported fishing.[31] It finds that current illegal and unreported fishing losses worldwide are between $10 billion and $23.5 billion annually (mean value of $16.75 billion, or 20.55 percent of declared import value), representing between 11 and 26 million tons. The study uses primary data sources from several key composite studies supplemented by country-specific studies. In the source studies, a number of different methods have been used to estimate the level of illegal fishing, including surveillance data, trade data, stock assessments based on survey data, and expert opinion. The price data used were those reported by the Food and Agriculture Organization of the United Nations (UN).

The frequently cited global estimates ranging from $5 to $20 billion have no scientific basis. According to TRAFFIC, "Comprehensive trade data only exist for species covered by CITES—and even here problems with the accuracy of CITES trade reporting mean that trade data are indicative rather than actual."

Illegal Waste. A report from the European Union Network for the Implementation and Enforcement of Environmental Law presents a data-driven assessment of illegal waste in which 72 percent of the quantity (93 percent of declarations) of toxic waste exported to non-OECD countries in 2003 potentially constituted illegal trafficking,[32] as the transporters did not have an approved notification. It nevertheless "highlights the lack of data on the scale and breadth of illegal waste movements." Separately, an "International Crime Threat Assessment" prepared by the U.S. Government in 2000 estimated that criminal organizations earn $10 to $12 billion per year for dumping trash and hazardous waste materials.[33] The source of this frequently cited estimate is not mentioned.

Defining the Problem

The UN Convention against Transnational Organized Crime, which came into effect in late 2003, is a reasonable starting point to develop a definition. The convention covers offenses with a transnational component as well as so-called serious crimes such as money laundering and corruption, which "are made in order obtain, directly or indirectly, a financial or other material benefit."[34] An offense is transnational if (1) it is committed in more than one state, (2) it is committed in one state but planned from another state, (3) it involves a criminal group that has criminal activities in more than one state, or (4) it has substantial effects in another state.

Paradoxically, as noted in UNODC's threat assessment on transnational organized crime, "the Convention contains no precise definition of transnational organized crime," and "the implied definition of 'transnational organized crime' encompasses virtually all profit-motivated criminal activities with international implications." It is further recognized that such a broad definition "leaves the exact subject matter rather vague." While the absence of a definition may be explained by the desire to integrate political sensitivities over illegal markets, it makes it more difficult to use the convention for measurement purposes.

In considering whether the definition of the problem should be articulated around the concept of transnational organized crime, one must note that many profit-driven criminal activities can occur within borders. Whether international ramifications are found does not necessarily affect the dynamics of the market or the motivations of its actors. For example, the counterfeiting of locally produced and consumed goods has become an important problem for China, forced labor and even human trafficking do not necessarily involve crossing borders, interstate cigarette smuggling is predominant in the United States, and cannabis is often consumed in the country of production, as it is bulky and can be grown nearly anywhere in the world. Ultimately, in a globalized world one might wonder whether being transnational is a helpful discriminant. The point here is not to minimize the importance of addressing the transnational component of organized crime, but whether this is a key characteristic of the problem.

Still, according to UNODC's threat assessment, if "failure to identify the market-driven dimension of organized crime, and particularly of transnational organized crime, is one of the reasons these problems can prove so intractable," a definition focused on the concepts of market and profit might be more relevant. Here, Thomas Naylor's theory of profit-driven[35] crimes may offer a useful framework. He suggests a typology for profit-driven crimes, which includes three types of primary offenses: predatory crimes, market-based crimes, and economic crimes. The three types of offense are differentiated by 1) the nature of the transfer (wealth versus goods or service); 2) the basic act by which it is transferred (robbery, trafficking, or legal market); and 3) the method which is used (force, voluntary transfer, or fraud):

+ Predatory crimes involve a transfer of wealth through an illegal act (robbery) and by using an illegal method (force or guile).

+ Market-based crimes involve production and/or distribution of new goods and services that are inherently illegal. Market-based crimes can be subdivided into those involving evasion of regulations, taxes, and prohibitions.

+ Commercial crimes are committed by otherwise legitimate entrepreneurs, investors, and corporations in a normal business setting. They involve the application of illegal methods to the production of inherently legal goods and services that would otherwise be produced by someone else using legal methods.[36]

Table 1 summarizes the characteristics of primary offenses.[37]

Table 1. Characteristics of Profit-driven Crimes

Type	Transfer of	Basic Act	Method
Predatory	Wealth	Illegal (theft)	Illegal (force or guile)
Market-based	Illegal goods and services	Illegal (trafficking)	Legal (market-exchange)
Commercial	Legal goods and services	Legal (market sale)	Illegal (fraud)

Table 2 shows examples of the three types of profit-driven crimes.

Table 2. Examples of Profit-driven Crimes by Type

Type of Profit-driven Crime	Examples
Predatory crime	Cargo theft, maritime piracy, cyber theft or scams
Market-based crime	Illegal drugs, counterfeiting, human trafficking, excise goods, illegal trade in wildlife
Commercial crime	Avoiding taxes on legal goods, commercial tax evasion, using illegal methods to reduce costs or obtain contracts, violation of regulation on environmental, labor or safety standards, criminal negligence

The advantage of Naylor's classification of profit-driven crimes is that it can accommodate different views of the problem:

+ concerned exclusively with market-based crimes where an illegal form of supply meets an illegal form of demand in order to avoid prohibitions, regulations, or taxations.

+ covers predatory and market-based crimes as long as they have a transnational component.

+ the standard definition of makes it essentially equivalent to illegal markets. A broader definition, such as the one given in the introduction, includes economic crime as well: in this case, a legitimate demand is met by illegitimate forms of supply, involving seemingly legitimate actors who in fact behave illegally. Depending on the understanding of "trade," it may be limited to illicit flows of goods, or it may include people and services as well (for example, gambling), or it may include illicit financial flows.

+ while we have not found an explicit definition, it apparently covers any type of profit-driven crime as long as the opportunity for the crime can be connected to the flows of globalization, defined as the cross-border integration of economic activity.[38] Harmful economic activities are also covered if they would be considered illegal according to Western norms. As there is a fair level of subjectivity in deciding whether an illegal economic activity is related to globalization or not, and even more so in deciding whether a legal economic activity violates Western norms, we will consider for the sake of definition and measurement that deviant globalization covers all profit-driven crimes as long as they are transnational. We note that deviant globalization and the broader definition of illicit trade are quite similar overall.

Table 3 summarizes the different names given to the problem, the underlying view of the problem in a nutshell, the types of offenses, and whether certain conditions apply.

Table 3. Comparison of Problem Definitions

Name of the Problem	Nutshell View of the Problem	Predatory Crimes	Market-based Crimes	Commercial Crimes	Condition
Global Black Markets	Imbalance between supply and demand	No	Yes	No	None
Transnational Organizational Crime	Organized crime benefits from globalization	Yes	Yes	No	Transnational crimes only
Illicit Trade (Narrow)	Imbalance between supply and demand	No	Yes	No	Can be restricted to flows of goods
Illicit Trade (Broad)	Globalization and the legal business infrastructure are enablers of illicit activities	No	Yes	Yes	None
Deviant Globalization	Globalization and the legal business infrastructure are enablers of illicit activities	Yes	Yes	Yes	Transnational

For measurement purposes, the most practical definition is the one used for Global Black Markets, which considers only one category of profit-driven crimes without any condition. The condition of transnationality used for Transnational Organized Crime and Deviant Globalization increases the complexity of the classification and measurement as well. One could take the view that a majority of profit-driven crimes have one or more transnational features, in which case the condition can be essentially ignored. However, this is clearly not true for predatory crimes, and predatory crimes with an obvious transnational component have received much attention lately. While the rest of the chapter essentially focuses on global markets, we note that commercial crimes in particular deserve more attention.

While there are many subtleties to and implications of Naylor's theory that are outside the scope of this chapter, let us make a few remarks. The different categories of offenses can be connected. "In some cases explicitly criminal acts and inherently legal ones are embedded together in a matrix of economic activity to such a degree that the two, while theoretically distinct, are mutually interdependent." For example, illicit markets can be used to monetize the proceeds of predatory crimes, such as cargo theft, and certain offenses may overlap different categories: for example, intellectual property crime involves the misappropriation of intellectual capital by the counterfeiter (predatory crime), the sale of an illegal good (market-based crime), and a commercial crime if the IP-infringing good is sold to an unsuspecting customer. Therefore, a distinction is made between primary offenses, which is the primary act generating the money, and secondary offenses. However, as the complexity of global supply chains allows legitimate business to interact with criminals and more generally to participate in illegal markets, the frontier between market-based and commercial crimes is blurred.

Measuring Impacts

The usual way to measure the global impact of illicit activities is to sum up their respective market size. If the market size is not available, in general any monetary measurement associated with the market will be used. This allows a straightforward comparison between illicit markets, while a simple addition yields a global estimate. One must be cautious though, since as we have seen before, some of the illicit markets that claim the largest market value have the most questionable statistics.

The deeper problem, however, is that market values, even if they are correctly estimated, are only one indicator of the impacts, which does not necessarily provide a fair metric for comparison. Compared to the market in counterfeit goods and illegal drugs, the market in endangered species might be one to two orders of magnitude less, and the market in stolen organs three times smaller. However, one may question whether the harm they are causing is proportionally less significant, and therefore focusing on the illicit market size can be misleading. This is not to say that market size is not a useful figure. As we will see, it is a helpful starting point for measuring the degree of nuisance of organized crime, its corrosive impact on society and institutions, and the overall indirect impacts on trade. However, it is not necessarily a relevant indicator of general social, economic, or environmental harm. Therefore, in pursuing the goal of finding a global measurement of the scale and impact of illicit trade, we should consider adjusted methods for measuring the harm caused by different illicit activities.

Any illicit market can have impacts in multiple categories. For example, counterfeiting creates costs and lost revenues for both businesses and government, causes job loss, and involves health and safety risks. It also creates intangible costs, such as a general loss of trust in institutions, and political tensions between states. And nearly any illicit market involves criminal activities that impact the security of citizens, undermine the authority of states, erode the social fabric, criminalize society, and generate an overall cost of crime that must be borne by society. As the types of impacts can be numerous, it is convenient to use the compact "direct/indirect impacts" typology from UNDOC.[39] It is described in the following way:

> The real threat of organized crime cannot be reduced to the violence associated with criminal markets. Rather, it is best described under two headings: Direct impacts, which are essentially the reasons each criminal activity was prohibited in the first place; Indirect impacts, in particular the ways organized crime as a category undermines the state and legitimate commercial activity.

According to this typology, direct impacts are specific to the direct externalities of a profit-driven crime, excluding the crime-related consequences.

Connecting Illicit Market Size to Indirect Impacts

Crime-related consequences, or indirect impacts, are essentially generic to any illicit market. They include all costs that criminals inflict on society. While these are numerous and include social impacts such as the erosion of social fabric and the criminalization of society, they do not lend themselves to measurement. However, with no pretension to be exhaustive, it is possible to estimate the economic cost of many consequences of criminal activities. A study from the Office of National Drug Control Policy (ONDCP) found that in 2002, the societal costs of illicit drug use in the United States totaled $181 billion.[40] The study divides costs into three categories: healthcare cost, productivity losses, and other effects (such as the criminal justice system), with each category containing multiple categories for which the costs are detailed. Subcategories of costs are also aggregated as being crime-related, or non-crime-related. For example, the category with largest impacts, productivity losses ($128 billion), has the following subcategories: premature death, drug abuse–related illness, institutionalization and hospitalization, productivity loss of victims of crime, incarceration, and crime careers. The first three categories are (essentially) non-crime-related, while the last three are obviously direct consequences of crime. The report shows that of the $181 billion in costs of drug abuse, about $108 billion (59.7 percent) are crime-related impacts.

This number can be put in perspective with an estimate on the market size of $64 billion for the year 2000[41] (we have not found estimates after that year), which come from the same agency. For that year, crime-related costs were $98.7 billion. Combining the two estimates yields the conclusion that each dollar spent on the illegal drug market generated crime-related costs, or indirect impacts, of $1.54 (that is, 98.7 divided by 64). In the absence of in-depth studies on the economic impact of other illegal markets (at least to our knowledge), we will make the assumption that, as a first approximation, this ratio of 1.54, rounded to 1.5, can be used for other illegal markets. In other words, if an illegal market generates revenue of $10 billion, it is in effect having an indirect impact valued at $15 billion.

Applying the Methodology

In this section we apply the proposed methodology to estimate the overall cost of different illicit markets by referencing the data sets and studies we rely on while giving and justifying the assumptions that must be made to infer the estimates.

Illegal Drugs

As a basis for calculation, we will use, instead of the previously mentioned ONDCP 2002 report, a more recent U.S. Government report that used a similar methodology to obtain an estimate of $193 billion.[42] This represents the total cost of drugs in the United States including direct and indirect impacts. Unfortunately, as noted in the ONDCP 2002 report, there have been very few studies on the total impact of drug abuse across the world. The UNODC World Drug reports provide country data on annual prevalence of drug abuse and on treatment demand, but the aim is to estimate the impact of drug abuse in other nations.

A simpler approach is to use gross domestic product (GDP) levels. The cost estimate of $181 billion in 2002 (about $650 per capita) is roughly equivalent to 1.7 percent of GDP. As the United States has a more severe drug abuse problem than most nations, we must use a different percentage for other nations. The ONDCP report cites different studies that obtained a cost estimate of 1.8 percent for the United Kingdom, 1.0 percent for Australia, 0.4 percent for Germany, and 0.2 percent for Canada. If we take as a very rough approximation an average worldwide cost outside the United States of 0.85 percent (the nonweighted average of these four countries), and considering that worldwide GDP outside the United States was $48.5 trillion in 2011, the global cost of drug abuse outside the United States is estimated at $412 billion. The worldwide total cost is therefore estimated at $605 billion.

We may alternatively make the assumption that the national cost of drug abuse is proportional to market size. UNODC's World Drug Report 2005 made a regional breakdown of the global illicit drug, and North America was estimated to account for 44 percent of the world's total drug sales at the retail level. Considering that the United States represents 85 percent of North America's GDP, we make the rough approximation that it accounts for 85 percent of North America's drug sales, and consequently for 37.4 percent of total worldwide drug sales. Therefore, the worldwide economic cost of drugs would be $516 billion (that is, $193 billion divided by 37.4 percent).

As UNODC numbers are more recent and come from a systematic study of worldwide drug abuse, we will keep this estimate for the table. Using the ratio of 59.6 percent from the ONDCP 2002 report for the crime-related, indirect impacts, the economic cost of drugs is split into $308 billion that is crime-related, versus $208 billion that is not crime-related.

Counterfeiting

Many of the economic costs of counterfeiting are difficult to capture. Furthermore, we showed that the current estimates of the global market in counterfeit goods are not reliable. Even if they were, they do not measure the actual illegal market size and revenues of traffickers since counterfeits can be sold at a considerably lower price than genuine goods. A study commissioned by BASCAP (Business Action to Stop Counterfeiting and Piracy from the International Chamber of Commerce)[43] attempts to measure the global social and economic impact of counterfeiting and piracy. However, a fair fraction of the overall estimate builds on the OECD estimate that we consider unreliable. In the absence of any valid number, a lower boundary can be inferred from what companies and institutions pay to protect themselves from counterfeiting: the market size is estimated in 2011 at approximately $15 billion in 2011 for security printing[44] and $2 billion for brand protection devices,[45] for a total of $17 billion. While it is obvious that counterfeiting imposes a much larger cost on society, the lack of reliable data and studies makes it very difficult to derive an overall cost.

We may attempt to develop an estimate of the direct impacts of counterfeit medications. Artesunate is the only affordable and effective medication that can cure parcifarum malaria, a particularly deadly form of the disease found in Southeast Asia. A large survey in the region by the WHO, Interpol (popular name of the International Criminal Police Organization), and a coalition of scientists and health practitioners found that 49.9 percent of medications sold were counterfeit and ineffective.[46] Such data can be used to estimate the number of people who may have died because they have used ineffective counterfeit medications. A report uses a rough estimate of the percentage of counterfeit medications and the probability that untreated sick people will die from a given disease, to infer that for malaria, 250,000 people would die per year from counterfeit medications.[47] The same report estimates that 450,000 people die because of ineffective treatment caused by counterfeit or substandard tuberculosis medication. While we stress that this is a rough estimate and the overall impact of counterfeit and substandard medications is far from being fully understood, those 700,000 deaths remain a reasonable calculation based on the available evidence.

We may infer an economic cost for these deaths by using market estimates for the value of a statistic life (VSL). Typical VSL values in the United States are in the range of $5 million to $12 million (median of $7 million).[48] Although VSL estimates are less well known for low-income countries, an elasticity of 1.0 is commonly used.[49] The GDP per capita in the United States is around $50,000 per year, and the GDP per capita for low-income countries where most of these deaths occur is on the order of $2,000 per year. Therefore, using the median VSL of $7 million in the United States and an elasticity of 1.0, the VSL for low-income countries is estimated at $280,000. For this VSL, the total economic value of these deaths would be $196 billion.

Human Trafficking

As we have seen earlier, human trafficking generates revenue at $32 billion per year. Using our cost of crime ratio of 1.5, this revenue pocketed by organized crime causes indirect impacts valued at $48 billion. Regarding direct impacts, a lower bound would be to consider the revenue of $32 billion of which victims of human trafficking are deprived. But a more accurate economic valuation of the harm might be to consider that victims are deprived of one year of quality life. Recent research shows that one year of quality life is worth on average $129,000 in the United States, that is, 2.58 times the U.S. GDP per capita of $50,000.[50] Assuming that trafficked people are deprived of 100 percent of a year of quality life, and that a year of quality life can be valued proportionally to GDP per capita, a rough economic valuation of those lost years of quality life would be given as follows: $32 billion x 2.58 = $83 billion.

Excised Goods

Each illicit package corresponds to a direct and precisely known loss of tax revenue. Government losses can be directly inferred from statistics on the number of illicit products on the market. This involves an addressable direct impact of $31 billion. The main other direct impact is related to public health: a report on tobacco taxation, using a transparent model that incorporates price elasticity of demand, calculates that if the illicit trade in tobacco was eliminated, consumption would drop by 2 percent. This would translate into 160,000 lives saved per year from 2030 onward. Given the estimated $193 billion per year in healthcare expenditures and productivity loss from smoking in the United States,[51] a 2 percent drop in consumption would be worth close to $4 billion per year. On a worldwide basis, 2 percent of an estimated $500 billion each year in healthcare expenditures, productivity losses, and other costs[52] would be worth $10 billion per year. The total direct impacts are therefore estimated at $41 billion. Regarding indirect impacts, we assume that organized crime captures 50 percent of the $50 billion in lost excise taxes (that is, illicit cigarettes are sold at half the regular price), and therefore makes a revenue of $25 billion. Using the cost of crime ratio of 1.5, we infer a loss of $37 billion.

Environmental Crime

To estimate the indirect impacts, one may add the estimates we reviewed for illegal logging (9 percent of $190 billion, or $17 billion), illegal fishing (20.55 percent of $81.5 billion, or $17 billion), other wildlife trade (roughly $5 billion), and the $11 billion for illegal toxic wastes.[53] Acknowledging that the last two estimates are not based on detailed studies, this places the illegal market size at $50 billion. Using the cost of crime ratio, this results in crime-related impacts of $75 billion.

Calculating direct impacts of human activity on the environment, whether resulting from illicit activities or not, is a daunting task. To obtain an order of magnitude of these impacts, we suggest a generic approach that does not get into the specifics of a given environmental crime. It consists of using a first estimate for the total yearly value brought by ecosystem services, a second estimate for the rate of depletion of these ecosystems, and a third estimate on the proportion of that depletion caused by environmental crimes. The multiplication of these three values provides a rough approximation of the direct permanent losses caused by one year of environmental crimes. Yet ecosystem losses are permanent; therefore, we need to integrate future losses as well. For that we use standard economic techniques for the current value of future losses. More details are given below:

+ Value of ecosystem services: a paper which was very influential in mainstreaming the economics of nature calculated that the annual value of ecosystem services is in the range of $16–54 trillion with an average of $33 trillion per year, or 183 percent of the global GDP.[54] Global GDP was $18 trillion per year at the time of the study, and put in perspective with the current global GDP of $63 billion, this represents an annual value of ecosystem services of $115 trillion.

+ Rate of depletion: While there is no such thing as a universal measurement of the rate of depletion of ecosystem resources, we may use as a first approximation the rate of depletion of forests, which are a fundamental pillar of our global ecosystem. The UN Food and Agriculture Organization publishes an annual report on the state of the world's forests. The report of 2011 states that the annual change rate of forest area was −0.13 percent during the period 2000 to 2010 (versus −0.2 percent for 1990 to 2000).[55] We will make the approximation that this rate of depletion generally applies to other ecosystem services.

+ Our review of various environmental crimes suggests that, according to the market, they represent approximately 10 percent to 25 percent of the total market activity. Based on that, we will consider that illicit trade contributes to 10 percent of the total depletion of ecosystem services. Therefore, each year $115 trillion × 0.13 percent ×

> 10 percent = $14.95 billion of ecosystem services are that permanently destroyed by illicit trade.

- Using a discount factor of 3 percent per year for future losses, the net present value of the total economic losses in the future is 33.33 times this value, or $498 billion.

In the previous calculation, the choice of the discount rate is critical. The Economics of Ecosystems and Biodiversity study states that "There are no purely economic guidelines for choosing a discount rate. Responsibility to future generations is a matter of ethics, best guesses about the well-being of those in future, and preserving life opportunities."[56] Yet a discount factor of 3 percent values the services provided by nature to future generations rather conservatively.

Conflict Resources

When considering civil wars in which resources were used for funding, what matters is certainly not the value of the trade, which would probably be comparatively small anyway, but rather whether a link can be established between the illicit trade and the prolongation of the conflict. If one of the belligerents had no other source of funding, the link can be established and it seems reasonable to assume that the impact of illicit trading is the entire cost of war. If only a fraction of the war budget comes from the illicit trade, then a ratio of the value of the illicit trade to the total war budget can be applied. The costs of a civil war can also be estimated using econometric techniques, which provide a number from which the illicit trade ratio can be applied. Collier and Hoeffler estimate that it takes on average 21 years to get back to the GDP level that would have prevailed without the conflict.[57] The total loss amounts to 105 percent of initial GDP, plus 18 percent for increased military spending. Including costs for neighboring countries as well, the average civil war costs around 250 percent of GDP. With the average GDP of $20 billion for low-income countries affected by civil war, a rough estimate of the cost of a single war is around $50 billion, with a yearly loss of around $2.5 billion. A very rough approximation of the global economic cost caused by conflict resources can be made by enumerating the number of conflicts driven by resources, something we will not attempt here.

Summarizing Impacts

Table 4 summarizes direct and indirect impacts for five illegal markets. Direct impacts are valued at slightly more than $1,043 billion, while indirect impacts are valued at $468 billion. Environmental crimes and illegal drugs have the highest impacts, but their cost structure is very different, as a much higher proportion of the costs for the latter are crime-related. What makes environmental crimes so costly is that they have long-term future effects that are captured in the economic valuation.

Table 4. Impacts of Illegal Markets (billions of dollars)

Illegal Market	Direct Impacts		Indirect Impacts		Total
Illegal Drugs	Non-crime-related economic cost for the United States normalized by its contribution to the global drug market	208	Crime-related economic cost for the United States normalized by its contribution to the global drug market	308	516
Human Trafficking	Global loss of one year of quality life	83	Revenue of human traffickers multiplied by the cost of crime ratio	48	128
Excised Goods	Lost revenue to government plus health costs of illicit trade	41	Revenue of traffickers multiplied by the cost of crime ratio	37	78
Environmental Crimes	Net present value of total future losses in ecosystem services caused by one year of environmental crimes	498	Sum of market estimates multiplied by the cost of crime ratio	75	568
Counterfeits	Counterfeit medications in low-income countries, value of statistical life expenditure for protection against counterfeiting	213	Unknown		213
Total		1,043		468	1,501

Conclusion

After arguing the importance of developing a better estimate of the global scale and impact of illicit trade, this chapter reviewed statistics on some of the major illegal markets, dissected the way they were constructed, and highlighted some of their limitations. The review uncovered different conceptual and empirical issues that must be addressed when defining, measuring, and sampling illegal markets. This helped to prepare the ground for the more ambitious task of developing a global estimate of illicit trade.

What is not properly defined cannot be measured. Most terms used to define the problem, including transnational organized crime, illicit trade, and illicit or deviant globalization, lend themselves to rather vague and open-ended definitions. We proposed a framework adapted from Naylor's theory of profit-driven crimes, in which different definitions or perceptions of the problem can be compared. Illegal markets appear as the common denominator of these different definitions, although we questioned the notion that being transnational is a key defining characteristic.

We then focused on illegal markets and adopted UNODC's direct/indirect classification of impacts. Direct impacts correspond to the fundamental reasons why a given trade is being prohibited, taxed, or regulated. Indirect impacts include effects that are more diffuse or difficult to quantify (for example, erosion of the social fabric and the role of the state). We therefore decided to focus on crime-related effects for which means of quantification exist and that probably integrate at least some part of the intangible costs. Using the results of studies on the economic cost of drug abuse, we inferred that each dollar spent on the illicit drugs generates an overall cost of crime of slightly more than $1.50. As we do not dispose of equivalent studies for other illegal markets, we conjectured that this cost of crime over market size ratio could be applied to other markets. When estimates on market size exist, we can therefore estimate their indirect impact using this cost of crime ratio.

Measuring direct impacts requires making a number of assumptions regarding economic valuation of the harm, although in most cases we can rely on studies and data sets to support them. The method is applied to the five illegal markets studied at the beginning of the chapter, and also for conflict resources although a global estimate was not provided. Going through the calculations illustrated the various choices that need to be made on the data sources, as well as the assumptions that are necessary, especially when data are lacking and must be supplemented by data on a different region or a different illegal market. Furthermore, the data and statistics that were fed into the model count themselves on multiple data sources and assumptions, which in most cases were selected for their scientific legitimacy or because they were coming from the organizations that have the best access to field data. The outcome is an estimated impact on the order of more than $1.5 trillion per year. While this number does not claim precision and relies on spelled-out assumptions, it should be taken as suggestive of the order of magnitude of the impacts. Another observation is that the total market size of the five studied illegal markets used in the application of the model is a bit less than $300 billion. This is five times less than the estimated impacts, indicating that, indeed, the effects of illicit markets matter more than their size.

Acknowledgments

The author would like to acknowledge that much of the intellectual input and stimulus for this chapter comes from discussions held at the World Economic Forum Global Agenda Council on Illicit Trade. Special thanks to Mariya Polner, Sandeep Chawla, Richard Danziger, Helena Leurent, and the editors for reading and providing extremely useful feedback on earlier versions.

Notes

[1] Whack-a-mole is an arcade game in which moles pop up from their holes at random, and the aim of the game is to force the individual moles back into their holes by hitting them directly on the head with a mallet; no matter how adept the player is at whacking the moles, they keep reappearing. The image of a "Whack-a-mole problem" is used when enforcement actions seem to have no lasting impact in addressing the problem. See, for example, *Illicit Tobacco: Various Schemes Are Used to Evade Taxes and Fee*, U.S. GAO Report GAO-11-313 (March 2011), available at <www.gao.gov/new.items/d11313.pdf>.

[2] Bruce Schneier, "The Global Illicit Economy," Schneier on Security blog, September 8, 2009, available at <www.schneier.com/blog/archives/2009/09/the_global_illi.html>.

[3] *Strategy to Combat Transnational Organized Crime: Addressing Converging Threats to National Security* (Washington, DC: The White House, July 2011), available at <www.whitehouse.gov/sites/default/files/microsites/2011-strategy-combat-transnational-organized-crime.pdf>.

[4] "Havocscope Black Market Ranking," and "Havocscope Black Market Country Ranking," Havocscope Global Black Market Information, October 16, 2011, available at <www.havocscope.com>.

[5] Ibid.

[6] Ibid.

[7] Peter Andreas and Kelly M. Greenhill, eds., *Sex, Drugs, and Body Counts: The Politics of Numbers in Global Crime and Conflict* (New York: Cornell University Press, 2010).

[8] Peter Andreas, "Illicit Globalization: Myths, Misconceptions, and Historical Lessons," available at <www.brown.edu/Departments/Political_Science/people/documents/Illicit_Globalization.pdf>.

[9] Ibid., 4.

[10] R.T. Naylor, *Wages of Crime Black Markets, Illegal Finance, and the Underworld Economy* (New York: Cornell University Press, 2005), 4.

[11] Ibid., 5.

[12] Raymond Baker and Eva Joly, "Illicit Money: Can It Be Stopped?" in *Deviant Globalization: Black Market Economy in the 21ˢᵗ Century*, ed. Nils Gilman, Jesse Goldhammer, and Steven Weber, 233 (New York: The Continuum International Publishing Group, 2011).

[13] Asif Efrat, *Governing Guns, Preventing Plunder: International Cooperation Against Illicit Trade* (Oxford: Oxford University Press, 2012).

[14] World Economic Forum, "Global Agenda Council on Illicit Trade 2012," available at <www.weforum.org/content/global-agenda-council-illicit-trade-2011>.

[15] United Nations Office on Drugs and Crime (UNODC), "World Drug Report 2011," available at <www.unodc.org/documents/data-and-analysis/WDR2011/WDR2011-ExSum.pdf>.

[16] UNODC, "World Drug Report 2005," available at <www.unodc.org/pdf/WDR_2005/volume_1_ex_summary.pdf>.

[17] Francisco Thoumi, "The numbers game: let's all guess the size of the illegal drug industry," *The Journal of Drug Series* (2005), available at <www.tomfeiling.com/archive/Thoumi_on_the_Numbers_Game.pdf>.

[18] Ibid.

[19] "Estimating the size of the global drug market: A demand-side approach," *RAND Technical Report TR711*, available at <www.rand.org/content/dam/rand/pubs/technical_reports/2009/RAND_TR711.pdf>.

[20] Ibid., 9.

[21] Ibid., 21.

[22] UNODC, "World Drug Report 2006," available at <www.unodc.org/pdf/WDR_2006/wdr2006_volume1.pdf>.

[23] P. Reuter, "The Mismeasurement of Illegal Drug Markets," *Exploring the Underground Economy* (Kalamazoo, MI: Upjohn Institute, 1996), 63–80.

[24] Ibid., 195.

[25] Business Software Alliance, "Global Piracy Study," available at <http://portal.bsa.org/globalpiracy2010/>.

[26] *Human Trafficking: Better Data, Strategy, and Reporting Needed to Enhance U.S. Antitrafficking Efforts Abroad*, U.S. GAO Report GAO-06-82, available at <www.gao.gov/new.items/d06825.pdf>.

[27] "ILO Minimum Estimate of Forced Labor in the World," International Labor Organization 2005, available at <www.ilo.org/wcmsp5/groups/public/---ed_norm/---declaration/documents/publication/wcms_081913.pdf>.

[28] "A Global Alliance Against Forced Labor," International Labor Organization 2005, available at <www.ilo.org/public/english/standards/relm/ilc/ilc93/pdf/rep-i-b.pdf>.

[29] American Forest & Paper Association, "Illegal Logging and Global Wood Markets: The Competitive Impacts on the U.S. Wood Products Industry," available at <www.illegal-logging.info/uploads/afandpa.pdf>.

[30] "Sustaining Forests: A Development Strategy," World Bank 2004, available at <http://siteresources.worldbank.org/INTFORESTS/Resources/SustainingForests.pdf>.

[31] D. Agnew et al., "Estimating the Worldwide Extent of Illegal Fishing," *PLOS ONE*, available at <www.plosone.org/article/fetchObjectAttachment.action;jsessionid=980C237B77A322A650864BF5A9D3ECDB.ambra01?uri=info%3Adoi%2F10.1371%2Fjournal.pone.0004570&representation=PDF>.

[32] "The Illegal Shipment of Waste Among Impel Member States," *IMPEL-TFS Threat Assessment Project 2005*, available at <http://impel.eu/wp-content/uploads/2010/02/2006-x-Threat-Assessment-Final-Report.pdf>.

[33] U.S. Government Interagency Working Group 2000, "International Crime Threat Assessment," available at <www.fas.org/irp/threat/pub45270index.html>.

[34] "United Nations Convention against Transnational Organized Crime," available at <www.unodc.org/documents/treaties/UNTOC/Publications/TOC%20Convention/TOCebook-e.pdf>.

[35] R.T. Naylor, "Towards a General Theory of Profit-Driven Crimes," *British Journal of Criminology* 43 (2003), 81–101.

[36] Ibid.

[37] Ibid.

[38] Nils Gilman, Jesse Goldhammer, and Steven Weber, eds., *Deviant Globalization* (New York: The Continuum International Publishing Group, 2011).

[39] "United Nations Convention against Transnational Organized Crime," available at <www.unodc.org/documents/treaties/UNTOC/Publications/TOC%20Convention/TOCebook-e.pdf>.

[40] Executive Office of the President, Office of National Drug Control Policy, "The Economic Costs of Drug Abuse in the United States, 1992–2002," available at <www.ncjrs.gov/ondcppubs/publications/pdf/economic_costs.pdf>.

[41] Executive Office of the President of the United States 2001, "What America's Users Spend on Illegal Drugs," available at <www.abtassociates.com/reports/american_users_spend_2002.pdf>.

[42] United States Department of Justice National Drug Center 2011, "The Economic Impact of Illicit Drug Use on American Society," available at <www.justice.gov/ndic/pubs44/44731/44731p.pdf>.

[43] International Chamber of Commerce BASCAP, "Estimating the global economic and social impacts of counterfeiting and piracy," available at <www.iccwbo.org/uploadedFiles/BASCAP/Pages/Global%20Impacts%20-%20Final.pdf>.

[44] "The Future of Global Security Printing to 2013," *Pira International*, available at <www.pira-international.com/The-Future-of-Global-Security-Printing-to-2013.aspx>.

[45] "The Future of Anti-Counterfeiting, Brand Protection and Security Packaging VI," *Pira International*, available at <www.pira-international.com/future-of-anti-counterfeiting-brand-protection-and-security-packaging-vi.aspx>.

[46] "Paul Newton et al., "A collaborative epidemiological investigation into the criminal fake artesunate trade in South East Asia," PLOS Medicine 2008, available at <www.plosmedicine.org/article/info:doi/10.1371/journal.pmed.0050032>.

[47] Julian Harris et al., "Keeping It Real; Combating the spread of fake drugs in poor countries," *International Policy Network 2009*, available at <www.policynetwork.net/sites/default/files/keeping_it_real_2009.pdf>.

[48] National Bureau of Economic Research, "The Value of a Statistical Life: A Critical Review of Market Estimates Throughout the World," available at <www.nber.org/papers/w9487.pdf>.

[49] J.K. Hammit, "The Income Elasticity of the Value per Statistical Life: Transferring Estimates between High and Low Income Population," *Journal of Benefit-Cost Analysis 2011*, accessed at <www.bepress.com/cgi/viewcontent>.

[50] "The Value of a Human Life: $129,000," *Time*, May 20, 2008, available at <www.time.com/time/health/article/0,8599,1808049,00.html>.

[51] *Illicit Tobacco: Various Schemes Are Used to Evade Taxes and Fee.*

[52] Campaign for Tobacco-Free Kids, "Toll of Tobacco Around the World," available at <www.tobaccofreekids.org/facts_issues/toll_global/>.

[53] Berend H. Ruessink and Gerhard J.R. Wolters, "Combating Illegal Waste Shipments Through International Seaports," available at <http://inece.org/conference/9/papers/RuessinkWolters_Netherlands_Final.pdf>.

[54] R. Costanza et al., "The value of the world's ecosystem services and natural capital," *Nature Magazine*, 1997, available at <www.coralreef.gov/mitigation/costanza_et_al_nature_1997.pdf>.

[55] UN Food and Agriculture Organization, "State of the World's Forests," 2011, available at <www.fao.org/docrep/013/i2000e/i2000e.pdf>.

[56] John Gowdy, "Discounting, ethics, and options for maintaining biodiversity and ecosystem integrity," available at <www.teebweb.org/LinkClick.aspx?fileticket=WIzKzKAFsuE%3D&tabid=1018&language=en-US>.

[57] Paul Collier and Anke Hoeffler, *Civil War* (Oxford, 2006), available at <http://users.ox.ac.uk/~econpco/research/pdfs/Civil-War.pdf>.

Part II.
Complex Illicit Operations

Chapter 4

The Illicit Supply Chain

Duncan Deville

In the age of globalization, the increased interconnectedness of the world allows individuals and groups to come together around common causes across national boundaries in order to form networks.[1] No longer restricted by physical, geographical, or political borders, networks increasingly operate outside of traditional government controls. In some instances, these networks form with the purpose of challenging traditional government authority. These challenges are not always bad. Some networks develop for the purpose of promoting the advancement of society. In doing so, they put pressure on governments to address political and social issues that are otherwise ignored or overlooked.[2]

Although some networks can be ultimately beneficial to the state, others have emerged with the capacity—and often even the objective—to be detrimental to it. When the end goal of a network is the advancement of its own parochial agendas at the expense of society, it can become a danger to society and the state. Among the most dangerous networks are those comprised of individuals and organizations engaged in criminal activities. Due to the globalization of the marketplace, illicit products and activities are no longer localized problems confined to individual states. Instead, the scope of criminal activities has expanded to macroeconomic levels, infecting the global marketplace via illicit networks. These networks function outside of the traditional state framework, spreading their operations across multiple organizational and geographical boundaries. Accordingly, illicit networks have no allegiance to any one government, state, or society at large. Instead, these networks are driven by market forces; their loyalty is reserved only for whatever increases their profits.

The damage that illicit networks inflict on states stems from their extensive use of market mechanisms to advance parochial and often antisocial goals at the expense of society and the public good. Crime flourishes in unstable and chaotic environments. Thus, in order to expand unchallenged in the marketplace, the more powerful networks are not satisfied with merely evading government and law enforcement detection. Instead, they seek and encourage direct challenges to—and ultimately the erosion of—state authority. Where these challenges are successful, networks are better positioned to expand criminal enterprises without effective government interference. When they gain enough power, they may threaten the overall stability of the state, its region, and even the world at large.

Operating Beyond the Traditional State Framework

In the past, government and law enforcement responses to crimes such as drug trafficking tended to focus on local criminal organizations with rigid hierarchical structures. In recent years, they have shifted their attention to global crime networks. Although their structures vary, these networks are driven more by market forces than the whims of individuals or specific groups. Therefore, even the successful targeting of individuals or specific groups may not have a significant impact on the larger network. The criminal activities will likely continue as long as the market and its incentives remain intact.[3]

While illicit activities are not typically localized to one state, the network's central operations are usually set up in states where government authority can be easily evaded or undermined. Conflict, postconflict, and underdeveloped countries tend to be preferred locations for illicit network operations. In states currently in some form of ongoing conflict, illicit networks often choose to align themselves with antiregime elements, providing funding to antigovernment or terrorist groups in order to prolong or sustain the conflict environment in which they flourish. For example, opium production and trafficking operations in Afghanistan provide revenue and material support to insurgents in exchange for protection of the traffickers' illicit activities.[4] As a result of this alliance, the continued instability in Afghanistan's primary poppy cultivation regions in the south and southwest has made it difficult for the government and law enforcement to devote any significant effort to containing drug-trafficking operations.[5]

In postconflict and underdeveloped countries with severely weakened government infrastructures, illicit networks may seek to infiltrate the government itself. In the most extreme cases, such target states may be degraded into sovereign criminal enterprises; states captured by narcotics-trafficking networks can devolve into "narco-states." For example, after law enforcement agents made progress in restricting the flow of drugs to the European market through the Caribbean, Latin American drug barons moved their trafficking operations to the West African country of Guinea-Bissau. As one of the poorest countries in the world,[6] Guinea-Bissau provided an ideal location. The weak government infrastructure made it easy for drug barons to infiltrate its ranks to the point where, in April 2010, the United States named two of the country's senior military officials as drug kingpins.[7] According to a senior U.S. Drug Enforcement Administration (DEA) official, "A place like Guinea-Bissau is a failed state anyway, so it's like moving into an empty house. . . . You walk in, buy the services you need from the government, army and people, and take over."[8] The Latin American drug-trafficking organizations did essentially that, effectively transforming the country into the world's first narco-state.[9]

Illicit Networks as Commercial Supply Chains: The Mexican Example

As market-driven enterprises, if government and law enforcement efforts begin to threaten their profits, illicit networks will often take steps to better protect their activities including moving their operations to other states or regions. For instance, prior to 1990, Latin America–U.S. cocaine trafficking activities (production and transport) were centralized in Colombia. However, in the 1990s, Colombians began outsourcing most of their U.S. cocaine business (that is, cocaine smuggling across the U.S. border and sales inside the United States) to Mexican illicit networks, more commonly known as cartels.[10] During this transition, the Colombians retained production in Colombia and moved transport operations completely into Mexico. This change resulted in part from increased U.S.-Colombian law enforcement and military cooperation that began in the 1990s. Colombian cartel leaders especially began to fear extradition to the United States, the fate that befell Gilberto "the Chess Player" Rodríguez (then head of the Cali Cartel) and "Don Diego" Montoya (then head of the Norte Del Valle drug cartel). By shifting transport to Mexicans, the Colombians could keep a lower profile in the eyes of U.S. authorities. Although this new business model resulted in a revenue loss for the Colombians, the enormous profit margins to be made from coca plant growing, processing, and exporting to Mexico still allowed for sizable revenue.

The key lesson from this transition, however, is that while law enforcement agents were successful in curbing the activities of Colombian criminal organizations, cocaine trafficking in Latin America did not cease. As long as the market and its incentives remain, the larger illicit network will find a way to move its products to the marketplace. In this particular case, the major trafficking activities were moved to Mexico, where they continue to the present day.

Since market forces primarily drive illicit networks, the networks function and can be best understood as supply chains. In the traditional commercial world, a supply chain is a system that moves a product from the supplier to the customer. It typically involves people, organizations, processes, technology, and so forth. Wal-Mart is a commonly cited example of a successful traditional commercial supply chain. Its model revolves around keeping prices low for its consumers by purchasing goods directly from suppliers and bypassing all intermediaries. In the case of the Mexican illicit networks, the methodology used to transport the product to the market and the subsequent illicit proceeds back to Mexico is constantly evolving in response to government and law enforcement activities. However, while the methodology may change, the larger supply chain model remains the same. The illicit networks employ their "staff" members to maintain distinct roles within the supply chain in order to enable key processes and functions to operate without interference from law enforcement. As long as these fundamental roles are not disrupted, the illicit network is able to adapt its activities to ensure the supply chain is not disrupted.

Like traditional commercial supply chains, illicit supply chains tend to be secure, redundant, and resilient to disruption. They are often designed with an extensive compartmentalization of operational knowledge throughout the layers of the various organizations involved. Furthermore, the end-to-end supply chain is designed with redundant nodes and simplified roles to limit the potential negative impact that any one individual or group of individuals can have on the success of the overall operation. In the Wal-Mart model, the organization limits negative impact by controlling virtually all aspects of the movement of goods from supplier to customer, bypassing the need to rely on any intermediaries that might slow down the system or add to the cost of its operations. In May 2010, Wal-Mart announced plans to become its own freight forwarder, which would further lower prices in its stores by cutting the cost it took for third parties to transport goods to its warehouses.[11] By examining operations using a supply chain model, we can identify the interrelationships, logistics, and communications used by successful illicit networks. This model not only allows us to examine how illicit networks function, but also leads us to a better understanding of how protection of these illicit supply chains undermines government efforts to contain them, as well as the larger detrimental effect illicit networks have on the state.

The supply chains of global criminal networks are bidirectional; that is, one chain conveys contraband from supplier to demander while a reverse chain conveys payment from demander to supplier. In the case of Mexico, narcotics smuggling—which fuels corruption, weapons trafficking, kidnappings, and violent crime—has risen to disturbing levels over the past few years. Mexican cartels ship indigenous drugs (marijuana, methamphetamine, and others) and, as described above, imported drugs (principally Colombian cocaine) into the United States. According to the 2009 National Drug Intelligence Center (NDIC) *National Drug Threat Assessment*, bulk cash is the primary method used by Mexican drug-trafficking organizations to transport their proceeds from the U.S. market back to Latin America. The U.S. Department of Homeland Security estimates between $18 billion and $29 billion in cash is smuggled from the United States to Mexico annually, nearly all from drug sales.[12] The primary recipients of illicit U.S.-generated gross drug revenue are the nodes along the supply chains of Mexican illicit networks. These proceeds are essential to sustain their operations, purchase additional drugs and supplies, and protect their activities in other ways including bribes to government and law enforcement officials.[13] Without these proceeds, illicit networks would be forced to identify alternative methods to fund their operations, greatly reducing their power. The movement of bulk cash across the border is arguably the most important part of the Mexican illicit network supply chain.

Cash makes up the majority of drug revenue for obvious reasons; it is convenient and virtually untraceable. Cash is the preferred method of payment for the product on the street for the same reasons. To a large extent, cash facilitates an illicit network's operations, making it more difficult for governments to detect and track the flow of illicit funds. Reliance on cash allows illicit networks to function in a shadow economy, operating in an unregulated sphere outside the global marketplace. In 2010, the Mexican government took steps to combat illicit networks' use of cash to fund operations by introducing new regulations that limited the amount of U.S. currency allowed to be deposited into Mexican financial institutions. However, it is too early to determine whether such regulations will have a significant impact on drug-trafficking operations. Prepaid access cards have emerged as a potential alternative to both transporting and laundering money, and a 2010 report by the Department of Homeland Security noted that these cards might offer more advantages than cash due to their small size and lack of reporting requirements. As of this writing, however, neither Mexican nor U.S. law enforcement agents have observed wide use of prepaid cards by Mexican illicit networks.[14]

Moving such large amounts of bulk cash poses a complex logistical challenge for illicit networks. To transport this cash, network managers compartmentalize the activities, assigning discrete management responsibilities to different cells within the supply chain. Each cell has a fixed role or function it must perform that is integral to the overall success of the network. Roles include consolidating and counting proceeds at centralized counting houses, packing cash into vehicles, moving cash across the border, providing intelligence related to law enforcement whereabouts, and so forth.[15]

For cash consolidation and the U.S. transport phase of the supply chain, network members take the drug proceeds to a regional centralized counting house. These houses are generally located in major U.S. cities such as Atlanta, Chicago, Boston, and Los Angeles.[16] In October 2011, Federal agents seized $200,000 in cash from a network of stash houses in Phoenix operated by members of the Sinaloa Cartel.[17] During this phase, in addition to money counting, members typically convert smaller bills to $100 or $50 notes to reduce the bulk of the cash.[18] The money remains in the counting house until it is ready to be transported across the border.

The next part of the supply chain is the preparation of the cash for transport into Mexico. After being converted to larger bills, the cash is vacuum-sealed into plastic bags and concealed in vehicles.[19] Illicit networks typically conceal cash in the wheel walls, panels, and spare tire compartments. Sometimes cash is hidden in tractor-trailer trucks, which are also used to transport drugs into the United States from Mexico.[20] In March 2011, during a routine traffic stop, Indianapolis police seized $500,000 from the hidden compartment of a truck's ceiling. The cash was bound for a truck stop in McAllen, Texas.[21] Drivers typically deliver the cash to a safe house along the border (as was the case in March 2011) or drive directly into Mexico.

The next node in the illicit cash supply chain is transporting proceeds across the border from the United States back to Mexico. To guard against seizure by Federal agents, network members frequently rotate cash transport vehicles. They also only ship a portion of their proceeds in any vehicle. Most vehicles will only carry between $150,000 and $500,000. Therefore, if a vehicle is stopped and cash is seized, the loss is not sufficient to render the overall operation unprofitable.[22] The network typically has a group of rotating drivers responsible for driving the cash to a designated border town close to the drug-trafficking center of its organization.[23] Times and locations of border crossings are chosen based on intelligence provided by lookouts (called *halcones*, or hawks). These halcones are responsible for carefully watching border traffic to spot any unusual patterns or searches. If they observe anything unusual, they will reroute the vehicles to another crossing.[24] The illicit networks hold the advantage in this step of the chain since U.S. law enforcement has primarily focused on stopping the northward flow of contraband. According to U.S. and Mexican officials, although 10 percent of southbound vehicles are supposed to be stopped for a secondary screening, the actual number is far less.[25] In a March 2011 report, the Government Accountability Office noted that the U.S. Customs and Border Protection (CBP) seized $67 million in illicit bulk cash at the border over the past 2 years.[26] Since NDIC estimates between $18 and $29 billion in cash is smuggled out of the country each year, the amount seized by U.S. officials is quite low at less than 1 percent. According to the same report, CBP efforts are limited by lack of staffing, effective infrastructure, and technology.[27] As one U.S. official in El Paso admitted, "We are simply not configured to deal effectively with southbound traffic."[28]

After the cash has been moved into Mexico, the final link in the supply chain is to break it down into smaller quantities to meet payroll, pay bribes, and compensate Colombian suppliers. This is estimated to take about half of the cash. The remainder is then integrated into the financial system using a variety of tactics. Often, the bulk cash is deposited into the Mexican banking system through *centros cambiarios* (currency exchange businesses), banks, retail stores, casinos, real estate purchases, and a variety of other money-laundering techniques. When needed, funds are repatriated back to the U.S. banking system through correspondent banking and banknote sales to U.S. institutions.[29] Cartel accountants supervise this process. One common tactic used by illicit networks is to launder drug proceeds through a system called the Black Market Peso Exchange (BMPE). In the BMPE, drug revenues are smuggled back into Mexico to be exchanged for pesos at a discounted rate. The peso brokers then use the dollars to buy products in the United States and ship them back to purchasers in Mexico. For example, in November 2011, U.S. Federal agents arrested Vikram Datta of Laredo, Texas, for laundering millions of dollars for the Sinaloa Cartel through his perfume business as part of a BMPE. Datta sold millions of dollars worth of perfume to corrupt buyers in Mexico since 2009.[30] According to prosecutors, he told an undercover agent that his responsibility was "just washing the whole money."[31] Over a 2-year investigation, Datta and other coconspirators deposited more than $25 million in American currency in dozens of bank accounts.[32]

These supply chain steps and roles are fairly standardized; however, the methods used to conduct the required activities change frequently. One example of their adaptability is their increasing use of emerging communication technologies (for example, Voice Over Internet Protocol, satellite technology, cell phones, and two-way radios) in the intelligence step of the chain. [33] By utilizing new modes of communication, cell members can anticipate law enforcement efforts and adjust their activities. In a 2010 interview, a U.S. official admitted that the cartels "have very good intelligence on our operations. We are always one step behind."[34] This advantage is largely rooted in the networks' reliance on the supply chain model to effectively organize and preempt threats to their chains.

In the case of Mexican illicit networks, the collection and movement of bulk cash drug proceeds function as a supply chain that operates in opposition to both the Mexican and the U.S. governments. By adopting and adapting the structure of commercial supply chains for their own illicit activities, illicit networks have been able to position their operations in opposition to the state.

Damage to the State

Mexican illicit networks have largely replaced their Colombian counterparts as the primary drug-trafficking organizations in Latin America. As a result, Mexican cartels have enjoyed a significant increase in their profit share. To continue increasing profits, these networks have devoted greater and greater resources to protecting their market share and supply chain. For the supply chain to function effectively, each process step or role must be protected by the network. In the case of Mexico, enforcement typically takes the form of violence and/or corruption— both of which severely undermine governance and the stability of the state.

Over the past 5 years, there has been a significant escalation in drug-related violence in Mexico. In 2010 alone, there were 11,583 drug-related murders (compared to 6,587 in 2009).[35] In the past 4 years, officials estimate that almost 40,000 have been killed in connection to drug-trafficking operations.[36] This escalation may be strongly linked to the government's efforts to curb illicit network activities.[37] As market-driven enterprises, Mexican illicit networks have a strong incentive to protect their profits. Violence is an effective means to that end. The majority of drug-related murders stem from conflicts over control of smuggling routes (both cash and product) between rival cartels.[38] Any progress law enforcement agencies make in targeting smuggling routes typically leads to increased competition—usually violent—among illicit networks for control over the remaining routes. Mexican illicit networks increasingly resort to military tactics, utilizing heavy weaponry such as sniper rifles, grenades, and rocket-propelled grenades in attacks on rival cartel members as well as government and law enforcement officials.[39]

Government and law enforcement officials are also increasingly being targeted in retaliatory killings. For example, in June 2009, 12 federal police agents were tortured and killed and their bodies dumped in retaliation after the Mexican police arrested a high-ranking member of a cartel.[40] Such activities demonstrate the lengths that illicit networks are willing to go to protect their market interests. When avoiding government detection is no longer an option, illicit networks openly challenge government efforts to contain them, as is occurring in Mexico today.

Violence is not the only tactic utilized by Mexico illicit networks to protect the effectiveness of the supply chain. Another way they ensure their activities are successful is infiltrating and corrupting the government itself. Much of this corruption is still enforced by violence. Many customs agents may collaborate with these networks at the border out of fear for their safety. Network agents have also succeeded in infiltrating local law enforcement units. In 2010, the Mexican government dismissed 3,200 members of its police force (10 percent of the total) due to corruption-related concerns.[41] The city of Torreon fired 1,200 police officers in 2010 once it was discovered that the city's entire force had been infiltrated by cartels. According to Mayor Eduardo Olmos, cartel members "bribed, threatened and recruited [the police] and were able to use their radios, vehicles, weapons, bulletproof vests, everything. . . . The police relaxed their ethics and discipline and just gave in. In the end they weren't working for them. They *were* them."[42] The more powerful Mexican illicit networks have expanded their scope beyond drug trafficking. In recent years, cartels have increasingly engaged in human smuggling and trafficking, prostitution, and kidnapping. Such activities allow illicit networks not only to further utilize their existing drug trafficking supply chain but also to further increase their political influence as well as their profits.

U.S. and Mexican law enforcement officials have observed that drug cartels are increasingly moving into human-smuggling operations, forcing immigrants to act as "mules" to transport drugs and cash proceeds across the border.[43] According to the United Nations Office on Drugs and Crime (UNODC) 2010 transnational crime threat assessment report, 90 percent of migrants who are smuggled into the United States have the help of a professional smuggler.[44] While data are not available regarding the percentage of these migrants who are aided by cartels, the report notes that, since cartels control the majority of Mexican border towns, there is the high likelihood that a significant portion of these migrants must rely on the cartels to cross the border. Working with cartels, however, does not come without a price. Many illicit networks take advantage of the vulnerability of migrants by forcing them to work as a part of the drug-trafficking network. Even migrants who do not directly choose to work with the cartels may still find themselves the victims of cartel brutality. Approximately 18,000 migrants are kidnapped each year with the aim of forcing them into servitude or extorting ransom payments from relatives.[45] When migrants do not comply with the cartels, they are often executed, which was the case in August 2010 when 72 migrants were allegedly killed by armed gangs in Tamaulipas.[46]

In addition to using human smuggling in support of their drug-trafficking operations, cartels have launched separate human-trafficking and prostitution enterprises to diversify their business and create additional profit lines. In July 2011, a U.S. Government official pointed to this shift in the cartels' original business model, noting, "They realize [human trafficking] is a lucrative way to generate revenue, and it is low-risk."[47] With government efforts still largely focused on drug trafficking, human-trafficking convictions in Mexico remain rare.[48] The diversification of cartel activities in Mexico reveals just how extensive their influence and power have become. Their drug-trafficking activities, which still fund the majority of their operations, are so well protected that they can expand into other illicit activities without fear of state interference.

Both violence and corruption have obvious impact on governance and the stability of a state. Regardless of how they choose to protect their activities, illicit networks are clearly undermining government authority across Mexico. One result is the growing exodus of Mexicans from their country to escape the increasing drug-related violence. Since 2006, an estimated 125,000 middle- and upper-class Mexicans have fled their country for the United States.[49]

Collateral Costs

The damage inflicted by illicit networks can extend far beyond the borders of the state in which their main operations or headquarters are located. Although the damage may be greatest in the home state, in the age of globalization, no state is immune from criminal networks no matter where they are located. The convergence of criminal organizations to form illicit networks is a threat not only to political and economic stability in individual states, but also to stability across the globe.

While the most dramatic impacts of illicit network supply chains and operations in Mexico are felt in Mexico itself, their impacts are increasingly being felt in other states of the region, including the United States. In addition to targeting Mexican agents and police officials, illicit networks are often working to corrupt, and more frequently targeting, U.S. law enforcement agents at the border. At a Senate subcommittee hearing in June 2011, a CBP official testified that 127 U.S. customs agents had been indicted for crimes including cocaine trafficking and money laundering since 2004.[50] In 2009, U.S. Customs Agent Luis Alarid was sentenced to prison after it was discovered that he had accepted over $200,000 in cash along with vehicles and electronics from cartel members in exchange for allowing vehicles carrying drugs to cross the border.[51]

In addition to corrupting customs agents, Mexico's illicit networks have used violence against U.S. law enforcement officials and citizens to protect their activities from U.S. intervention, although so far not as much—at least not yet—as many fear. The number of U.S. law enforcement and citizen deaths and disappearances related to Mexico's drug trafficking has increased in the past 5 years. According to State Department figures, "at least 106 U.S. residents were victims of 'executions' or 'homicides' directly related to drug battles in Mexico in 2010, compared to 79 in 2009 and 35 in 2007."[52] U.S. border towns have seen a sharp increase in kidnappings and home invasions over the past few years. Moreover, Mexican illicit networks have begun to directly target U.S. law enforcement agents who threaten to impede their activities, instructing their members "to shoot and kill American border agents using AK-47 assault rifles."[53] These tactics were on display in February 2011 when one U.S. immigration agent was killed after cartel members targeted and attacked his vehicle while he was driving on assignment in Mexico.[54]

Illicit networks may threaten stability in multiple states when they adapt their operations in response to law enforcement activities. For example, in response to the aggressive crackdown on drug trafficking by the Mexican and U.S. governments in Mexico itself, Mexican illicit networks have begun to move some of their operations into Central American countries where law enforcement and government capacity is substantially weaker.[55] Honduras is one country that has been impacted by the shift. The 2009 military coup there facilitated the movement of some cartel activities into that country. The movement to Honduras was largely successful. Honduras was placed on the U.S. list of major illicit drug transit or major illicit drug producing countries for the first time in 2010.[56]

Cooperation between Mexican cartels and Middle Eastern terrorist networks has recently become an area of concern. In testimony before Congress, the former chief of operations and intelligence for the DEA stated:

> *Collaboration between Latin American drug cartels and groups such as Iran's Quds Force and the Islamic terror group Hezbollah is growing far faster than most policy-makers in Washington, DC, choose to admit. This increasing cooperation means that Iran's al-Quds Force and its proxy Hezbollah have more opportunities to leverage the transportation, money laundering, arms trafficking, corruption, human trafficking and smuggling infrastructures of the Colombian and Mexican drug trafficking cartels.*[57]

Similarly, former Ambassador to the Organization of American States Roger Noriega stated, "There are two parallel terror networks growing at an alarming rate in Latin America. One is operated by Hezbollah and its collaborators and the other is managed by a cadre of Quds operatives."[58] For example, in September 2011, a Mexican cartel was accused of cooperating with an Iranian Revolutionary Guard member plot to assassinate the Saudi ambassador in Washington. Earlier that year, Manssor Arbabsiar, working with Iranian Revolutionary Guard member Gholam Shakuri, through friends and relatives allegedly approached a member of the Zetas Cartel (who was actually a DEA informant). Arbabsiar and the informant are alleged to have worked out a deal under which Arbabsiar was to pay $1.5 million to the Zetas to kill the Saudi ambassador at a Washington restaurant.

In December 2011, a Lebanese drug kingpin was charged with working with the Mexican Zeta Cartel and Hizballah to launder large amounts of cash. Ayman Joumaa (also known as "Junior") and his associates allegedly shipped an estimated 85,000 kilograms of cocaine into the United States and laundered more than $850 million in drug money coming out of Mexico from the Zeta Cartel through front companies and the Lebanese Canadian Bank. The two groups cooperated: Hizballah received cartel cash and protection in exchange for expertise in money laundering and explosives training for the Zetas.

The White House unveiled its *Strategy to Combat Transnational Organized Crime* in July 2011. The report underscored the seriousness of the risk that illicit networks pose to international security, noting the damage that networks are capable of if they gain enough power to penetrate or corrupt state institutions. To guard their market interests, illicit networks "insinuate themselves into the political process" through bribery and/or infiltrating financial and security industries.[59] In doing so, these networks "build alliances with political leaders, financial institutions, law enforcement, foreign intelligence, and security agencies . . . exacerbating corruption and undermining governance, rule of law, judicial systems, free press, democratic institution-building, and transparency."[60]

Conclusion

The challenge that increasingly global and diversified criminal networks pose to states and to the international system of states stems from the extensive and successful use of market mechanisms to serve parochial and antisocial ends at the expense of society and the public good. These networks function outside of the traditional state framework, working often in direct opposition to governments to protect their profits. When these networks gain sufficient power, they can infiltrate and corrupt governments. Meanwhile, states do not in most cases have the tools necessary to protect themselves from these corrupting agents. As UNODC spokesman Walter Kemp asked in February 2010, "How can an international system, created to deal with tensions between states, confront non-state actors who become rich, powerful, and dangerous by respecting neither laws nor borders?"[61]

To successfully counter the spread and cancerous impacts of illicit networks, governments must cooperate. As modern markets transcend individual state boundaries, so illicit networks operate increasingly *beyond* individual state boundaries and cannot be defeated ultimately within any individual state. Eradicating criminal organizations in one state is simply not enough to contain the larger illicit networks and their activities. Instead, law enforcement agencies must work together across state boundaries, sharing information and resources to break down supply chains and cut off network access to the global marketplace.

Acknowledgment

The author thanks Rachele Byrne for her assistance in preparation of this chapter.

Notes

[1] National Intelligence Council, *Global Trends 2025: A Transformed World* (Washington, DC: Government Printing Office, November 2008).

[2] Ibid.

[3] Ibid.

[4] Department of State, *International Narcotics Control Strategy Report 2011* (Washington, DC: Department of State, 2011).

[5] Ibid.

[6] In 2010, Guinea-Bissau was ranked 175 out of 177 countries on the United Nations Human Development index.

[7] "U.S. Names Two Guinea-Bissau Military Men 'Drug Kingpins,'" BBC News, April 9, 2010, available at <http://news.bbc.co.uk/2/hi/africa/8610924.stm>.

[8] Ed Vulliamy, "How a Tiny West African Country Became the World's First Narco-State," The Guardian, March 9, 2008, available at <www.guardian.co.uk/world/2008/mar/09/drugstrade>.

[9] Kevin Sullivan, "Route of Evil," The Washington Post, May 25, 2008, available at <www.washingtonpost.com/wp-dyn/content/article/2008/05/24/AR2008052401676.html>.

[10] The term cartel is used commonly to describe Mexican narcotics illicit networks, and this practice is adopted herein. In actuality, a cartel is a combination of independent organizations that unite to limit competition or fix prices (for example, the Organization of the Petroleum Exporting Countries). Mexican cartels are ruthlessly fighting one another for control of drug supply routes, so technically the term is incorrect. The term drug-trafficking organization also falls short in that it fails to capture the other criminal activities the organization sometimes pursues (for example, human trafficking and prostitution).

[11] Chris Burritt, "Why Wal-Mart wants to take the Driver's Seat," Bloomsburg Businessweek, May 2010, available at <www.businessweek.com/magazine/content/10_23/b4181017589330.htm>.

[12] National Drug Intelligence Center (NDIC), National Drug Threat Assessment 2009, Product No. 2008-Q0317-005 (Washington, DC: Department of Justice, 2008).

[13] Douglas Farah, "Money Laundering and Bulk Cash Smuggling: Challenges for the Merida Initiative," Working Paper on U.S.-Mexico Security Cooperation, The Woodrow Wilson International Institute for Scholars, Washington, DC, May 2010.

[14] Ibid.

[15] Ibid.

[16] Ibid.

[17] "70 tied to Mexico drug cartel busted in Arizona," Reuters, October 31, 2011, available at <www.msnbc.msn.com/id/45100583/ns/us_news-crime_and_courts/t/tied-mexico-drug-cartel-busted-arizona/>.

[18] Farah.

[19] Ibid.

[20] Ibid.

[21] Charlie Morasch, "Indiana drug seizure reveals Mexican cartels' infiltration in states," Land Line, October 28, 2011.

[22] Farah.

[23] Ibid.

[24] Ibid.

[25] Ibid.

[26] Government Accountability Office, Testimony Before the Senate Caucus on International Narcotics Control, "Moving Illegal Proceeds: Opportunities Exist for Strengthening the Federal Government's Efforts to Stem Cross-Border Currency Smuggling," March 9, 2011.

[27] Ibid.

[28] Farah.

[29] See Drug Enforcement Administration (DEA) Web site.

[30] Julian Aguilar, "Money-Laundering Case Speaks to Border Fears," The New York Times, November 12, 2011, available at <www.nytimes.com/2011/11/13/us/a-money-laundering-case-in-laredo-speaks-to-the-fears-of-texas-leaders.html?pagewanted=1>.

[31] Ibid.

[32] Ibid.

[33] Farah.

[34] Ibid.

[35] Department of State.

[36] "Mexico town's police force resigns over drug attacks," BBC News, August 4, 2011, available at <www.bbc.co.uk/news/world-latin-america-14411672>.

[37] Ibid.

[38] NDIC, *National Drug Threat Assessment 2010*, Product No. 2010-Q0317-001 (Washington, DC: Department of Justice, February 2010).

[39] Department of State.

[40] United Nations Office on Drugs and Crime (UNODC), *The Globalization of Crime: A Transnational Organized Crime Threat Assessment, 2010* (Vienna: UNODC, 2010).

[41] "Mexico fires thousands of police to combat corruption," Reuters, August 10, 2010, available at <www.reuters.com/article/2010/08/30/us-mexico-drugs-police-idUSTRE67T52D20100830>.

[42] Rory Caroll, "Mexico's drugs war: in the city of death," *The Guardian*, September 10, 2010, available at <www.guardian.co.uk/world/2010/sep/16/mexico-drugs-war-massacre-in-torreon>.

[43] Josh Meyer, "Mexico's Drug War: Drug Cartels Raise the Stakes on Human Smuggling," *The Los Angeles Times*, March 28, 2009.

[44] UNODC.

[45] Human Rights Watch, *World Report 2011* (New York: Seven Stories Press, 2010), available at <www.hrw.org/world-report-2011>.

[46] Ibid.

[47] Anne-Marie O'Connor, "Mexican cartels move into human trafficking," *The Washington Post*, July 27, 2011, available at <www.washingtonpost.com/world/americas/mexican-cartels-move-into-human-trafficking/2011/07/22/gIQArmPVcI_story.html>.

[48] Ibid.

[49] "Affluent Mexicans Flee Violence, Move to United States," *The Dallas Morning News*, August 9, 2011, available at <www.pittsburghlive.com/x/pittsburghtrib/news/nation-world/s_750616.html>.

[50] "U.S. goes after agents who moonlight for Mexican cartels," *Globalpost*, July 10, 2011, available at <www.globalpost.com/dispatch/news/regions/americas/mexico/110708/us-border-agents-cartels-corruption>.

[51] Randall C. Archibald, "Hired by Customs, but Working for Mexican Cartels," *The New York Times*, December 16, 2009, available at <www.nytimes.com/2009/12/18/us/18corrupt.html>.

[52] Sevil Omer, "More Will Die: Mexico Drug War Claims U.S. Lives," *MSNBC.com*, April 22, 2011, available at <www.msnbc.msn.com/id/42232161/ns/us_news-crime_and_courts/t/more-will-die-mexico-drug-wars-claim-us-lives/>.

[53] Ibid.

[54] Barry Leibowitz, "Jaime Zapata, U.S. Immigration Agent, Shot Dead in Mexico in Apparent Ambush," CBS News, February 16, 2011, available at <www.cbsnews.com/8301-504083_162-20032296-504083.html>.

[55] Randall C. Archibald, "Drug Wars Push Deeper into Latin America," *The New York Times*, March 23, 2011, available at <www.nytimes.com/2011/03/24/world/americas/24drugs.html?pagewanted=1&_r=1&ref=americas>.

[56] Ibid.

[57] Michael A. Braun, former chief of operations and intelligence for DEA, testifying before the House Foreign Affairs Committee, February 2, 2012.

[58] Ambassador Roger F. Noriega, visiting fellow at the American Enterprise Institute for Public Policy Research, testifying before the House Committee on Homeland Security's Subcommittee on Counterterrorism and Intelligence, July 7, 2011.

[59] *Strategy to Combat Transnational Organized Crime: Addressing Converging Threats to National Security* (Washington, DC: The White House, July 2011).

[60] Ibid.

[61] Walter Kemp, "Organized crime: a growing threat to security," *SIPRI.org*, February 10, 2010, available at <www.sipri.org/media/newsletter/essay/feb10>.

Chapter 5

Fixers, Super Fixers, and Shadow Facilitators: How Networks Connect

Douglas Farah

The multibillion-dollar illicit trade of commodities (from cocaine to blood diamonds, weapons, and human beings) has many complexities, from creating or extracting the product, to moving the product to international markets, to delivering payments. The return cycle is equally complex including the types of payments used to acquire the commodities, from cash to weapons and other goods, the seller may need. This cycle relies on a specific group of individuals who act as facilitators in connecting different facets of the criminal and/or terrorist networks of state and nonstate actors. This chapter addresses this crucial role of a cohort of actors—"fixers," "super fixers," and "shadow facilitators"[1]—in empowering social networks that operate within illicit commodity chains.

These commodity chains often span significant geographic space and require multiple steps, in multiple countries, to be successfully completed. One individual, or even one criminal and/or terrorist group, seldom has the capacity to operate throughout this complex landscape. Instead they must turn to specialized individuals, often primarily motivated by economic incentives rather than ideology, who can navigate specific links in that chain. These individuals are the crucial bridges among different worlds that do not usually overlap. The often isolated and relatively unsophisticated procurers of commodities (for example, coca leaf growers and cocaine laboratory operators in The Chapare, in central Bolivia, or warlords who control lucrative mines in the Democratic Republic of the Congo) do not have direct access to the markets for those commodities in Europe, the United States, and elsewhere. Those who can move the products often have no direct access to the money laundering, procurement, and transportation networks for the profits for those commodities. Payments are made not only with cash, but also with sophisticated weapons, chemicals, and other materials that need to be transported back to the source region or elsewhere.[2]

This chapter looks at two case studies of the "fixer" chain, using a model that has applicability in many parts of the world in building an understanding of illicit networks. The value of identifying these players is that this serves to identify crucial points of vulnerability in disrupting multiple criminal activities simultaneously because the super fixers and shadow facilitators usually deal with more than one criminal network at a time. Eliminating them or removing them from operations can not only wound several networks with one blow, but also offer insight into the operations of multiple networks from an intelligence perspective. Evidence of this is that of the 43 foreign terrorist organizations listed by the Department of State, the Drug Enforcement Administration (DEA) states that 19 have clearly established ties to drug-trafficking organizations (DTOs), and many more are suspected of having such ties.[3] As discussed below, in many of these cases the networks rely on the same super fixers and/or facilitators.

The first case is the relationship of Russian weapons trafficker Viktor Bout with Charles Taylor in Liberia, and the commodities-for-weapons trade in a highly criminalized state that had a significant impact on the conflicts in Sierra Leone and Liberia. The second is the enduring Cold War network of weapons supplies and suppliers consisting of Communist Party operatives in El Salvador and the Revolutionary Armed Forces of Colombia (FARC) guerrillas, which is now primarily a commercial rather than ideological enterprise. This network enjoys growing support from the rapidly criminalizing Bolivarian states in the region, primarily Nicaragua and Venezuela.[4]

The term *criminalized state* refers to states where the senior leadership, on behalf of the state, is aware of and involved with—either actively or through passive acquiescence—transnational criminal enterprises; where transnational organized crime (TOC) groups are used as an instrument of state power; and where levers of state power are incorporated into the operational structure of one or more TOC groups. The benefits may be for a particular political movement, theocratic goals, terrorist operations, or personal gain of those involved, or a combination of these factors.[5] Few states are wholly criminalized. Most in this category operate along a continuum. At one end are strong but "criminal" states, with the state acting as a TOC partner or an important component of a TOC network. On the other end are weak and "captured" states where certain nodes of governmental authority, whether local or central, have been seized by TOCs, who in turn are the primary beneficiaries of the proceeds from the criminal activity—but the state, as an entity, is not part of the enterprise.

This differs in important ways from the traditional look at "weak" or "failed" states, which assume that a government that is not exercising a positive presence and fulfilling certain basic functions (public security, education, infrastructure) does not represent a functioning state. In fact, such states can be highly efficient at the functions they choose to perform, particularly if they participate in an ongoing criminal enterprise. By choice, their weakness exists in the fields of positive state functions that are traditionally thought of as essentially sovereign functions, but not in other important areas.

One can understand the complexities of illicit trade best by viewing the pieces or nodes as part of a series of recombinant chains with links that can merge and decouple as necessary, rather than by looking at the purchase or exchange of commodities for cash or other goods as a series of individual transactions.[6] Many illicit goods pass through the same physical space, same border crossings, and same specialized groups of handlers for specific jobs. A ton of cocaine or a load of AK-47 assault rifles or 100 Chinese being smuggled to the United States will move through the same pipeline and the same choke points when moving from Latin America across the southern border of the United States.

The flow of goods and cash for those goods is not linear, but circular (figure 1). Nor is the flow usually limited to a single commodity. In Liberia, for example, timber and diamonds flowed out through interrelated networks to different markets, while weapons, uniforms, food, and fuel flowed back through a trusted group of super fixers.

In the cocaine trade, the drugs flow from South America to markets in the United States and Europe, often through the same channels as illicitly moving human beings, contraband, and other drugs such as marijuana. The return flow brings bulk cash, weapons, precursor chemicals, aircraft, and other products necessary to keep the business functioning.

The transient and sometimes fungible nature of these pipelines means they are often in flux and easily rerouted when obstacles arise in any part of the chain.[7] This makes them almost impossible to track at ground level in real time. Investigations usually offer a snapshot of what has taken place rather than a moving picture of how they are evolving.

Figure 1. Circular Flow of Goods and Cash

The "Fixer" Chain

Illicit networks often develop in times of conflict or in the absence of a positive state presence,[8] where multiple porous borders and disdain for the often predatory and/or corrupt state have led to smuggling routes that have often endured for generations. These historic routes, in turn, engender the accompanying "cultures of contraband"[9] —particularly in border regions—that often lead to violence and the acceptance of illicit smuggling activities as a legitimate livelihood. For example, one of the primary routes to move cocaine across Honduras into El Salvador for onward movement to Mexico is controlled by the *Cartel de Texis*, named for the town of Tex-istepeque, where some of its leaders come from. Its operational territory along the Honduran border includes the once famous *ruta del queso* or "cheese route" used to smuggle Honduran cheese into El Salvador at the turn of the 20th century, a smuggling route that has endured for a hundred years. [10]

Within specific geographic spaces, usually controlled by nonstate armed actors, producers or extractors of illicit commodities rely first on "local fixers," often traditional local elites or outsiders with significant business holdings in local areas, to supply the criminal state or nonstate armed actor with connections to the market and financial networks needed to extract and sell the commodity out of the area of production. [11] For example, in Sierra Leone, the Revolutionary United Front (RUF) relied on Lebanese diamond traders near their zones of control and authorized by Charles Taylor in neighboring Liberia to collect and pay for the diamonds that were mined. [12] Demonstrating the enduring nature of these networks, the structure of this relationship between local miners and Lebanese diaspora diamond-smuggling networks was described by Graham Greene in his book *The Heart of the Matter*. [13]

These local fixers knew the terrain and how to operate in conditions that would have been extremely hostile to outsiders, but they knew little beyond the world they inhabited. So they, in turn, had to rely on "super fixers" to move the diamonds to the international market. As will be examined below, Taylor had a small core of trusted super fixers who could move in the broader world, operate bank accounts, and seek out connections but did not have the capacity to actually provide and move the products he wanted.

For this they rely on international "shadow facilitators" who can move weapons and commodities, launder money, and obtain the fraudulent international documents such as end user certificates, passports, business registrations, shipping licenses, and other needed papers.

Many of the local social networks are surprisingly fungible *and* durable because they offer services critical to any incoming regime and are usually nonideological in their network-building. Driven largely by economic imperatives, the groups have proven skillful at adapting to new political realities and exploiting them. They flourish even in times of violence because they offer services that are essential in moving the commodities to market and ensuring in return that the regime (or nonstate actor) acquires the resources to enrich itself and maintain its government (or nonstate) hold on power in specific regions. These geographic regions necessarily include the "honey pot" regions, or regions where illicit goods such as cocaine or heroin can be produced or commodities such as diamonds and timber can be extracted. That honey pot is crucial for financing the ongoing conflict and for geographic control. [14]

Although usually not broadly known outside of local areas of operation, the internal balance of political power often shifts when a state or armed group loses control of its local fixer networks or the networks challenge the state for power. These networks are often made up of diaspora communities who remain at least somewhat separate from the local population and maintain strong ties to their homeland. While they often cannot aspire to direct political power in the countries where they live, they often, because of their external contacts, control key import-export sectors of the economy and have a range of contacts that are coveted and necessary to the regime. [15]

For example, in Liberia many of the most powerful businessmen, particularly from the Lebanese diaspora, who were supportive of the traditional political elites from the earlier regimes, worked for Samuel Doe after his 1981 coup despite his regime's execution of many of the traditional political leaders. [16] From Doe, they migrated to the Taylor structure in the early 1990s once it was clear that he was going to be the most powerful force in the civil war and post–civil war era. Taylor kept the businessmen's loyalty until, in their rational cost-benefit analysis, they concluded that a change in government would benefit their own interests.

This adaptability of the transnational networks is one of the reasons postconflict expectations of deep social and economic changes are often left unmet. Postconflict regimes, particularly if controlled by forces that were traditionally nonstate armed actors, frequently have little governance experience and seldom have the knowledge or experience to operate without the help of these networks or simply find it easier to rely on them than to antagonize them.

The U.S.-led occupation of Haiti in 1994 to restore President Jean-Bertrand Aristide provides a clear example. The occupying forces contracted almost exclusively with the families and networks that funded Aristide's initial ouster because they had the port, fuel, and housing facilities the United States needed for its forces. With the new regimes hamstrung by their dependence on the old network, the promised economic reforms never materialized.[17]

But local elites do not survive in isolation. Rather they are—or become—part of a complex web of relationships that reach both backward (to the armed actor) and forward (to the international market). While local fixers can make themselves indispensable to the new regime and know how things operate internally in the region during or after conflict, they often lack the expertise to supply the honey pot controller, either a criminalized state or nonstate armed actor, with the more specialized services required.

While a local node of the network may know how to deliver rice or oil in adverse conditions or how to navigate the world of local militias, the same node may not know how to acquire a shipment of AK-47 assault rifles. Those who can provide those rifles from the international market and who are adept at breaking embargos or moving through the "gray market" arms bazaars want to sell the weapons but often do not know the lay of the local land and political structures.

It is in this niche that one finds a small group of super fixers (figure 2) who operate as intermediaries among these different groups, across multiple countries, at a handsome profit. These individuals—through family ties, successful smaller ventures in the past, personal charm, or a combination of these attributes—have ties to numerous elites across the region and to suppliers from the outside world. In the case of Liberia, many of these individuals were identified and can be traced. While this small group knows how to access local power brokers and make contacts in certain difficult markets, its members may not themselves have access to the specialized markets of weapons, helicopters, and the necessary paperwork or banking facilities to make deals happen.

A separate but sometimes overlapping ring of actors I refer to as shadow facilitators then comes into play to acquire the specialized equipment and paperwork on the world's gray or black market, create or activate shell companies, and make the logistical arrangements to sell the commodity on the international market. The paperwork includes end-user certificates, which allow a government to appear to legally purchase weapons that may end up elsewhere; front companies to handle freight loading and delivery; offshore bank accounts to make the money untraceable; falsified flight routes to justify the time in the air; and air operations certificates to show that aircraft are certified airworthy.

Figure 2. The "Fixer" Chain

Few of these facilitators work exclusively in illegal activities. Of particular interest was Viktor Bout, a Russian weapons merchant convicted in November 2011 on multiple counts of attempting to supply a designated terrorist entity with weapons to kill American troops. Yet in previous years his air services were used by an array of governments and institutions even as they denounced his illicit activities. These included the U.S. Government for services in Iraq, the United Kingdom (UK) government for aid work in Africa, and the United Nations (UN) for its peacekeepers and World Food Program operations in the Democratic Republic of the Congo and elsewhere.[18]

Bout is also noteworthy because his contacts spanned such a wide array of conflicts. Not only did he supply warlords and criminal states with illicit arms shipments, but also terrorist networks from the Taliban in Afghanistan, to the FARC in Colombia, to Muammar Qadhafi in Libya, to Hizballah in Lebanon.[19]

In sub-Saharan Africa, he relied on several super fixers, depending on his geographic location. One of those relationships will be described below. In United Arab Emirates, other local fixers and super fixers helped him establish a presence in the free trade zone, repair facilities for his aircraft, and housing for his crews.

The Taylor Liberian Fixer Networks

Charles Taylor of Liberia, currently standing trial at the Special Court for Sierra Leone in The Hague for crimes against humanity, offers a useful example of the symbiotic relationship among armed groups, a criminalized state, and those in the fixer chain.[20]

Several individuals had overlapping roles in different parts of the fixer chain. Among them were local fixer Gus Kouwenhoven, super fixer Sanjivan Ruprah, and shadow facilitator Viktor Bout operating along the lines in the model outlined above.

All of the above have been named by the Liberian Truth and Reconciliation Commission as responsible for economic crimes.[21] They have also been named in various UN Panel of Experts reports and placed on the UN Security Council travel ban and asset forfeiture lists for being "threats to regional peace."[22] Because of the extensive documentation available on this network, it offers one of the most comprehensive looks at how the fixer chain works.

As a rebel warlord, Taylor gained control over forested territory along the country's borders and used commodities to fund his insurgents, the National Patriotic Front of Liberia (NPFL), against the Liberian state (1989–1997). After he became president in 1997 and managed a remarkably efficient resource-extraction system, he continued to use revenue from commodity sales and bribes in exchange for concession allocations to fund his war against subsequent insurgencies against his rule (1997–2003).

From December 1989 to 1997, Taylor waged a brutal and destructive war against the Liberian government and other warlords. In 1991, he helped to establish and train the RUF, a rebel group and proxy army in neighboring Sierra Leone. From the start he largely funded and pursued the wars through a desire to control the region's lucrative natural resources, including timber, diamonds, iron ore, and rubber, though his two main sources of income were diamonds and timber.

Taylor always relied on illegally extracting and selling commodities to fund his armed efforts and viewed control of natural resources as a means of funding "Greater Liberia," a territory he saw encompassing the bauxite region in neighboring Guinea and the diamond fields of neighboring Sierra Leone.[23]

Within a year of launching the insurgency, Taylor controlled most of the vital economic regions of Liberia and was taking in millions of dollars to buy weapons and pay his troops.[24] In 1996 the U.S. Government estimated that from 1990 to 1994 Taylor had "upwards of $75 million a year passing through his hands," largely through selling timber and other commodities.[25]

Even after winning elections in 1997, Taylor was, in effect, not president of a country but was controlling what Robert Cooper has called the "pre-modern state," meaning territory where

> *chaos is the norm and war is a way of life. Insofar as there is a government, it operates in a way similar to an organized crime syndicate. The pre-modern state may be too weak even to secure its home territory, let alone pose a threat internationally, but it can provide a base for non-state actors who may represent a danger in the post-modern world . . . notably drug, crime and terrorist syndicates.*[26]

This "pre-modern state" did not preclude the development of sophisticated illicit networks. Indeed one could argue that the absence of functioning state institutions greatly empowered these networks because almost the entire basis of control depended on them.

The extensive financial, military, and political networks that Taylor established before and during his time as president were impressive. His network for acquiring weapons ranged from the Balkans to Central America, from Bulgaria to the Islamic Republic of Iran. His inner circle of financial advisers, local fixers, international facilitators, sanction busters, and weapons purchasers included American, Belgian, Dutch, Israeli, Lebanese, Libyan, Russian, Senegalese, and South African citizens. His access to these international criminal networks greatly increased the resources with which to prolong the carnage that his troops and his allies could inflict on the region.[27]

Taylor, like Doe before him, relied on the same essential economic groups partly, as William Reno states, because "they were so integrally connected to the exercise of violence." These groups had been both privileged in their economic operations and cut out of the political power structure. This "selective access to rights to profits," which Taylor could arbitrarily revoke if he wished, kept the franchise operators constantly beholden to him.[28]

Beginning in the early 20th century, Lebanese businessmen became the region's primary diamond buyers from the alluvial mining regions largely because they were willing to run the risk of living in the bush and transporting their valuable cargo to diamond markets, primarily in Antwerp, Belgium. They also knew how to move in the world of trade and finance, skills that few Liberians had at the time. Most of the mining activity took place in Sierra Leone, but as the war there gained momentum the financial center of power for the trade shifted to Monrovia.

The Lebanese family and clan structure in West Africa has endured in part because its members work hard to retain ties to their homeland and in part because, as outsiders to both the indigenous cultures and the elite Liberian-American culture,[29] they banded together for the creation of reliable business networks. Much of this revolves around the import and export of commodities. For example, in the Liberia diaspora businessmen handled the importation of petroleum products, rice, vehicles, and chicken, as well as the diamond trade and much of the timber industry.

Lansana Gberie, in one study of the Lebanese diaspora in West Africa, noted their role:

> The Lebanese in West Africa, even those born there, remained and continue to remain intensely aware of events in Lebanon. . . . In Lebanon they are referred to as "Africans." Their loyalty, however, remains with the Middle East, and many have made regular contributions to factions in that region's never-ending conflicts.[30]

Gus Kouwenhoven, a Dutch national, operated a logging company known as TIMCO on land controlled by Taylor's NPFL insurgents during Liberia's civil war, from which Taylor's forces profited directly. This put him in the category of local fixer and local elite.[31] Although not a member of the Lebanese diaspora, Kouwenhoven was part of the broader though small community of long-time foreign residents who had carved out different economic niches in the chaos of the nation's civil wars. When Taylor became president, Kouwenhoven, who also owned the Hotel Africa in Monrovia, was given a privileged position in Taylor's financial circle.

On July 28, 1999, Kouwenhoven established the Liberian Forest Development Company in Monrovia, which was owned by two other companies, OTC and Royal Timber Corporation (RTC). While OTC's owners were listed as three Indonesian-Chinese businessmen, RTC's owners were listed as Kouwenhoven and Taylor.[32] Robert Taylor, the president's brother, as head of the Forest Development Authority in 1999 granted the company a logging concession of over 3 million acres, the largest concession by far in Liberia. OTC tripled Liberia's timber exports from 1999 to 2000 and committed numerous violations by cutting undersized trees and by clear cutting.[33]

Taylor did not hide his interest in OTC, publicly dubbing the company his "pepperbush," a local expression meaning something that is dear to one's heart and profitable.[34] Liberians joked that OTC stood for "Only Taylor Chops" because of the fierceness with which he defended the timber company and attacked those who questioned its operations.

The OTC concession was illegal because it was not approved by the Liberian legislature as required by law. When the concession was called into question, Taylor simply had the legislature pass the Strategic Commodities Act of 2000, which granted him the sole power to give and maintain concessions over all the nation's natural resources.[35]

This is a central point in the relationships between local fixers and the government or insurgent warlord where they operate, whether state or nonstate: while providing that governing body with economic benefits, the fixer gets privileged access that enhances his own economic standing. It is a symbiotic relationship.

The way the OTC concession was granted shows an additional benefit of dealing with a criminalized state that is not subject to the rule of law or to normal checks and balances among the different branches of government. Operators received the concession and began to operate with no fear of legal sanction even though the concession was against Liberian law. When it became politically necessary to legitimize it, Taylor simply wrote a new law—certain to be passed by a legislature he controlled—in such a way as to legalize virtually anything he did with any commodity on behalf of the Liberian state.

Various investigations by the UN, nongovernmental organizations, and the Truth and Reconciliation Commission have established that OTC and other logging companies both paid and fed Taylor's militias and former combatants and helped Liberia evade UN sanctions. These militias, hired as security forces for logging companies, were often commanded by notorious NPFL commanders, and many of their members were charged with serious human rights abuses.[36]

These crosscurrents illustrate the costs for local businessmen in dealing with a criminal state. While reaping the benefits of such a system, they are also subject to its demands, which can change and become more excessive over time. There is no legal recourse.

Kouwenhoven, during his subsequent trial in Holland, described Taylor's escalating demands for OTC to pay more cash and provide other services:

> I did have contact with Taylor about OTC matters.... If he needed anything he would call me.... Most of the time it had to do with financial requests. After we had concluded the OTC agreements and he was prepared to give concessions to OTC to make it a profitable company, we were asked to make an advance payment of $5,000,000 for future taxes.... So he would make all kinds of requests. Apart from that he asked us to send a number of tractors to his farm, or he said that he wanted a road, that he needed electricity and he would ask me if I could advance the money. He also simply asked for payments.... He called for me and told me that it was understood that the Liberian government and OTC had an official tax relationship. The regular government budget was not sufficient and they would ask businessmen for financial aid. He would receive 50 percent of royalties I received from OTC.[37]

Kouwenhoven later explained to a Dutch newspaper:

> The president is like the top God. If it turns out that he needs money at the end of the year to pay his civil servants, he just calls at various businesses and asks them for an advance on next year's tax. And in that way you help pay the civil servants' salaries. This is not a "Kouwenhoven system"—everyone is involved in it.[38]

Despite this increase in demands for cash and services by Taylor, Kouwenhoven remained directly active in the weapons-for-commodities pipelines. While the Truth and Reconciliation Commission documented at least eight weapons shipments—six by ship and two by air—it is the air shipments that most clearly show how local fixers such as Kouwenhoven, with Taylor's approval, reached out to transnational super fixers such as Sanjivan Ruprah, who in turn tapped into the world of shadow facilitators.

Ruprah, a Kenyan of Indian descent, was one of the people documented as directly receiving payments for weapons from Borneo Jaya Pte Ltd., OTC's parent company.[39] He continues to operate and is a well-known super fixer in sub-Saharan Africa who has surfaced repeatedly in criminal investigations across the continent.[40]

Ruprah had worked with several private military companies and mining interests in the Democratic Republic of the Congo and was married to the sister of a leader of one of that country's main Rwandan-backed military factions. Described as an "arms broker" in numerous UN reports,[41] Ruprah had also directed the Kenyan office of Branch Energy, a company that in the early 1990s negotiated to obtain control of the diamond-mining rights of Sierra Leone. Branch Energy, through Ruprah, also introduced Executive Outcomes to the government of Sierra Leone, which used them to fight the RUF because its own forces were in such disarray. Executive Outcomes, made up largely of white, former special forces operatives from South Africa and Zimbabwe, pioneered the idea of hiring themselves out as mercenaries in exchange for extensive concessions in natural resources such as diamonds and timber.[42]

Ruprah, by his own admission, met Taylor in the mid-1990s in Burkina Faso, before Taylor was president. He was seen more frequently in Monrovia from 1999, often staying at Kouwenhoven's Hotel Africa, where Taylor housed his privileged guests. Recognizing how valuable his services were (or could be), in 1999 Taylor issued Ruprah a diplomatic passport under the name of Samir M. Nasr. He also gave him the title of Deputy Commissioner of Maritime Affairs.

When Taylor was in desperate need of weapons and scrambling to acquire them from a variety of external sources, he, through Kouwenhoven, asked Ruprah for help, particularly in obtaining attack helicopters.[43]

Through his business dealings in the Democratic Republic of the Congo (DRC), Ruprah had made a connection that put him in touch with the next circle—the shadow facilitators.

His contact was Viktor Bout, who was already active in Angola, the DRC, Rwanda, South Africa, and elsewhere. Dubbed the "Merchant of Death" by a senior British official following the discovery that Bout was arming multiple sides of several conflicts in Africa, Bout had made his mark by building an unrivaled air fleet that could deliver not only huge amounts of weapons but also sophisticated weapons systems and combat helicopters to armed groups. From the mid-1990s until his arrest in Thailand in 2008, Bout, a former Soviet intelligence officer, helped arm groups in Africa, Afghanistan, Colombia, and elsewhere.[44]

Ruprah introduced Bout to Taylor's inner circle, a move that fundamentally altered the supply of weapons to both Liberia and to the RUF in Sierra Leone by giving both groups access to more sophisticated weapons, in greater volume and at lower cost. One of the favors Ruprah and Taylor offered Bout was the chance to register several dozen of his rogue aircraft in Liberia.

Ruprah had taken advantage of operating in a criminal state and used his access to Taylor to be named the Liberian government's Global Civil Aviation Agent Worldwide to further Bout's goals. This position gave Ruprah access to the aircraft and possible control of it.[45] "I was asked by an associate of Viktor's to get involved in the Aviation registry of Liberia as both

Viktor and him wanted to restructure the same and they felt there could be financial gain from the same," he has stated.[46]

Bout was seeking to use the Liberian registry to hide his aircraft because the registry, in reality run from Kent, England, allowed aircraft owners to obtain online an internationally valid airworthiness certificate without having the aircraft inspected or disclosing their name.[47]

Within months, mutually beneficial transactions were flowing. As the UN Panel of Experts reported, on August 26, 1999, OTC's parent company paid $500,000 for weapons deliveries to Taylor. The money was paid to San Air, a Bout company operating in the Sharjah, United Arab Emirates, "through Sanjivan Ruprah."[48] In this circular pipeline, the timber flowed out and the cash flowed in, then the cash flowed out as the weapons flowed in.

The personal relationship between Ruprah and Bout, and then of those two with Taylor, was a crucial part of the network that was built on a level of personal trust as well as mutual financial interests. Something repeatedly said about Bout's operations was that he personally closed deals with the heads of state or warlords with whom he dealt.

The benefits, like the goods and profits, flowed in a circular manner. Taylor, operating a criminal state, could offer more than just cash benefits to national and international facilitators. The local fixer could acquire premium concessions, control of a port, and official protection in a violent and uncertain context. He, in turn, could bring in a super fixer who, besides cash and official protection, could acquire a diplomatic passport, which allowed him to travel around the world without being searched or detained.

The super fixer could then use his access to both Taylor and the international facilitator to bring a beneficial relationship to all involved. Finally, the international facilitator could enter a lucrative market and access an aircraft registry that allowed him to hide his aircraft from scrutiny for several years. Indeed so vital was Bout to Taylor that he dubbed the arms merchant another of his pepper trees—untouchable. "Taylor would say that Bout was the root of his pepper tree, and without the root the tree dies," said one Taylor confidant.[49]

Ramiro, the FARC, and the Bolivarian Revolution

On March 1, 2008, Colombian military and police forces launched a lightning strike into neighboring Ecuador, targeting a senior command post of the FARC guerrillas[50] just inside Ecuadoran territory. While the attack caused a significant diplomatic rupture between Ecuador and its Bolivarian allies and Colombia, it achieved its main objective of killing Raúl Reyes,[51] the FARC's second in command and international spokesman. In addition to killing Reyes, the raid yielded an unexpected dividend—several hundred gigabytes of data stored on computers and flash drives. The contents were an historic archive of FARC internal and international communications going back almost a decade, offering unprecedented insight into the internal workings of the secretive group. Interpol, given access to the hard drives, verified they had not been tampered with by Colombian officials.

The documents outline the support the FARC has received from the Chávez government in Venezuela, including financing, joint business operations, and the establishment of political front groups to defend the FARC internationally. They also show the FARC gave several hundred thousand dollars to the presidential campaign of Ecuadoran president Rafael Correa and had direct dealings with senior members of Correa's national security cabinet. The archives also divulge the FARC's dealings with members of Evo Morales' government in Bolivia, Daniel Ortega in Nicaragua, and various other nonstate armed actors in Latin America and Europe.[52]

Also found in the cache were emails among different FARC commanders about the help they were receiving in acquiring sophisticated weapons from an individual identified as "Ramiro" in El Salvador, identified by Salvadoran, Colombian intelligence, and U.S. officials as the nom de guerre of José Luis Merino.[53]

Merino was a senior commander of the Communist Party (PC) during El Salvador's civil war (1980–1992). The PC was the smallest of the five organizations that made up the Marxist-led Farabundo Martí National Liberation Front (*Frente Farabundo Martí Para la Liberación Nacional*—FMLN). While the other four factions of the FMLN demobilized after the war, a significant cadre of the PC did not. They maintain arms caches, safe houses, and relationships with nonstate armed groups around the world to this day.[54]

As one report on the documents noted,

> The (FARC) documents also show that the FARC has an international support network stretching from Madrid to Mexico City, Buenos Aires to Berne. Merino, the documents suggest, is a key link in that international chain, the FARC's man in El Salvador, and one of the architects of an arms deal that includes everything from sniper rifles to ground-to-air missiles.[55]

It is this network where one can again see the vital role played by the fixer chain in moving illicit weapons to a designated terrorist nonstate group deeply involved in global cocaine trafficking.

Merino's PC is now the dominant faction of the FMLN coalition. The FMLN-backed candidate, Mauricio Funes, won the presidency in 2010. The former guerrilla group's electoral triumph, the first since the signing of historic peace accords in 1992 that ended the 12-year conflict, was widely viewed as a key milestone in national reconciliation.

Merino, a former urban commando trained in Moscow, had maintained a low public profile during the campaign but emerged as a key power broker in the new government. His leadership position is derived largely from being the point man for receiving and distributing subsidized gasoline shipments from Venezuela's Hugo Chávez. The sale of the gasoline, given by Venezuela at a steeply discounted price then resold at party-owned gas stations, is a major source of income for the FMLN.

While Merino has steadfastly refused to discuss the issue of weapons sales to the FARC, Funes said he had spoken to Merino about the emails and had been assured Merino had no involvement in weapons sales to the FARC.[56]

During the war, Merino commanded an elite urban commando unit that carried out several spectacular hits, including the 1989 assassination of attorney general Roberto García Alvarado. The assassination was carried out by men on a motorcycle who strapped an explosive to the roof of García's car, killing him while the assassins sped away unhurt.[57] The Salvadoran Truth Commission, which investigated the worst human rights abuses of the civil war as part of the peace agreements, did not name Merino but concluded the assassination was carried out by the armed branch of the PC.[58]

The spectacular rise in crime immediately after the war, particularly kidnappings, led to deep concern. A major investigation of postconflict armed groups in 1994 found that the "illegal armed groups" operating after the war had "morphed" into more sophisticated, complex organizations than had existed during the war and that, as self-financing entities, they had a strong economic component as well as political aspect to their operations.[59]

This PC group commanded by Merino is suspected of carrying out several high-profile kidnappings at the time, including the 1995 abduction of 14-year-old Saul Suster, the son of a prominent businessman and politician. Suster was kept in a tiny underground water tank for almost a year before being freed in exchange for $150,000.[60] While Merino, who was widely reported to have planned the operation, was never charged with the crime, Raul Granillo, a former senior commander of the PC and Merino aide, along with one of his PC former lieutenants, were convicted of it. Granillo escaped arrest and remains at large.[61]

The relationship between Merino and the FARC appears to go back to the 1980s when the PC was in charge of the FMLN's international relations and Merino regularly met with the FARC as members of the broad international Marxist revolutionary front.[62]

After the war, the relationship appears to have revolved around mutually beneficial economic arrangements, many of them illegal. Glimpses of the broader network involved in these activities can be seen in the Reyes documents, although in many cases the final results of the discussion are not known. While the overall results of the efficiency of the network are inconclusive, the documents clearly show the role of the FARC fixers interacting with Merino, the super fixer able to reach into the outside world. In this case the identity of the shadow facilitator or facilitators is not clear, but Merino was reaching out to others to carry out the deal, in one case at least to a group of Australians with significant access to sophisticated weapons and means of transporting them.

In one 2003 email, Reyes suggests the FMLN and FARC carry out a joint kidnapping for $10 million to $20 million in Panama to raise money for the FMLN's 2004 electoral campaign, although there is no record of such a kidnapping having occurred. The profits, the email said, would be "split down the middle."

A 2004 missive describes a meeting of the FARC leadership with "Ramiro" and a Belgian associate in Caracas to discuss the FARC and FMLN obtaining Venezuelan government contracts through front companies to operate waste disposal and tourism industries. Interestingly, Merino then acquired a major interest in a waste disposal company, Capsa, which operates in 52 Salvadoran municipalities.[63] In another email, Reyes recounts a 2005 meeting with "Ramiro," who reportedly boasted that he had gained full control of the FMLN and was reorienting the party toward "the real conquest of power."

But the email that caused the most concern in El Salvador is from September 6, 2007, written by Iván Ríos, a member of the FARC secretariat, to other secretariat members in which he lays out the multiple negotiations under way for new FARC weapons, including highly coveted surface-to-air missiles:

1. *Yesterday I met two Australians who were brought here by Tino, thanks to the contact made by Ramiro* [Salvador]. *We have been talking to them* [the Australians] *since last year.*

2. *They offer very favorable prices for everything we need: rifles, PKM machine guns, Russian Dragunovs with scopes for snipers, multiple grenade launchers, different munitions … RPGs [rocket propelled grenades], .50 [caliber] machine guns, and the missiles. All are made in Russia and China.*

3. *For transportation, they have a ship, with all its documents in order, and the cargo comes in containers. The crew is Filipino and does not know the contents, with the exception of the captain and first mate. They only need a secure port to land at.*

4. *They gave us a list of prices from last month, including transportation. They offer refurbished Chinese AKs that appear as used, but in reality are new, and were not distributed to the Chinese army, which developed a new line of weapons, for $175. AK 101 and 102, completely new, for $350. Dragunovs, new with scopes, $1,200. RPG launchers for $3,000 [and] grenades for $80. They say they have a thermobaric grenade that destroys everything in closed spaces (like the bombs the gringos use against Alqaeda hideouts) for $800. Chinese missiles (which they say are the most up-to-date at this time) with a 97 percent effective rate, $93,000, and $15,000 for the launchers. They say it is very easy to use, and they guarantee the training. If one of these missiles were identified inside Colombia it would cause them a lot of problems, but if, on the side, they include old Russian SA-7s [shoulder-fired surface-to-air missiles], it would serve to confuse, mislead or at least give the impression that the guerrillas have weapons of different types, not just Chinese. The ammunition for AKs is 21 cents a round, but if we buy more than 3 million rounds, the price drops to 9 cents a unit.*

5. *They promised to give us an exact price on other material. Two months ago they sent me a price list (very favorable, for example, a used .50 [caliber] machine gun for $400, new for $3,000), but I didn't take the list to the meeting place.*

6. *They do the purchasing without the need of a down payment, but when the merchandise is on the ship, they want 50 percent. When it is delivered they collect the other 50 percent. The money moves through a bank in the Pacific, in an independent country where they can move money without any questions being asked. Once the cargo is shipped it can take one month, or a month and a half to arrive in Venezuela. They said we could have a representative, it doesn't matter what nationality, on board the ship while it sails to its final destination.*[64]

The above missive clearly lays out the different roles of the fixer community. The FARC leaders contact a trusted colleague who brings in extraregional actors who have access to the desired product as well as ways of moving the payment through international channels. As the final paragraph shows, the expected destination was to be an allied nation where few questions would be asked.

Subsequent emails on November 12 and November 23 indicate that the Australians arrived in Venezuela to consummate the deal and Ríos said a person identified as "El Cojo" (The Cripple) was in charge of paying the first quota and logistics for whatever was to be delivered.

While the end result of what was delivered is unknown, as we will see, this network appears to be part of a regional pipeline moving illicit products and connecting parts of the PC to the FARC and elements of the former Sandinista intelligence structure in Nicaragua that remain active. In 2006 and again in 2011, Daniel Ortega, the leader of the triumphant Sandinista revolution in 1979, was reelected president of Nicaragua.

Since at least the early 1990s, Ortega has maintained a cordial relationship with the FARC leadership. The alliance was first formed in the 1970s when Ortega led the Sandinista revolution that succeeded in taking power in Nicaragua in 1979. Throughout his first presidency (1979–1990), Ortega was at the center of revolutionary movements active throughout Latin America, including the FARC, in part because Nicaragua was the one state besides Cuba in the hemisphere where a Marxist revolution had triumphed.

In 1998, Ortega, as the head of the Sandinista party, awarded the Augusto Sandino medal, his party's highest honor, to Manuel Marulanda, the commander in chief of the FARC, while visiting FARC headquarters in Colombia. Various emails in the Reyes documents are addressed to Ortega directly and indicate that, at least until he returned to the presidency, he maintained regular contact with the FARC leadership.[65]

In 2000, Ortega attended an international convention in Libya organized by Qadhafi for "political parties, revolutionary movements, liberation movements and progressive forces."[66] This meeting was significant because senior leaders of the FARC were also in attendance. The timing is important because it was at a point when the FARC was seriously looking to purchase surface-to-air missiles in order to have more effective defenses against the U.S.-supplied helicopters that were beginning to arrive in Colombia for the police and military as part of Plan Colombia.[67]

A September 4, 2000, email from Reyes' computer addressed to Qadhafi and signed by Reyes offered a bold alliance and laid out the strategy:

> *Comrade Colonel Muammar Qadhafi, Great Leader of the World Mathaba, receive our revolutionary and Bolivarian greeting. We want to express our gratitude for the invitation that you gave us to visit your country and the hospitality you showed our delegation during the recent Summit of Heads of State, Governments, Parties and Organizations of the World Mathaba. We want to let you know that the FARC continues its struggle for the conquest of political power to govern Colombia. . . . As a member of the [FARC] high command I have been asked by the commander in chief to request from you a loan of 100 million dollar, repayable in five years. Our strategic objectives and circumstances of our war oblige us to seek weapons with greater range to resist our enemy's advances. One of our primary needs is the purchase of surface-to-air missiles to repel and shoot down the combat aircraft. Our strategy is to take power through the revolutionary armed struggle.[68]*

Apparently, Qadhafi was not immediately responsive to the request, so the FARC tried to follow up on the request, this time through Daniel Ortega. In a February 22, 2003, note "From the Mountains of Colombia," that was hand-delivered to Ortega, Reyes requested help:

> Dear compañero Daniel, this is to send you my warm and effusive revolutionary greeting, and that of commander Manuel Marulanda. We also are writing to see if you have any information on the request we made to our Libyan comrades, which was made in writing in the name of the secretariat of the FARC, and which I signed. The Libyans said they would answer us, but we have not yet received any information. . . . While we were in Libya they explained to us that the political responsibility for carrying out Libya's policies in the region [was] in the hands of Daniel Ortega. For that reason, we are approaching you, in hopes of obtaining an answer. . . . I also want to reaffirm to you that the primary priority of the FARC, in order to achieve greater success in its military operations against enemy troops, in order to take political power in Colombia, is acquiring anti-aircraft capabilities, in order to counteract the efficiency of the Colombian and U.S. aircraft against our troops.[69]

It is not clear what the result of the discussions were. According to documents obtained by La Prensa, Nicaragua's most important newspaper, the key envoy between Ortega and the FARC is Luis Cabrera, the Nicaraguan ambassador in Cuba. Cabrera is a nationalized Argentine who has maintained a friendship with Ortega since the Sandinista revolution.

Citing other documents from Reyes's computer, the newspaper said FARC leaders had met numerous times with Cabrera, who promised the rebels any help within Nicaragua's ability. He even offered to let the FARC establish a link on the Sandinista Web site in order to make it easier for people to get in touch with the organization. Reyes wrote to the FARC representatives in Havana not to forget to "visit the Nicaraguan ambassador with great frequency. Take him documents and express our appreciation for the stimulating statements of comandante Ortega."[70]

In 2008, following the bombing of the Reyes camp, Ortega granted political asylum to Nubia Calderón, a leader of the FARC's International Commission, who was wounded in the attack. In addition, FARC emissary Alberto Bermúdez was issued a Nicaraguan identity card by a member of Nicaragua's Supreme Electoral Council. Bermúdez, known as "El Cojo," is believed by Salvadoran and Colombian authorities to be the person mentioned above who received the FARC weapons shipment arranged by Merino.[71]

In this case, the lines among the different types of fixers are harder to draw, as is the case in most chains where roles are seldom as clearly delineated as they are conceptually.

While the identities of all the players in this case are not known, the model is useful in understanding the overall structure. Merino, with his international contacts, falls into the super fixer category, while the FARC leadership falls more into the local fixer category. The shadow facilitators are not well known, although it is clear from Reyes documents that Chávez in Venezuela has played that role with the FARC and in hosting Merino's meetings with the Australian providers and offering his services in setting up businesses and other opportunities that gave the FARC a broader international reach. Nicaragua's Ortega also played a key role as a shadow facilitator, empowering both the PC and the FARC to carry out business through Ramiro's structure.

Conclusions

In order for transnational criminal pipelines and networks to function, different parts of the network need to communicate with each other and be able to supply mutually beneficial services and commodities. The fixers, or links, among different parts of the network provide this vital service.

Local fixers, often members of the local elites and/or entrenched diaspora communities, help move the illicit product from production in conflict zones, criminalized states, or territories not governed by the state to the internal market. Super fixers, those with broader international connections, can then move the product to the international market and provide some financial services. However, they rely on shadow facilitators to acquire the more complex and sophisticated equipment—often weapons or high-technology components that are not readily accessible through normal market channels.

The shadow facilitators usually deal with more than one criminal or terrorist network, providing rare and valuable services to multiple actors who may even be in conflict with each other. As the Viktor Bout case shows, if a dealer provides valuable enough services that are highly specialized, he can arm multiple sides of the same conflict and greatly increase his profits.

Because shadow facilitators' international contacts and structures are so specialized and difficult to replace and operate for multiple groups, targeting them by law enforcement and intelligence communities should be a high priority. The DEA, in particular, has focused on this group. The high profile cases of Viktor Bout and Monzer al-Kassar,[72] two of the world's most elusive and powerful brokers in recent decades, show that operations can be undertaken and prosecuted successfully.

The impact in both cases was significant. Viktor Bout's arrest crippled and grounded an air fleet that was supplying weapons across Africa, including Somalia, and seeking inroads into Latin America. Replacing items outside that network was possible but costly both financially and in time. Almost 4 years after his arrest (he was convicted in November 2011),[73] no one has successfully put the pieces of the enterprise back together. In the al-Kassar case, the network also collapsed. In both cases, there are indications that other individuals have picked up pieces of the network, but in neither case has capacity emerged to offer anything like the full range of services these two super facilitators did.

However, much of the conventional strategy for combating TOC groups relies on conceiving of illicit networks as hierarchical organizations with a strong top-down structure. Rather than focusing on the links (individuals providing specific services) and mutual usefulness of different small, relatively flat networks that overlap and rely on many of the same specialists, the focus remains the kingpins.

In addition, the focus is almost exclusively on nonstate actors rather than on the role criminalized states also play in empowering these networks, particularly in the world of shadow facilitators.

A more node- or fixer-centric understanding of transnational organized crime and terrorist networks can help identify and disrupt key elements of the pipelines that often deal with more than one group simultaneously. This means, as was clearly visible in the case of Viktor Bout, that targeting one shadow facilitator disrupted multiple supply chains in which he played a key role. While not eliminating the networks, such targeting makes them far less efficient, disrupts trusted relationships by forcing new actors to meet and interact, and raises the cost of the criminal networks for doing business. Targeting fixers is not a silver bullet, but it is one of the most efficient methods for weakening and disrupting illicit networks.

Notes

[1] *Shadow facilitators* was first used as a term of art by the U.S. Drug Enforcement Administration to describe those individuals deemed to be controlling central points of interaction among drug traffickers and other criminal organizations, such as weapons merchants. See Michael Braun, "Drug Trafficking and Middle Eastern Terrorist Groups: A Growing Nexus," presented to the Washington Institute for Near East Policy, Washington, DC, July 25, 2008.

[2] For a more extensive look at how this chain operated in the Liberian timber and diamond trades, see Douglas Farah, "Transnational Crime, Social Networks and Forests: Using Natural Resources to Finance Conflicts and Post Conflict Violence," *Forests, Fragility and Conflict: Overview and Case Studies*, Program for Forests, World Bank, June 2011, 67–112.

[3] Braun.

[4] The self-declared Bolivarian states are led by Venezuela under president Hugo Chávez, and include Bolivia under Evo Morales, Ecuador under Rafael Correa, Nicaragua under Daniel Ortega, and several small Caribbean states. This bloc endorses a concept of socialism for the 21st century and strongly anti-U.S. positions, while working hard to build long-term alliances with Iran, China, and Russia to replace U.S. influence.

[5] For a more comprehensive look at the concept of criminalized states, see Douglas Farah, "Terrorist-Criminal Pipelines and Criminalized States: Emerging Alliances," *PRISM* 2, no. 3 (June 2010), 15–32.

[6] Douglas Farah, "The Criminal-Terrorist Nexus and Its Pipelines," Nine Eleven Finding Answers (NEFA) Foundation, presentation delivered to U.S. Special Operations Command Tampa, FL, January 14, 2008.

[7] Ibid.

[8] The descriptions of "positive" and "negative" sovereignty are drawn from Robert H. Jackson, *Quasi-states: Sovereignty, International Relations and the Third World* (Cambridge: Cambridge University Press, 1990). Jackson defines *negative sovereignty* as freedom from outside interference, the ability of a sovereign state to act independently, both in its external relations and internally, toward its people. *Positive sovereignty* is the acquisition and enjoyment of capacities, not merely immunities. In Jackson's definition, it presupposes "capabilities which enable governments to be their own masters." The absence of either type of sovereignty can lead to the collapse or absence of state control.

[9] For an examination of the "cultures of contraband" and their implications in the region, see Rebecca B. Galemba, "Cultures of Contraband: Contesting the Illegality at the Mexico-Guatemala Border," Ph.D. dissertation, Brown University Department of Anthropology, May 2009.

[10] Sergio Arauz, Óscar Martínez, and Efren Lemus, "El Cartel de Texis," *El Faro*, May 16, 2011, available at <www.elfaro.net/es/201105/noticias/4079/>.

[11] Richard Snyder and Ravi Bhavnani, "Diamonds, Blood and Taxes: A Revenue-Centered Framework for Explaining Political Order," *The Journal of Conflict Resolution* 49, no. 4 (2005), 563–597.

[12] This description of the diamonds-for-weapons trade is drawn largely from the author's fieldwork in Sierra Leone and Liberia as a correspondent for *The Washington Post*, and the investigations contained in Douglas Farah, *Blood from Stones: The Secret Financial Network of Terror* (New York: Broadway Books, 2004).

[13] Graham Greene, *The Heart of the Matter* (New York: Penguin Classics, republished 2004).

[14] See William Reno, *Warlord Politics and African State* (Boulder, CO: Lynne Rienner, 1999), 221–223, for a look at how accidents, luck, and an individual's learning ability all help shape how the armed actor or warlord responds to internal and external factors.

[15] See Snyder and Bhavnani for a description of how the Lebanese diaspora community in Sierra Leone was both favored in diamond mining and marginalized socially for the regime's benefit and how the balance of coercive power eventually shifted in favor of Lebanese merchants.

[16] Reno, 82.

[17] See Douglas Farah, "Haitian Elite Set to Profit from U.S. Presence," *The Washington Post*, September 24, 1994, A4.

[18] Douglas Farah and Stephen Braun, *Merchant of Death: Money, Guns, Planes and the Man Who Makes War Possible* (New York: J. Wiley and Sons, 2007).

[19] Ibid.

[20] The Special Court for Sierra Leone was established to try those most responsible for crimes against humanity committed in relation to the war in that country, but not for crimes committed in Liberia. Given Taylor's role in arming and supplying the rebel RUF in Sierra Leone, he was indicted by the court.

[21] Truth and Reconciliation Commission of Liberia, *Title III: Economic Crimes and the Conflict, Exploitation and Abuse*, vol. 3 of Final Report (Monrovia, 2009), 6–9.

[22] See United Nations Security Council 2005a and 2005b. Kouwenhoven was arrested in Holland and convicted in June 2006 on charges of supplying weapons to the Taylor government via OTC in 2000–2003 and sentenced to 8 years in prison. An appeals court overturned the conviction in March 2008. See Agence France-Presse 2008. The Supreme Court, in turn, overturned the appeals court ruling and ordered a new trial, which began in December 2010. See <http://allafrica.com/stories/201012230695.html>.

[23] For a comprehensive look at how Taylor used commodities to fund his insurrection and presidency, see Arthur G. Blundell, "The Financial Flows That Fuel War," *Forests, Fragility and Conflict: Overview and Case Studies*, Program for Forests, World Bank, June 2011, 113–156.

[24] Stephen Ellis, *The Mask of Anarchy: The Destruction of Liberia and the Religious Dimension of an African Civil War* (New York: New York University Press, 1999), 90.

[25] William Twadell, "Foreign Support for Liberian Factions," Testimony before the International Relations Committee, U.S. House of Representatives, Washington, DC, June 26, 1996.

[26] Robert Cooper, "The Post-Modern State," in *Re-Ordering the World*, ed. Mark Leonard, 11–20 (London: The Foreign Policy Centre, 2002), 18.

[27] Coalition for International Justice, *Following Taylor's Money: A Path of War and Destruction* (Washington, DC: Coalition for International Justice, 2005).

[28] Reno, 98.

[29] The term "Americo-Liberian" refers to the former American slaves who settled in Liberia. Most did not originally come from that region in Africa, and they were culturally and linguistically distinct from the native groups already living in what became Liberia. The Americo-Liberians dominated political and cultural life in the nation. For a closer examination, see Ellis.

[30] Lansana Gberie, *War and Peace in Sierra Leone: Diamonds, Corruption and the Lebanese Connection*, Occasional Paper No. 6 (Ottawa: Partnership Africa Canada, Diamond and Human Security Project, 2003).

[31] Global Witness, *Logging Off: How the Liberian Timber Industry Fuels Liberia's Humanitarian Disaster and Threatens Sierra Leone* (London: Global Witness, 2002), 17.

[32] Truth and Reconciliation Commission of Liberia, 13.

[33] Ibid.

[34] "Nobody Plays with Taylor's Pepperbush," *The Perspective*, June 27, 2001, Monrovia, Liberia.

[35] Global Witness,17.

[36] The most comprehensive documentation of both the human rights abuses and weapons trafficking can be found at Truth and Reconciliation Commission of Liberia, 19; Global Witness; and Douglas Farah, "Liberian Leader Again Finds Means to Hang On: Taylor Exploits Timber to Keep Power," *The Washington Post*, June 4, 2002, A01.

[37] Truth and Reconciliation Commission of Liberia, 18.

[38] Marian Husken and Harry Lensink, "All This Misery Stems from One Source," *Vrij Nederland*, March 10, 2007.

[39] United Nations Security Council, *Report of the Panel of Experts on Liberia Submitted in Accordance with Resolution 1343*, UNSC S/2001/1015, New York, October 26, 2001.

[40] Farah and Braun, 156.

[41] The most relevant report is United Nations Security Council, *Report of the Panel of Experts Pursuant to Resolution 1343*, paragraph 19, "Concerning Liberia," 92 and following.

[42] Farah and Braun, 155–159.

[43] United Nations Security Council, *Report of the Panel of Experts Pursuant to Resolution 1343*, 92 and following.

[44] Farah and Braun. In November 2010 Bout was extradited to the United States to stand trial for allegedly planning to sell weapons to a designated terrorist organization.

[45] United Nations Security Council, *Report of the Panel of Experts Appointed Pursuant to Security Council Resolution 1306*, S/2000/1195, paragraph 19, in relation to Sierra Leone, New York, December 20, 2000.

[46] Ruprah email to author for Farah and Braun 2007, 159.

[47] Farah and Braun, 159; UN Security Council 2000a, paragraphs. 142–143.

[48] UN Security Council 2001, paragraph 349.

[49] Farah and Braun 2007, 162.

[50] The Revolutionary Armed Forces of Colombia (*Fuerzas Armadas Revolucionarias de Colombia*—FARC) is the oldest insurgent group in the Western Hemisphere. Founded in 1964 as an offshoot of Liberal Party militias, the FARC is now considered by U.S. and Colombian officials to be one of the largest drug-trafficking organizations in the world. The FARC was designated a terrorist entity by the United States in 1997, a designation ratified by the European Union in 2001. See "FARC Terrorist Indicted for 2003 Grenade Attack on Americans in Colombia," Department of Justice Press Release, September 7, 2004, available at <www.usdoj.gov/opa/pr/2004/September/04_crm_599.htm>; Official Journal of the European Union, Council Decision of December 21, 2005, available at <http://europa.eu.int/eurlex/lex/LexUriServ/site/en/oj/2005/l_340/l_34020051223en00640066.pdf>. For a more detailed look at the history of the FARC, see Douglas Farah, "The FARC in Transition: The Fatal Weakening of the Western Hemisphere's Oldest Guerrilla Movement," NEFA Foundation, July 2, 2008, available at <www.nefafoundation.org/miscellaneous/nefafarc0708.pdf>. For a look at the details of the raid and a complete analysis of the documentation found in the computers of Raúl Reyes, see "The FARC Files: Venezuela, Ecuador and the Secret Archives of 'Raúl Reyes,'" *An IISS Strategic Dossier*, International Institute for Strategic Studies, May 2011.

[51] Reyes' real name was Luís Edgar Devia Silva, but he was known throughout his long militancy in the FARC as Reyes. He was the first member of the FARC's general secretariat to be killed in a combat operation after more than four decades of combat.

[52] "The FARC Files: Venezuela, Ecuador and the Secret Archives of 'Raúl Reyes.'" The Colombian government gave the International Institute for Strategic Studies, a respected London-based security think tank, complete access to the Reyes files, which it then catalogued, organized, and printed. The quotations and references to these documents come from this work.

[53] José de Córdoba, "Chávez Ally May Have Aided Colombian Guerrillas: Emails Seem to Tie Figure to a Weapons Deal," *The Wall Street Journal*, August 28, 2008.

[54] Author interviews with current and former PC members, San Salvador, September 2011.

[55] José de Córdoba, "The Man Behind the Man," *Poder 360 Magazine*, April 17, 2010.

[56] Córdoba, "Chávez Ally May Have Aided Colombian Guerrillas."

[57] For details of Merino's alleged ties, see Córdoba, "The Man Behind the Man"; Douglas Farah, "The FARC's International Relations: A Network of Deception," NEFA Foundation, September 22, 2008; Córdoba, "Chávez Ally May Have Aided Colombian Guerrillas."

[58] Belisario Betancur et al., "From Madness to Hope: The 12-year War in El Salvador," Report of the Commission on the Truth for El Salvador, March 15, 1993, available at <www.usip.org/files/file/ElSalvador-Report.pdf>.

[59] The investigation was carried out by a special commission formed in 1992 and composed of the nation's human rights ombudsman, a representative of the United Nations Secretary General, and two representatives of the Salvadoran government. The commission was formed by a political agreement among all the major parties due to a resurgence in political violence after the signing of the historic peace accords. See "Informe del Grupo Conjunto Para la Investigación de Grupos Armados Ilegales con Motivación Politica en El Salvador," El Salvador, July 28, 1994, available at <www.uca.edu.sv/publica/idhuca/grupo.html>.

[60] For a more complete look at Merino and the role of the Communist Party, both with the FARC and with criminal activities, see Córdoba, "Chávez Ally May Have Aided Colombian Guerrillas."

[61] Juan O. Tamayo, "Left-Right Distrust Festers in a 'Pacified' El Salvador," *Miami Herald*, November 12, 1997.

[62] Author interviews with current and former FMLN leaders in San Salvador, El Salvador, September 2011.

[63] Córdoba, "The Man Behind the Man."

[64] The Reyes documents cited are in possession of the author.

[65] Following Ortega's disputed electoral triumph in November 2011, the FARC published a congratulatory communiqué lauding Ortega and recalling their historically close relationship. "In this moment of triumph how can we fail to recall that memorable scene in Caguán when you gave the Augusto Cesar Sandino medal to our unforgettable leader Manuel Marulanda. We have always carried pride in our chests for that deep honor which speaks to us of the broad vision of a man who considers himself to be a spiritual son of Bolivar." Available at <http://anncol.info/index.php?option=com_content&view=article&id=695:saludo-a-daniel-ortega&catid=71:movies&Itemid=589>.

[66] José de Córdoba, "Ortega's Old Friendships are Liabilities as Election Nears," *The Wall Street Journal*, October 19, 2001.

[67] Plan Colombia is a multibillion dollar program through which the United States has funded the expansion, retraining, and arming of the Colombian national police and military. Begun in 1999, at the end of the Clinton administration, it has continued with bipartisan congressional support through the Bush years.

[68] Letter to Qadhafi, in possession of NEFA Foundation. It can also be seen at Octavio Enriquez, "Ortega, Puente Entre Gaddafi y las FARC," *La Prensa*, June 27, 2008.

[69] Letter to Daniel Ortega, in possession of the NEFA Foundation. It can also be seen at Octavio Enriquez, "Ortega, Puente Entre Gaddafi y las FARC."

[70] Octavio Enriquez, "Contacto Está en La Habana," *La Prensa*, June 28, 2008, available at <laprensa.com.ni/archivo/2008/junio/28/noticias/nacionales/268624_print.shtml>. In addition to this support, *La Prensa* reported that a well-known Nicaraguan painter, Genaro Lugo, helped a senior FARC official identified as Alberto Bermudez obtain Nicaraguan identity papers, something Lugo admitted to doing. This is interesting because the incident was described in Reyes' papers.

[71] U.S. Department of State, *Country Reports on Terrorism 2008: Nicaragua*, April 30, 2009, available at <www.unhcr.org/refworld/docid/49fac6b5c.html>, and author interviews.

[72] For a look at the story of Monser al-Kassar's long career as a shadow facilitator and weapons dealer, as well as his arrest and conviction, see Patrick Radden Keefe, "The Trafficker: The Decades-long Battle to Catch an International Arms Broker," *The New Yorker*, February 8, 2010.

[73] Noah Rosenberg, "Guilty Verdict in Russian Arms Trial," *The New York Times*, November 2, 2011.

Chapter 6

The Geography of Badness: Mapping the Hubs of the Illicit Global Economy

Patrick Radden Keefe

Suppose, for a moment, that you have come into possession of a small quantity of weapons-grade uranium. Perhaps you filched it from one of the casually-guarded facilities scattered across the former Soviet Union. You are desperate, or just unscrupulous, so you decide to take this precious bit of lethal contraband and sell it on the open market. Once you have devised a method for transporting the material without tipping off the authorities or exposing yourself to incapacitating levels of radiation, you will find yourself confronted with a further predicament: where do you transport it *to*? You are told that there is robust international demand for fissile material and that well-financed terrorist groups, select state governments, and a multitude of shady middlemen and brokers are all seeking precisely the sort of product you have to sell. But where are they? How do you connect with them? In the global black market for the raw ingredients of a nuclear bomb, where do buyers and sellers meet?

The answer, as often as not, is Turkey. Situated at the crossroads of Europe and Asia, with ready access both to the former Soviet Union and to the Middle East, Turkey has been a vital conduit of global trade, licit and otherwise, for centuries. "Turkey is the world's grand bazaar," William Langewiesche remarks, "[and] it is hardly surprising that in recent years people have gone there looking to sell their nuclear goods."[1] With its advantageous location, official corruption, weak penal code, and porous frontiers, the country offers an appealing logistical way station for smugglers of various stripes. By some accounts, as much as 75 percent of the narcotics entering Western Europe either originates in or transits through Turkey.[2] So it should come as no surprise that in the murky underworld of clandestine nuclear transactions, Turkey has emerged as perhaps the dominant venue of exchange. There were over a hundred reported incidents of nuclear smuggling in the country during the first decade after the Cold War alone[3] (though in many instances, the material exchanged did not ultimately turn out to be dangerous). "It is a well-known rumor," Georgia's chief nuclear investigator Archil Pavlenishvili observes, "that you can sell anything in Turkey."[4]

If, as Phil Williams contends, organized crime is "best understood as the continuation of commerce by other means," and transnational criminal groups are "the illicit counterparts of multinational corporations," then it stands to reason that like their legitimate cousins, underworld entrepreneurs might come to rely, in their global activities, upon certain especially convenient logistical hubs.[5] Whether your cargo is narcotics or light arms, undocumented people or dirty money, endangered species or highly enriched uranium, you will find that some geographic locales prove to be especially congenial and can function as a staging area or transshipment point or even a base of operations.

One fundamental insight in the emerging study of illicit networks is the notion that underworld enterprises are mobile, even migratory; as nimble, dispersed, adaptive networks, they can tailor their logistical footprints to the supply of raw goods and labor, the demands of a global market, the shifting regulatory atmosphere in any given region, and pressures both from competition and from law enforcement. Perhaps the most famous exemplar of this new brand of globetrotting criminality, the arms broker Viktor Bout, initially based his operations in Belgium. When authorities there began scrutinizing his munitions flights into war-torn African countries, Bout simply relocated to the United Arab Emirates, then to South Africa, and eventually to Moscow.[6]

Federico Varese has recently questioned the assumption that criminal organizations can easily relocate from one country to another, noting that "transplantation is fraught with difficulties."[7] And to be sure, we must be careful not to overstate the case: the very ad hoc, decentralized nature of contemporary criminal networks renders some of the more alarmist accounts of a rapacious campaign of coordinated expansion less than credible. But because Varese's critique focuses on traditional, geographically-rooted, hierarchical mafias whose primary revenue stream is the collection of extortion, it overlooks the very networked attributes that allow an enterprise like Viktor Bout's to migrate. In my own study of the "snakeheads" of southeast China, who in recent decades have smuggled hundreds of thousands of undocumented migrants to destinations around the planet, I discovered a pronounced tendency on the part of these immigration brokers to identify and exploit certain locations such as Bangkok and Guatemala.[8] Whether it is gunrunners like Bout, snakeheads moving migrants, or nuclear proliferators like A.Q. Kahn, some black market entrepreneurs have demonstrated an extraordinarily sophisticated ability to engage in what Williams calls "jurisdictional arbitrage."[9] Today, organized criminal groups from Nigeria alone are reportedly active in no fewer than 60 countries.[10]

If we can agree that transnational criminal networks *are* mobile, and that like legitimate multinationals they will migrate their operations to especially appealing locations, then it stands to reason that some of these jurisdictions will emerge as hubs in the underground economy. Given that by some estimates the illicit international economy may now account for as much as 15 to 20 percent of global gross domestic product (GDP), you would think that tallying the major geographic centers of this thriving shadow market would be an easy and noncontroversial exercise.[11] The existence of places that are especially susceptible to transnational organized crime has certainly been acknowledged, in a casual way, in the literature of illicit networks. Observers have referred to these locations, variously as "geopolitical black holes,"[12] "twilight zones,"[13] and "pseudo-states."[14]

But surprisingly, given the central role these hubs play as both engines and enablers of illicit global commerce, there has been little rigorous empirical scholarship on how a city, country, or region becomes a black market mecca.[15] While there is hardly room here for the kind of sustained, methodical inquiry that is ultimately required, I will seek in this chapter to initiate a discussion of the geography of the illicit global economy by exploring what features make some places congenial to underworld actors and others not.

The Role of Hubs in a Networked World

It took a while for observers to grasp the revolution that has transformed global crime over the last few decades, but the study of globalized crime and illicit networks has gradually come into focus. One of the bedrock notions at the heart of this emerging field is the idea that a new breed of illicit, transnational, nonstate actors has been very busy in recent years, and that whether they are ideological terrorist groups or profit-driven criminals, they share certain characteristics at the level of their organizational DNA. They are networks.[16]

A decade ago, John Arquilla and David Ronfeldt described a new kind of nonhierarchical adversary: "dispersed organizations, small groups, and individuals who communicate, coordinate, and conduct their campaigns in an internetted manner, often without a precise central command."[17] In the language of networks, each criminal actor came to represent a *node* in a dispersed associational map. The associations between these nodes can be described as *links*.[18]

Much of the early scholarship on illicit networks focused on the democratic character of these structures. Whereas traditional mafias, like nation-states, were entrenched and hierarchical, networks often had a flatter, more fluid, and adaptive leadership structure. This quality made them especially robust and difficult to attack because the removal of any one node in a criminal or terrorist network might have only a limited impact.

Still, even in a decentralized network, all nodes are not created equal. Physicist Albert-László Barabási has shown that some nodes serve as points of connection for a disproportionate number of links. These nodes come to function as hubs. Imagine the difference between a map of the highways in the United States and a map of the air traffic routes. On the road map, cities are the nodes and the highways connecting them are the links. Each major city has at least one link to the highway system, but there are no cities served by hundreds of highways. So most of the nodes are fairly similar, with roughly the same number of links. But an airline map looks different. In order to operate efficiently, airlines route flights through certain hubs. So on a map of air traffic, a few specific nodes, such as Houston, Denver, or New York, will stand out with an exponentially larger number of links as they connect the vast majority of the airports around the country.[19]

In the same way Continental Airlines, Amazon.com, and United Parcel Service construct their delivery routes around a handful of critical hubs, black marketeers come to rely on some locations more than others. In some instances, this may simply be a matter of cultivating a few contacts on the ground in the jurisdiction in question who can make the relevant payments to ensure that a shipment of contraband passes through unmolested. But in other cases, criminal actors will come to base a large part of their business in these hubs, staging their shipments there, sourcing otherwise hard-to-come-by documentation, or laundering their money.

For those seeking to study or combat illicit networks, the importance of these geographic hubs cannot be overstated. All it takes is one lax jurisdiction to create a wormhole that undermines the legitimate international system in its entirety. If one nation is willing to issue fake end-user certificates or genuine passports for a fee, or a handful of border guards are prepared to look the other way at a checkpoint, it won't matter how studiously the rest of the world honors national and international anti-smuggling regulations. If, for instance, word gets out that Guatemala is a back door into the United States—and we live in a world in which there are few barriers preventing the enterprising and mobile criminal or terrorist from simply making his way to Guatemala—then the laudable work of the U.S. Coast Guard off the Florida Keys and the Canadian Mounties at Boundary Bay is sadly beside the point. So the critical question, both in terms of understanding the present landscape of vulnerabilities and of developing some rudimentary method for predicting those vulnerabilities that might emerge tomorrow, is "Why Guatemala?"

Hubs versus Havens

"There are so many forgotten places, out of government control, too scary for investors and tourists," Antonio Maria Costa, head of the United Nations Office on Drugs and Crime (UNODC), told the UN Security Council in 2010. "These are precisely the places where smugglers, insurgents and terrorists operate."[20] Costa's general point is unassailable; in 2003, the Central Intelligence Agency identified some 50 "lawless zones" around the world that might be conducive to illegal activity.[21] But he paints with too broad a brush, a common tendency when describing criminal and terrorist networks, which has resulted in some degree of conceptual confusion. Costa's characterization of underworld hubs brings to mind the 9/11 Commission's description of terrorist "safe havens" as "areas that combine rugged terrain, weak governance, room to hide or receive supplies, and low population density with a town or city near enough to allow necessary interaction with the outside world."[22] But criminal hubs and terrorist safe havens are not necessarily the same thing.

Given some of the organizational similarities shared by terrorist and criminal networks, along with the fact that many terrorist groups use crime to finance their activities, there is an understandable tendency to discuss the two in the same breath. It is also unquestionably the case that jurisdictions such as Dubai have served as an instrumental facilitation point for criminals and terrorists alike. But really, there is little reason to believe that in seeking a base of operations or a key logistical hub, terrorists and smugglers would be looking for the same thing.[23] It has become something of a cliché over the last decade that weak or failed states are especially attractive to terrorists and criminal groups. Even in the particular case of terrorism, this has not always proven to be the case, but when it comes to crime, the assertion is often flat-out wrong. Whereas a terrorist group in Afghanistan or Somalia might benefit from the lawlessness and anonymity afforded by state failure, transnational criminal organizations rely for their very existence on some baseline level of infrastructure and services. In fact, many of the dominant hubs in the illicit global economy can be found in the major cities of relatively coherent states. As Stewart Patrick suggests in his book *Weak Links*—which is the most sustained examination to date of the geography of illicit transnational activity—"the relationship between state fragility and cross-border criminality is variable and complicated."[24]

A multitude of different factors will influence whether or not a given corner of the globe becomes a magnet for criminality, and the strength or weakness of the state in question is just one variable in a highly dynamic situation. Indeed, so many factors influence the level of cross-border criminality in a given country or region that the exercise of classifying the relevant vectors may degenerate into something akin to Justice Potter Stewart's test for identifying pornography: I know an illicit economic hub when I see one. Consider the description of post-Soviet Ukraine's emergence as a black market hub from former Deputy Assistant Secretary of State Jonathan Winer: "There's concentrated power, resources in very few hands, no oversight, no separate functioning judiciary, a huge porous border, huge inherited military facilities, lots of airstrips, a bunch of old planes." What made Ukraine so attractive to underworld elements, in Winer's view, was its role as "a one-stop shopping infrastructure for anyone who wants to buy anything."[25] He also notes the concentration of power—a feature that is not associated with the kind of state failure that often characterizes terrorist safe havens. Winer memorably characterized Ukraine as "the epicenter of global badness," and the country remains a major venue for numerous forms of illicit activity. In seeking to identify other critical locales in the illicit global economy, it should be possible to make some general observations about the characteristic features of a criminal hub.

Hallmarks of a Criminal Hub

In 2003, the Library of Congress published a study called *Nations Hospitable to Organized Crime and Terrorism*. The study feels somewhat dated today; it was a product of the immediate aftermath of the attacks of 9/11, perhaps not the most conducive environment for a sober assessment of the relevant threats. But it provides a thoughtful examination of some of the general characteristics that might render a country appealing to professional criminals:

> *The main domestic elements making a nation "hospitable" to transnational crime and terrorism are official corruption, incomplete or weak legislation, poor enforcement of existing laws, non-transparent financial institutions, unfavorable economic conditions, lack of respect for the rule of law in society, and poorly guarded national borders. . . . Such purely domestic factors often are exacerbated by a nation's geographic location (along a key narcotics trafficking route or in a region where arms trafficking is prolific, for example) or the influence of regional geopolitical issues such as a long-standing territorial dispute.*[26]

Notwithstanding the fact that the study has unhelpfully conflated terrorism and crime, this list of variables is a useful starting point, which I would adapt somewhat to comprise the items below. My own list is hardly meant to be exhaustive. Rather, it is a provisional effort to explore how these dynamics affect transnational crime in particular, which will, I hope, suggest some avenues for future research.

Partial State Degradation

From the perspective of a transnational criminal organization that is primarily in the business of making money by moving contraband products of one sort or another, the prospect of doing business in a failed state is a dubious one. These criminal entrepreneurs rely on access to the global marketplace, which, in turn, necessitates some baseline level of infrastructure. Geopolitical black holes can be "remote and relatively isolated from the main power centers," Moisés Naím suggests. He adds, "But surprisingly, they are extremely well connected to the markets in other continents where they place their lucrative illicit exports."[27] This is not to suggest that criminals are not attracted to some degree of state degradation. Because international commerce is such an important element of transnational organized crime, a failure or a refusal by the relevant authorities to police their own borders creates a decisive strategic advantage for smugglers. Likewise, a diminished willingness or capacity on the part of law enforcement to crack down on criminality can be a major enticement for transnational criminal actors. As UN Secretary-General Ban Ki-moon pointed out in 2010, "Criminal networks are very skilled at taking advantage of institutional weaknesses on the ground," and there is no doubt that this is the case.[28]

Because they are motivated primarily by profit rather than political or religious ideology, and they are often running extensive, high-margin operations, transnational criminal organizations will likely be repelled by a degree of state failure that results in isolation from the broad currents of global communications and commerce, or an unacceptable level of anarchy or uncertainty on the ground. According to UNODC, Colombian drug traffickers first began routing large quantities of cocaine via West Africa in 2004. Countries such as Guinea-Bissau, Guinea, and Mali provided an ideal transshipment point for narcotics bound for Europe. On the surface, West Africa would appear to be an ideal location for a criminal hub. Many states are barely functioning, so the cartels could channel vast quantities of product through unguarded ports and airstrips and onward to the lucrative markets of Europe. By 2008, Western officials estimated that $300 million worth of cocaine, the equivalent of Guinea-Bissau's entire GDP, was moving through the country each *month*.[29]

Then a funny thing happened. Just as international authorities were beginning to awaken to the emerging narcotics hub in West Africa, but before any concerned parties had the chance to conduct a sustained or vigorous crackdown, the flow of cocaine that had been stampeding through the country quite suddenly began to taper off. After 2008, "both maritime seizures and airport seizures on flights originating in West Africa virtually disappeared," according to UNODC. One speculative explanation offered by UN officials is that the corruption-corroded, coup-prone nations of West Africa may actually have proven *too* politically unstable for risk-averse Colombian traffickers.[30] Antonio Mazzitelli, head of UNODC programs in West Africa, reasoned that the region had grown so fragile that even drug runners were abandoning it. "They don't need a failed state," he says. "They need a weak state."[31]

This is a key distinction, and one which is often elided in discussions that conflate criminals and terrorists. The ideal locale for a transnational criminal organization is one which boasts many of the advantages of a strong, functioning state, such as modern infrastructure and communications, a banking system, and enough rule of law to make life generally predictable.

In fact, many of the major hubs in the illicit global economy are also hubs of the licit economy. Bangkok, for instance, has long been appealing to human smugglers, weapons traffickers, document forgers, and other transnational criminals precisely because it is a regional hub for legitimate transportation and a prominent shipping center with a major international airport. Because of the size of the licit economy in Thailand, Bangkok can absorb a high degree of criminality without becoming completely overrun by, or identified, with it. This is not always the case with more fragile states, and here we arrive at another key dynamic: while transnational criminals may not be attracted to outright state failure, once they arrive in a state that is already weak, their very presence can hasten the further degradation of sovereign control. Mexico is a large and prosperous enough country that even in the context of a raging drug war, the basic rule of law widely persists, the state continues to function up to a point, and the country continues to enjoy both tourism and foreign direct investment.

In recent years, Mexico's drug cartels have begun to expand southward into neighboring Guatemala, which is already arguably on the brink of state failure.[32] In recent years, Guatemala's murder rate has been a multiple of Mexico's; a 2007 study by the UN and the World Bank ranked it the third most murderous country in the world with 97 percent of its homicides unsolved.[33] Whereas Mexico struggles with drug cartels but for the most part maintains the basic functions of a viable state, sovereign authority in Guatemala, such as it is, will likely erode even further as the cartels become entrenched. Similarly, Kenya can assimilate the presence of East African criminal groups, whereas the existing disorder in Somalia is only exacerbated by their activities. Think of crossborder criminality as a kind of virus: in searching for a host, it naturally gravitates to weakness. But whereas a big and only somewhat weak state like Mexico or Kenya is robust enough to live with the virus without succumbing to it, an already demoralized state like Guatemala or Somalia is much more vulnerable. So corrosive is the influence of transnational organized crime that even if black market entrepreneurs aren't attracted to state failure, they may ultimately induce it. Occasionally, the virus simply kills its host.

Borderlands and Breakaway States

Importantly, these hosts are not always nation-states per se. Hubs of illicit activity often develop in areas of overlapping or contested sovereignty. Consider the Tri-Border Area, where Argentina, Brazil, and Paraguay meet. Though none of its constituent nations is itself a fragile state, the Tri-Border Area has long been a notorious hotbed of black market activity. It "provides a haven that is geographically, socially, economically, and politically highly conducive for allowing the activities of organized crime, Islamic terrorist groups, and corrupt officials," the Library of Congress report observed in 2003. "Those groups are supported by drug and arms trafficking, money laundering, and other lucrative criminal activities."[34] Paraguay's Ciudad del Este has long been a vibrant hub of licit commerce, generating some $13 billion in merchandise sales in 2000. But it also serves as a clearinghouse for counterfeit goods and has been the subject of longstanding rumors (though not as much in the way of verifiable information) about fundraising for Islamic extremism. The 2003 report identifies "Brazilian and Paraguayan mafias" that are active in the area, "as well as non-indigenous syndicates from Chile, Colombia, Corsica, Ghana, Italy, Ivory Coast, Japan, Korea, Lebanon, Nigeria, Russia, and Taiwan."[35]

A more recent study hypothesizes that beyond this famous triple border, other areas where the sovereign boundaries of three nations converge, like the point where Brazil meets Peru and Bolivia, or Colombia and Venezuela, are also conducive to organized crime.[36] It may be that simple geography dictates these conditions—that any location with ready access to three different countries will diminish accountability and increase the likelihood that criminal actors can find border authorities willing to take a bribe. It may be that a development disparity between neighboring countries will give rise to elevated levels of smuggling. Another explanation is that conditions of overlapping or contested authority are inherently congenial to transnational organized crime.

Another type of jurisdiction in which black market economies tend to flourish is the breakaway state. The former Soviet Union's answer to the Tri-Border Area in South America may well be the Republic of Moldova's Transnistria. An area the size of Rhode Island that broke from Moldova in 1990, Transnistria functions as an unrecognized state within the geographic borders of Moldova. You will not find Transnistria on a map, and according to international law, it does not exist. Yet during the Cold War, the area hosted Russia's Fourteenth Army, and today this unrecognized republic holds one of the most extensive arms stockpiles in all of Europe.[37] "During the last 15 years, the Transnistria area became a black hole," Oazu Nantoi, of Moldova's Institute of Public Policy, says. "There is smuggling of alcohol, tobacco, human beings, drugs, and . . . arms trafficking."[38] The breakaway Georgian republics of Abkhazia and South Ossetia have come to serve a similar function for transnational criminal activity. Residents of Abkhazia reportedly joke that it is "the world's biggest duty free shop."[39]

Corruption

If there is a sine qua non for transnational criminal activity, it is probably official corruption. Graft is the oxygen of organized crime, and kickbacks are a commonplace in most of the global hubs of the illicit economy. The connection between corruption and illicit activity has been fairly well explored elsewhere, so there is no need to dwell on it at length here, but if one appealing quality for a criminal organization on the move is a semifunctioning state, it follows that criminal actors will want to neutralize or co-opt those state functions which might impact their business through well-placed bribes. Depending on the level of corruption in the country in question, and the skill and resources of the briber, corrupt accomplices can often be secured at the very highest levels of government—indeed, even in the very organs of government whose mission it is to crack down on crime. In 2006, the head of Guatemala's drug enforcement agency, along with his deputy and another top counternarcotics official, pled guilty to charges of conspiring to ship cocaine to the United States.[40] In 2009, Guatemala's President, Álvaro Colom, fired the director general of the national police along with three other senior officials after a large quantity of cocaine and cash that was under their supervision went missing.[41]

The only scenario in which corruption does not need to be present for an area to become an illicit economic hub is one in which activity that might otherwise be regarded as illegal or corrupt has simply been legislated into respectability. Think of bank secrecy, for instance. In Switzerland and a host of offshore tax havens, the circumstances that allow for the illegal laundering of criminal proceeds are themselves actually legal. When laws differ from one locality to the next, particularly when the laws in question pertain to the ease and efficiency of money laundering, capital will move in large quantities, an illustration of jurisdictional arbitrage at its most efficient. A banking haven need not suffer from high levels of corruption to attract a criminal element. In these jurisdictions, the rule of law manages to accommodate criminality, perhaps because the corruption we find in these places does not manifest itself in a departure from the legal codes. The corruption is written into the law itself.

Poverty

Closely linked to corruption, poverty is often but not always a feature of areas that become criminal hubs. It seems beyond dispute that if the standard of living were higher and legitimate vocational prospects were more plentiful in southern Afghanistan, say, or in Somalia, we might expect fewer people to go into poppy cultivation or maritime piracy. But in part because the global demand for narcotics and other forms of contraband tends to emanate from prosperous developed nations, it would be a mistake to draw too strong a parallel between the poverty of nations and their appeal to cross-border criminal elements. In a destitute country, crime may appear to loom large because it is often one of the few viable revenue streams available. But in fact, most illicit transnational activity unfolds in wealthy countries, where it is simply obscured by the white noise of a booming legitimate economy. "In absolute terms, transnational organized crime is strongest in the richest countries," a recent UNODC report points out, "but its share of total economic activity there is so small that it does not rise to become a substantial threat."[42]

Here again we are confronted by the dynamic nature of criminal hubs: while poverty is only sometimes a prerequisite for transnational organized crime in a given region, if poverty *is* a problem in the place in question, criminal activity will often exacerbate it. Poverty may drive young Somali men into maritime piracy, but as piracy renders the Somali coast unsafe, it begins to diminish the likelihood that alternative lines of work, like fishing or shipping, might be an option. Piracy also frustrates efforts by the international community to deliver aid to Somalia by ship. These developments exacerbate the poverty, which, in turn, drives more people into piracy.

The dynamics can become self-perpetuating to the point where organized crime can actually derail international development programs. Epic quantities of aid money channeled into Afghanistan are simply diverted outright by corrupt officials and siphoned out of the country by cash couriers. Between 2007 and 2010, more than $3 billion in cash was openly flown out of Kabul International Airport, a sum so vast that even allowing for legitimate transfers by Afghans of honest cash and the profits drawn from drug-smuggling and other illicit activities, it surely reflects large amounts of international money that was designated for reconstruction.[43] In another example of the pernicious effects that organized crime can have on economic development, a 2005 report by UNODC concluded that one explanation for the relatively low levels of development investment in Africa is "the perception among investors that the rule of law does not prevail" there.[44] Thus, as with state weakness, the relationship between poverty and organized crime appears to be highly dynamic.

An Informal Economy

Of course, many African countries share with Afghanistan an especially vital feature in the facilitation of organized crime: they tend to have robust informal economies. In Afghanistan, as much as 90 percent of the economy remains effectively "off the books." A 2006 UNODC study found that 60 percent of fund transfers in the country were drug-related, while as many as 90 percent of the *hawala* money remittance brokers in Kandahar and Helmand provinces were involved in drug transactions.[45] (*Hawala* is an informal banking system widely used in Afghanistan and other parts of Asia.) Most of the cash that leaves Kabul International Airport each day goes to Dubai, which, with its largely unregulated financial markets, its gold souk, and its army of *hawaladars*, has emerged as a crucial nexus in the illicit global economy—a kind of anonymous transit lounge for capital in which clean and dirty money are liberally intermingled.

Dubai has labored in recent years to shrug off its post-9/11 image as a cashpoint and playground for terrorists and criminals, but the fact remains that funds for the 9/11 hijackers and the East African embassy bombers were transferred through the city. Dubai played a key role in A.Q. Khan's black market proliferation network, and half of all applications to purchase U.S. military equipment that originate from Dubai are reportedly from bogus companies.[46] The major figures of Indian organized crime have operated in recent years not from India, but from Dubai.[47] And Iran is, in the words of one U.S. official, "building a bomb through Dubai."[48] Efforts by U.S. authorities to get Dubai to crack down on cash couriers transporting large volumes of illicit funds into the United Arab Emirates have been unsuccessful. A recent proposal by U.S. law enforcement to create a training program to teach inspectors at the airport to spot suspicious couriers was blocked by Dubai's Central Bank.[49]

From Thailand to Transnistria, from Equatorial Guinea to Guatemala, an informal economy is as crucial an ingredient as corruption for an underworld economic hub because the fungibility of cash and other commodities in an underground marketplace allows criminal enterprises to launder their proceeds.

Tribes and Kinship Networks

One common theme that links a number of the key attributes of a criminal hub is a certain tendency to dismiss the prerogatives and institutions of state power as somewhat arbitrary, or even irrelevant. Just as porous border areas and breakaway republics signify the absurdity of the lines and boundaries of official cartography in the face of very different facts on the ground, a common feature in many criminal hubs is the existence of tribal or kinship networks, and the power structures and struggles associated with them, which have a tendency to trump the authority of the state. Much of the expansion of transnational criminal networks in recent decades was driven by loosely dispersed diasporic networks. From the former Yugoslavia to the Federally Administered Tribal Areas of Pakistan, transnational criminal actors may find a congenial base of operations in what the Library of Congress study characterized as "societies based on family and clan ties that have resisted the rule of law."[50] These societies themselves can provide a haven for certain types of criminality, but these kinship networks are also global, and whether it is Fujianese snakeheads in Bangkok or Lebanese traders in Ciudad del Este, ethnic and cultural networks often furnish the transnational links that enable global criminal enterprises to thrive.

State Capture, Rogue States, and the Criminal State

One final factor that is worth mentioning is the propensity for some states to become either so thoroughly corrupted or so alienated from the international system that they do not merely abet criminal activity but actively perpetuate or even authorize and control it. By some estimates, North Korea's senior leadership gleans roughly $1 billion each year from illicit activities. Government trucks reportedly transport the opium harvested in North Hamgyong province to a factory outside Pyongyang where it is processed by a government-owned firm.[51] Iran's smuggling industry, meanwhile, is estimated to generate some $12 billion a year, much of it controlled by the Revolutionary Guard.[52] When government authorities are not merely abdicating their role as guarantors of the rule of law, but deliberately engaging in the illicit global economy themselves, we are confronted with an especially powerful engine of black market commerce, the criminalized state.[53]

Naming, Shaming, Co-opting, and Further Study

The above list of factors hopefully can serve as a starting point for further research. To date, studies of illicit networks have tended to focus either on a kind of mile-wide and inch-deep perusal of the broad contours of the globalization of crime, or alternatively, on inch-wide and mile-deep examinations of the particular characteristics of specific countries or illegal trades. But as the study of illicit networks broadens and matures, the geography of the illicit global economy should merit further research. Identifying the attributes of criminal hubs, and distinguishing them from terrorist safe havens, can serve an important function for scholars (in clarifying our understanding of the dynamics of black markets and the migration of crime) and for practitioners (in suggesting law enforcement and diplomatic responses). A close study of illicit economic hubs may occasionally suggest shifting our focus from one problem area to another. Just as aid money disappearing from Afghanistan is funneled to the hub in Dubai, a close study of the financial networks of Somali pirates might lead not to the failed state of Somalia, but rather to Kenya where the ransom proceeds are often invested.

Nor should Americans be surprised if, in pursuing the links of the illicit economy, they find themselves confronted by major hubs that may be familiar: New York, Atlanta, or Los Angeles, for instance. Until quite recently, Wachovia Bank, which is now part of Wells Fargo, was serving as a major facilitator for Mexican drug cartels.[54] As Raymond Baker, director of Global Financial Integrity, wrote recently, "We cannot succeed in curtailing part of these flows while at the same time facilitating other parts of these flows."[55] Just as drugs move into the United States, assault rifles and laundered money move back out to Mexico. So in identifying and reforming the hubs of the shadow economy, we must be careful not to neglect our own.

Notes

[1] William Langewiesche, *The Atomic Bazaar: The Rise of the Nuclear Poor* (New York: Farrar, Straus & Giroux, 2007), 24–25.

[2] Library of Congress, *Nations Hospitable to Organized Crime and Terrorism* (hereinafter, *Nations* report), Federal Research Division, October 2003.

[3] Lyudmila Zaitseva, "Illicit Trafficking in the Southern Tier and Turkey since 1999: A Shift from Europe?" *The Nonproliferation Review* 9, no. 3 (Fall–Winter 2002).

[4] Lawrence Scott Sheets, "A Smuggler's Story," *The Atlantic*, April 2008.

[5] Phil Williams, "Crime, Illicit Markets and Money Laundering," in *Managing Global Issues: Lessons Learned*, ed. Chantal de Jonge Oudraat and P.J. Simmons (New York: Carnegie Endowment for International Peace, 2001).

[6] See Douglas Farah and Stephen Braun, *Merchant of Death: Guns, Planes and the Man Who Makes War Possible* (New York: J. Wiley and Sons, 2007).

[7] Federico Varese, *Mafias on the Move: How Organized Crime Conquers New Territories* (Princeton, NJ: Princeton University Press, 2011), 4.

[8] See Patrick Radden Keefe, *The Snakehead: An Epic Tale of the Chinatown Underworld and the American Dream* (New York: Doubleday, 2009); and Patrick Radden Keefe, "Snakeheads and Smuggling: The Dynamics of Illegal Chinese Immigration," *World Policy Journal* 26, no. 1 (2009).

[9] Phil Williams, "Transnational Criminal Networks," in *Networks and Netwars: The Future of Terror, Crime, and Militancy*, ed. John Arquilla and David Ronfeldt (Santa Monica, CA: RAND, 2001).

[10] Stewart Patrick, *Weak Links: Fragile States, Global Threats, and International Security* (New York: Oxford University Press, 2011), 138.

[11] Misha Glenny, *McMafia: A Journey Through the Global Criminal Underworld* (New York: Knopf, 2008), xv.

[12] Moisés Naím, *Illicit: How Smugglers, Traffickers, and Copycats Are Hijacking the Global Economy* (New York: Doubleday, 2006), 261.

[13] Timothy Luke and Gearóid Ó Tuathail, "The Fraying Modern Map: Failed States and Contraband Capitalism," *Geopolitics* 3, no. 3 (Winter 1998).

[14] John O'Loughlin, "Pseudo-States as Harbingers of a Post-Modern Geopolitics: The Example of the Trans-Dniester Moldovan Republic (TMR)," *Geopolitics* 3, no. 1 (Summer 1998).

[15] One conspicuous recent exception is Stewart Patrick's *Weak Links*.

[16] See Malcolm Sparrow, "The Application of Network Analysis to Criminal Intelligence: An Assessment of the Prospects," *Social Networks* 13, no. 3 (1991); Williams, "Transnational Criminal Networks."

[17] John Arquilla and David Ronfeldt, "The Advent of Netwar (Revisited)," in *Networks and Netwars: The Future of Terror, Crime, and Militancy*, ed. John Arquilla and David Ronfeldt, 6 (Santa Monica, CA: RAND, 2001).

[18] For a general introduction to the science of networks, see Duncan Watts, *Six Degrees: The Science of a Connected Age* (New York: Norton, 2003). Also see Patrick Radden Keefe, "Can Network Theory Thwart Terrorists?" *New York Times Magazine*, March 12, 2006.

[19] Albert-László Barabási, *Linked: How Everything Is Connected to Everything Else and What It Means for Business, Science, and Everyday Life* (New York: Penguin, 2003), 51.

[20] Louis Charbonneau, "Global Gangs Exploit Blind Spots for Trafficking: UN," Reuters, February 10, 2010.

[21] Patrick, 4.

[22] *The 9/11 Commission Report: Final Report of the National Commission on Terrorist Attacks upon the United States* (Washington, DC: Government Printing Office, 2004), 366.

[23] For an exploration of the types of features that make a given jurisdiction appealing for terrorist, as distinct from criminal, groups, see Rem Korteweg, "Black Holes: On Terrorist Sanctuaries and Governmental Weakness," *Civil Wars* 10, no. 1 (March 2008).

[24] Patrick, 13.

[25] Peter Landesman, "Arms and the Man," *New York Times Magazine*, August 17, 2003.

[26] Library of Congress, *Nations* report.

[27] "Illicit Networks Operate at the Frontiers of Globalization," interview with Moisés Naím, *Brown Journal of World Affairs* 16, no. 1 (Fall/Winter 2009).

[28] Secretary-General Ban Ki-moon, "An Agenda for Prosperity and Peace," Remarks to the Summit of the African Union, January 31, 2010.

[29] Patrick, 151.

[30] United Nations Office on Drugs and Crime (UNODC), *Crime and Instability: Case Studies of Transnational Threats*, February 2010, available at <www.unodc.org/documents/data-and-analysis/Studies/Crime_and_instability_2010_final_26march.pdf>.

[31] Patrick, 151.

[32] See "The Rot Spreads," *The Economist*, January 20, 2011.

[33] David Grann, "A Murder Foretold," *The New Yorker*, April 4, 2011.

[34] Library of Congress, *Nations* report.

[35] Ibid.

[36] Southern Pulse Report, "Ungoverned Spaces in the Americas, Part 1," July 8, 2011, available at <www.southernpulse.com>.

[37] Glenny, 91–96; Helen Fox, "Trans-Dniester's Surreal Life," BBC, September 2, 2005.

[38] Fox.

[39] Sheets.

[40] Drug Enforcement Administration (DEA), "Former Guatemalan Senior Anti-Narcotics Officers Plead Guilty to Conspiracy to Manufacture and Distribute Cocaine," press release, September 8, 2006.

[41] UNODC, *Crime and Instability*.

[42] Ibid.

[43] Matthew Rosenberg, "Corruption Suspected in Airlift of Billions from Kabul," *The Wall Street Journal*, June 25, 2010.

[44] UNODC, "Crime and Development in Africa," June 2005.

[45] UNODC, *Crime and Instability*.

[46] See Angela Giuffrida, "Dubai Labors Under Money Laundering Image," *The New York Times*, December 1, 2010; Patrick Radden Keefe, "Quartermasters of Terror," *New York Review of Books* 52, no. 2, February 10, 2005; "An Unlikely Criminal Crossroads," *U.S. News & World Report*, November 11, 2005.

[47] Library of Congress, *Nations* report.

[48] "An Unlikely Criminal Crossroads."

[49] U.S. Senate, Foreign Relations Committee, "Afghanistan's Narco War: Breaking the Link Between Drug Traffickers and Insurgents" (Washington, DC: Government Printing Office, 2009).

[50] Library of Congress, *Nations* report.

[51] Bill Powell and Adam Zagorin, "The Tony Soprano of North Korea," *Time*, July 12, 2007; also see Sheena Chesnut, "Illicit Activity and Proliferation: North Korean Smuggling Networks," *International Security* 32, no. 1 (Summer 2007); Paul Rexton Kan, Bruce Bechtol, Jr., and Robert M. Collins, *Criminal Sovereignty: Understanding North Korea's Illicit Activities* (Carlisle, PA: U.S. Army War College, April 12, 2010).

[52] Babak Dehghanpisheh, "Smugglers for the State," *Newsweek*, July 10, 2010.

[53] Robert Killebrew and Jennifer Bernal, *Crime Wars: Gangs, Cartels and U.S. National Security* (Washington, DC: Center for a New American Security, 2010).

[54] Ed Vulliamy, "How a Big U.S. Bank Laundered Billions from Mexico's Murderous Drug Gangs," *The Observer*, April 2, 2011.

[55] Quoted in Jeremy Haken, "Transnational Crime in the Developing World," *Global Financial Integrity*, February 8, 2011.

Chapter 7

Threat Finance: A Critical Enabler for Illicit Networks

Danielle Camner Lindholm and Celina B. Realuyo

Money alone sets all the world in motion.

—Publilius Syrus (100 B.C.), Maxim 656

Endless money forms the sinews of war.

—Cicero (106 B.C.– 43 B.C.), Philippics

Through globalization, the free flow of goods, services, capital, ideas, and technology has positively transformed our lives in an increasingly interconnected world. Simultaneously, these same catalysts have allowed bad actors to enrich themselves at the expense of the prosperity, security, and integrity of the global community. These actors, and the illicit networks in which they typically function, actively seek out governance gaps, socioeconomic vulnerabilities, and character weaknesses as openings to conduct their nefarious activities and expand their power and influence throughout the world. No region, country, or community is immune. While we know that illicit networks are enabled by multiple factors, financing is essential for anything they do and enhances their power. Money is the oxygen that brings the activities to life of any licit or illicit organization. Finances sustain these entities and provide a window—and perhaps a vulnerable entry point—into these shadowy organizations.

"Following the money trail" has become a critical crosscutting component for counter-crime, counternarcotics, and counterterrorism efforts against illicit networks. We define the financing of all these illicit activities as *threat finance*. In combating threat finance and associated networks, the international community must collaborate and employ all the instruments of national power to dismantle, degrade, disrupt, and deter illicit networks. These instruments include diplomatic, military, intelligence, information, law enforcement, economic, and financial tools that can be applied alone or in combination to counter national security threats including transnational crime, terrorism, and the proliferation of weapons of mass destruction. Such countermeasures have been successfully leveraged at the local, national, and international levels to combat and degrade illicit networks around the globe; however, these must be continually revamped to keep up with the resourcefulness of illicit actors as they adapt to and circumvent countermeasures.

This chapter focuses on financing as an enabler of all things illicit and the financial front of efforts to combat illicit networks. We will begin with an overview of the threat posed by illicit finance and how illicit actors raise, move, and hide their money. Subsequently, national and international strategies to combat threat finance through law enforcement and intelligence operations, public designations, international cooperation, and capacity-building programs will be examined. We will demonstrate how "following the money trail" enhances law enforcement investigations and intelligence analyses to root out illicit actors and their financiers. Finally, we will underscore the importance of recognizing threat finance as a key enabler of illicit networks and the critical need to combat illicit finance.

The Nature of Threat Finance

Threat finance encompasses money laundering and terrorist financing by illicit networks and endangers the integrity of financial systems around the world. Threat finance is employed by rogue nations, terrorist facilitators, weapons of mass destruction (WMD) proliferators, money launderers, drug kingpins, and others who endanger national security.

As the July 2011 *Strategy to Combat Transnational Organized Crime* (SCTOC) states, the political reach and financial power of transnational organized crime groups allow them to corrupt governments, undermine state stability and sovereignty, subvert and degrade democratic and financial institutions, and threaten strategic markets and the global financial system. The participation of organized criminals in licit markets undermines legitimate competition and market reliability and transparency. Their laundering activities and use of violence, fraud, and corruption create an unfair competitive advantage that drives out honest businesspeople while distorting and possibly destabilizing strategic markets. This is particularly threatening because of the entry of transnational organized crime (TOC)–linked businesses into sensitive markets such as energy, telecommunications, and precious metals.[1] Furthermore, terrorists and insurgents are increasingly turning to criminal networks or engaging in traditional criminal activities to generate funding and acquire logistical support. This relationship is frequently referred to as the growing "crime-terror nexus with groups like Hezbollah, Al Qaeda in the Islamic Maghreb, and the FARC [Revolutionary Armed Forces of Colombia] in Colombia." Transnational organized crime also undermines the integrity of the interconnected trading, transportation, and transactional systems that move people and commerce throughout the global economy and across our borders.[2]

Money Laundering. According to the U.S. Department of the Treasury, money laundering generally refers to financial transactions by which criminals, including terrorist organizations, disguise their identities and the proceeds, sources, and nature of their illicit activities. Money laundering facilitates a broad range of serious underlying criminal offenses and ultimately threatens the trustworthiness of the financial system.[3] It is the process of making financial proceeds from illicit activities appear legal through the following three stages:

In the initial stage of money laundering—*placement*—the launderer introduces his "dirty" illegal profits into the legitimate financial system. This might be done by breaking up large amounts of cash into less conspicuous sums that are then deposited directly into bank accounts, or by purchasing a series of monetary instruments (checks, money orders, etc.) that are then collected and deposited into accounts at other locations.

After the funds have entered the financial system, the second stage—*layering*—takes place. The launderer engages in a series of conversions or movements of the funds to distance them from their sources. The funds might be channeled through the purchase and sales of investment instruments, or the launderer might simply wire the funds through a series of accounts at various banks across the globe. This use of widely scattered accounts for laundering is especially prevalent in jurisdictions that do not cooperate in money-laundering investigations. In some instances, the launderer disguises the transfers as payments for goods or services, giving them a legitimate appearance.

Having successfully processed his criminal profits through the first two phases, the launderer then moves them to the third stage—*integration*—in which the now "clean" funds reenter the legitimate economy. The launderer might choose to invest the funds into real estate, luxury assets, or business ventures.[4]

Terrorist Financing. Terrorist financing refers to the processing of funds to sponsor or facilitate terrorist activity. A terrorist group, like any other criminal organization, builds and maintains an infrastructure to facilitate the development of sources of funding, channel those funds to the providers of materials and services to the organization, and possibly launder the funds used in financing the terrorist activity or resulting from that same activity.

Terrorist organizations derive income from a variety of sources, often combining both lawful and unlawful funding. The agents involved do not always know the illegitimate end of that income. The forms of financing can be grouped in two types:

Financial support comes in the form of donations, community solicitation, and other fundraising initiatives. It may originate with states and large organizations or individuals.

Revenue-generating activities provide income derived from criminal activities such as kidnapping, extortion, smuggling, and fraud. Income may also be derived from legitimate economic activities such as diamond trading or real estate investment.[5]

As demonstrated, criminal and terrorist financing are similar in that "they often exploit the same vulnerabilities in financial systems that allow for anonymity and non-transparency in the execution of financial transactions."[6] For the purposes of studying the financial activities of illicit networks, these similarities (and crossover via the crime-terror nexus) outweigh the differences.

Common Modes of Threat Finance

Financing is essential for any organization and its activities, and this axiom applies equally to illicit networks. Their activities can be categorized into *operational* and *support* activities. Operational activities include surveillance and reconnaissance, rehearsal, final preparations, and execution of the actual illicit activity (for example, a terrorist attack, cybercrime, or drug deal). Support activities entail security, propaganda/marketing, recruitment and retention of personnel, fundraising, procurement, transportation and travel, safe havens, multiple identities, communications, money services, and training. *All* of these activities require financing. In the case of terrorism, while the actual cost of a terrorist attack can be merely in the thousands of dollars, developing and sustaining a terrorist network requires millions of dollars. So how do illicit networks raise, move, and hide their money?

Over the years, illicit networks have relied on a broad spectrum of methods to fund their networks and operations. While money laundering entails disguising identities and funds obtained through illicit activities, terrorist financing does not always involve "dirty money," which complicates the challenges for the intelligence and law enforcement communities. The attacks of September 11, 2001, brought to light how al Qaeda and its affiliated groups exploited the international financial system to finance its preparations for and execution of the attacks. In the decade since, we have observed how terrorist groups have turned to a range of funding sources and mechanisms to move or conceal these funds to circumvent heightened government *and* private sector vigilance and the oversight of traditional banking. While these methods are frequently black-market opportunities tailored to meet illicit needs, they are more often adaptations of legitimate commercial tools (such as technologies designed to ease access to consumers) that are diverted or abused for nefarious purposes. With the good comes the bad, and as the following examples demonstrate, every innovation or convenience for the public provides a new opportunity for enterprising criminals. This list is not exhaustive by any means. The alternatives available to criminal organizations are limited only by their ingenuity.

Nongovernmental Organizations and Charities

Most charitable organizations and nongovernmental organizations (NGOs) are legitimate groups dedicated to their stated altruistic goals. However, due to this same generous mission and the often-opaque nature of their organizational structures and finances, they make attractive targets for illicit actors including terrorists, criminals, and corrupt politicians looking to move their assets. According to the Financial Action Task Force report *Money Laundering Typologies 2002–2003*, charities may have served as a cover for moving funds to support terrorist activities, usually on an international basis, in addition to serving as a direct source of income. For example, according to court documents, the Global Relief Foundation, an Illinois-based charity, sent more than 90 percent of its donations abroad, and, according to the Department of Justice (DOJ) the foundation has connections to and has provided support and assistance to individuals associated with Osama bin Laden, the al Qaeda network, and other known terrorist groups.[7] The Global Relief Foundation has also been linked to financial transactions with the Holy Land Foundation, another charity that has been accused of supporting terrorism. Similarly, the DOJ asserts that the Illinois-based Benevolence International Foundation moved charitable contributions fraudulently solicited from donors in the United States to locations abroad to support terrorist activities since the foundation had offices worldwide through which it could facilitate the global movement of its funds.[8] These activities occur with or without the donors' knowledge. The charities and NGOs can serve as "fronts" for the illicit purpose, or the donations that were intended by donors for legitimate activities may be diverted by charity executives for criminal uses.

Cash Couriers/Bulk Cash Smuggling

Cash is king! Despite newer methods to move and hide money, the exchange of paper currency is still regarded as the dominant and preferred mode of payment across the globe, especially for transnational criminal organizations like drug cartels. Cash couriers and bulk cash smuggling are an attractive mode of moving money and facilitating money laundering and terrorist financing. It usually occurs in U.S. dollars and euros, which are widely accepted as international currency and can always be converted. There is often no traceable paper trail and no third party such as a bank official to become suspicious of the transaction, and the terrorist or criminal can control the movement of that money. However, the costs of couriers and equipment, the risk of the courier stealing the money, the risk of informants within the network, and losses due to border searches or government inquiries that could compromise the network or mission are omnipresent vulnerabilities in the transnational organized crime supply chain. Physically, too, cash can be large, heavy, and difficult to conceal: 1 million dollar bills weigh just over 1 ton and $1 million in $100 bills weighs about 22 pounds.[9]

In the United States, bulk cash smuggling is a money-laundering and terrorism-financing facilitation technique that is designed to bypass financial transparency reporting requirements.[10] Often, the currency is smuggled into or out of the United States concealed in personal effects, secreted in shipping containers, or transported in bulk across the border via vehicle, vessel, or aircraft. According to the Federal Bureau of Investigation (FBI), some of the 9/11 hijackers allegedly used bulk cash as one method to transfer funds. Furthermore, in response to the 9/11 events, U.S. Customs initiated an outbound-currency operation, Operation *Oasis*, to refocus its efforts to target 23 identified nations involved in money laundering. According to the Department of Homeland Security's (DHS) Immigration and Customs Enforcement, between October 1, 2001, and August 8, 2003, Operation *Oasis* seized more than $28 million in bulk cash. However, while some of the cases involved were linked to terrorism, Homeland Security officials were unable to determine the precise number and the extent to which these cases involved terrorist financing. Bulk cash smuggling across the porous borders of Iraq has been used by al Qaeda in Iraq to finance foreign fighters.[11]

According to the U.S. National Drug Intelligence Center (NDIC), bulk cash seizures in the United States alone totaled $798 million from January 2008 through August 2010. These seizures were mostly related to drug-trafficking cases involving Mexican-based transnational criminal organizations. Demonstrating the size and significance of the bulk cash operations related to the illicit drug trade, however, is the negligible effect that these seemingly large seizures have had on stemming transnational criminal operations. NDIC stated in its *National Drug Assessment Report 2011* that "bulk cash interdiction efforts have not impacted overall TCO operations to a significant extent."[12] While U.S. counternarcotics strategies focus on drug interdiction efforts, they now track proceeds from narcotics trafficking that empower transnational criminal organizations to control global supply routes and corrupt government authorities. As a result, U.S. agencies including the Drug Enforcement Administration

(DEA) and the DHS continue to aggressively pursue the detection and disruption of bulk cash smuggling operations.

Alternative Remittance Systems

Illicit networks including terrorist groups use an informal banking system known as *hawala* to move their assets in some regions, owing to the system's nontransparent and liquid nature. A remittance is a transfer of money by a foreign worker to his or her home country. An informal banking system is one in which money is received for the purpose of making it, or an equivalent value, payable to a third party in another geographic location whether or not in the same form. Such transfers generally take place outside the conventional banking system through nonbank money services businesses or other, often unregulated and undocumented, business entities whose primary activity may not be the transmission of money. Traditionally, expatriates—traders and immigrant laborers—used informal banking systems to send money home from or to countries lacking formal and secure banking systems. Informal systems are still used by immigrant ethnic populations in the United States and elsewhere today. Such systems are based on trust and the extensive use of connections such as family relationships or regional affiliations. These systems also often involve transactions to remote areas with no formal banking system or to countries with weak financial regulations, such as in Somalia, where the Al Barakaat informal banking system moved funds for al Qaeda. Figure 1 provides an example of how a simple hawala transaction can occur.[13] It is believed that the perpetrators of the Mumbai attacks of November 2008 relied on hawala transactions to fund their operations.[14]

Trade-Based Money Laundering

Trade-based money laundering is the movement of illicit funds through commercial transactions and organizations that are or appear to be legitimate. This might involve practices such as padding or underreporting of the amounts of invoiced goods or services (for example, shipping more items than documented to allow the recipient to profit from resale); repeating invoices (for example, delivering one set of items but receiving payment for two sets); falsification of receipts (for example, the goods shipped are described as a less expensive item when they're really something more costly); and sales of commodities above or below market. The transactions can be simple (two parties colluding to use a commercial transaction to deflate the value of an exchange to benefit from the difference between the resulting understatement in cost and the value of the goods on the open market) or complex (involving multiple parties in numerous nations knowingly or unknowingly involved in the fraudulent aspects of the transactions). They may also involve trade that is hard to detect or accurately price, such as artwork or unusual services. The key is that these practices take advantage of commercial opportunities to hide their intended purpose and to mix illegally obtained funds into the legitimate financial system. A well-known version of this mode is the "black market peso exchange" (BMPE) popular with South American transnational organized crime. BMPE is regarded as the "most commonly used money laundering method among Colombian transnational criminal organizations" and plays a major role in the movement of funds for Mexican criminal organizations as well.[15]

Figure 1. A Simple Hawala Transaction

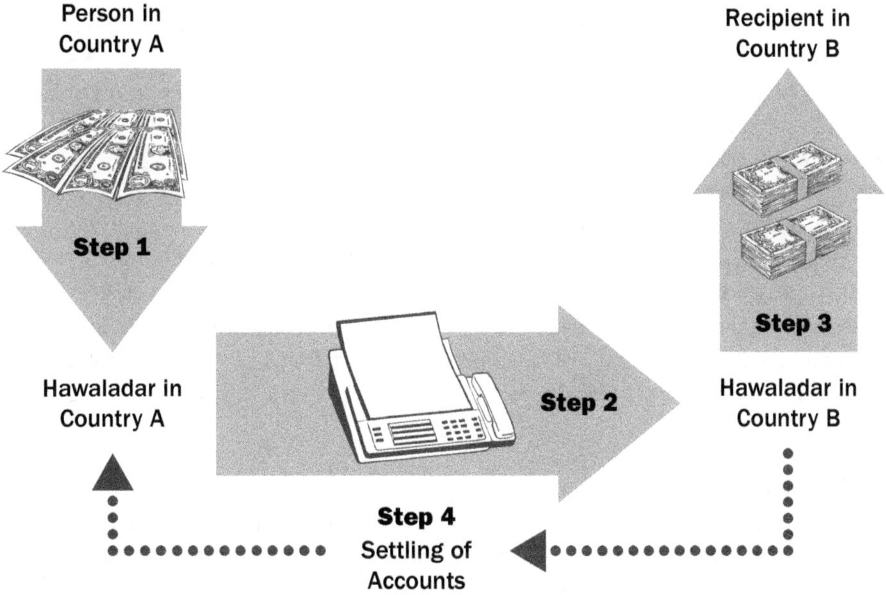

Step 1. A Person in Country A would like to send money to a recipient in Country B. The person in Country A contacts a hawaladar, a hawala operator, in Country A and gives the operator money and instructions to deliver the equivalent value to the recipient in Country B.

Step 2. The hawaladar in Country A contacts the counterpart hawaladar in Country B via fax, email, telephone, or other method and communicates the instructions.

Step 3. The hawaladar in Country B then contacts the recipient in Country B and through verification by some code passed from the person in Country A to the recipient in Country B, delivers the equivalent value (in foreign currency or some commodity), less a transaction fee, to the recipient in Country B.

Step 4. Over time, the accounts between the two hawaladars may become unbalanced and must be settled in some manner. Hawaladars use a variety of methods to settle their accounts, including reciprocal payments to customers, physical movement of money, wire transfer or check, payment for goods to be traded, trade or smuggling of precious stones or metals such as gold and diamonds, and invoice manipulation.

Sources: GAO, using information from the Department of the Treasury, the Financial Action Task Force on Money Laundering (FATF), Interpol, U.S. law enforcement, and other experts.

To transfer drug proceeds back to a cartel's home country, the funds earned internationally must be converted back to the preferred domestic currency. Nonbank institutions conduct legal currency exchanges for their customers, but through a series of both licit and illicit transactions, they can also move funds from "bad" to "good" for the benefit of the criminal client or owner. There are variations, but typically the BMPE involves the foreign-based money house matching traffickers selling drug dollars for pesos with Colombian merchants who are selling pesos for dollars to buy U.S. merchandise. As the middleman, the exchange house buys and sells the drug dollars at a profit. The string of transactions provides the cartel with the ability to move its proceeds without having to engage in the risky transnational transport of large quantities of bulk cash.[16]

Law enforcement agencies are cognizant of many of these methods, but given the total volume of international trade, legal and otherwise, it is impossible to screen and detect fraud in every transaction. In addition, the variations and opportunities for the fraud of over- and under-invoicing goods and services used to mask the illegitimately gained money on behalf of the illicit networks are ever-changing; once detected, the "whack-a-mole" game simply moves to another method or vehicle. Regulations demanding reporting as well as education and outreach to private-sector staff who might encounter such fraud are helpful but are neither foolproof nor a deterrent against this mode.

Prepaid Instruments

Prepaid cards have become fixtures in our daily lives, from the retail gift cards we receive during the holiday season, to the reloadable expense cards college students are given by their parents, to wages distributed via stored-value card to employees and soldiers. The total market for these cards was projected to reach nearly $550 billion by 2012.[17] Consumers appreciate prepaid cards as a convenient means of transferring and carrying funds without the hassles of the paperwork and background checks that are required for debit and credit cards. Issuers and retailers love them because they attract customers, particularly from the hard-to-reach unbanked segment[18] (that is, those without bank accounts often due to their lack of legal identification).

Criminals find them similarly appealing, especially due to the fact that many can be loaded and reloaded with minimal oversight and maximum anonymity. They can be carried legally across the border without being declared or seized as well.[19] According to the NDIC, for example, in 2009, Colombian drug traffickers in the Philadelphia area were found to be using prepaid cards to launder and carry their illicit proceeds. The cartels preferred the cards to the Black Market Peso Exchange because of the ease of movement and the more favorable rates once they exchanged the dollars for Colombian pesos.[20] Hackers from Eastern Europe also found prepaid payroll cards useful when they stole money from a Royal Bank of Scotland payment processing division. In only 12 hours, the criminals loaded the stolen account information onto cloned ATM cards; through their network, they withdrew money from 2,100 ATMs in 280 cities worldwide.[21]

Some of these cards' vulnerabilities may be limited soon due to new U.S. Federal regulations. One recently finalized rule requires reporting from some of the vendors within the prepaid instruments supply chain.[22] According to James H. Freis, Jr., Director of the Financial Crimes Enforcement Network (FinCEN), an agency of the Treasury Department, the regulation is designed to "provide a balance to empower law enforcement with the information needed to attack money laundering, terrorist financing, and other illicit transactions through the financial system while preserving innovation and the many legitimate uses and societal benefits offered by prepaid access."[23] There are plans in place to begin regulating the international transport of prepaid access products, as well, per a notice of proposed rulemaking issued on October 11, 2011.[24]

Mobile Payments

For many years, consumers worldwide have been using their mobile phones to make basic monetary exchanges through simple transactions such as trading phone minutes and downloading ringtones. The availability and variety of payment options and services are proliferating as smart phone usage with Web access increases (especially in developed countries with disposable income). Studies suggest that 35 percent of Americans currently own smart phones,[25] an increase from 17 percent of adults in 2009.[26] Whether one uses a phone as a mobile wallet that is preloaded with currency, or as a tool to access bank or phone accounts for payments, or as a vehicle for financial transactions via text/SMS, Web browsers, or applications ("apps"), the phone may become the consumer's financial settlement method of choice. Long available in countries like India, Mexico, Kenya, and Afghanistan (with varying models and degrees of use), the explosion of mobile services in the United States and globally is imminent. One projection sets the mobile payments market at U.S. $630 billion globally by 2014.[27]

The extent of criminal abuse of mobile payments is unknown at this time. What is certain is that where there is money, there are criminals. Phone-based transactions can be more secure than other means of financial transfer (for example, credit cards, which often leave the account holder for use and are widely targeted by criminals who copy or steal the account access). Nonetheless, the same malware that infects a desktop computer can and increasingly will affect smart phones, making transactions vulnerable to abuse. "Apps" can be infected, passwords can be stolen, and "phishing" emails and "smishing" texts can solicit personal information. (Antivirus software is not widely used on smart phones because, at this stage in product history, to run it constantly would drain the phone battery and limit the device's convenience.) A new study found that mobile malware for Android phones alone increased by 400 percent in the 18 months between the summer of 2010 and the report's release in May 2011.[28] The biggest vulnerability, however, may be the user who is unaware of the threat or is typically more relaxed with connections made via social-networking tools. For example, a SANS Technology Institute poll determined that 85 percent of smart phone users do not scan their devices to search for malware from downloads and other sources.[29] There is a concern, too, that monetary sums aside, some mobile transactions are not traceable or recordable to the extent needed by law enforcement or intelligence to disrupt or prosecute those committing illicit activities. Simply put, mobile payment systems can be considered the "Wild West" for savvy criminal organizations.

Virtual Payments

It may sound odd that virtual currency in a cartoon-like Internet space could have real-world monetary value. This is the case, however, with Linden Dollars, the monetary unit within the "Second Life" online community. Players in these worlds can earn or pass virtual dollars through activities online (for example, by creating a virtual bar at which your virtual customers pay you virtual dollars for your virtual food, drink, and entertainment) and can cash out for real-world currency. (As of October 29, 2011, for example, one online site was exchanging 286 Second Life Lindens [SLL] per U.S. dollar and 392 per euro.[30]) There are dozens of similar online-based financial transaction services available, many of which use the Open Metaverse Currency—OM¢—and which trade through online currency exchanges or even virtual ATMs within their multiuser virtual worlds.

Other Web-based sources of currency exchange such as bitcoin ("no prerequisites or arbitrary limits"[31]) and paysafecard ("personal details not required"[32]) can be a concern for law enforcement and intelligence agencies. These exchanges cater to those whose goal is to remain anonymous in their financial transactions, typically consumers who want to avoid the compliance checks common to traditional banking systems. These activities can be hard to trace, are unregulated or less regulated, and are thus popular with Internet-savvy criminals. (A Twitter post on July 27, 2011, from LulzSec, a hacker group, recommended to like-minded colleagues: "Get yourself a slice of MyBitCoin, Liberty Reserve, WebMoney, Neteller, Moneybookers, and start using prepaid credit and gift cards" to avoid detection.)

In the case of E-Gold, a digital currency firm, the U.S. Government alleged that the company's untraceable exchanges had become "a preferred means of payment for child pornography distributors, identity thieves, online scammers, and other criminals around the world to launder their illegal income anonymously."[33] In 2008, its directors pleaded guilty to charges of money laundering and running an unregistered money service business.[34] Assistant Director James E. Finch, of the FBI's Cyber Division, summed up the original charges by observing that, "the advent of new electronic currency systems increases the risk that criminals, and possibly terrorists, will exploit these systems to launder money and transfer funds globally to avoid law enforcement scrutiny and circumvent banking regulations and reporting."[35]

Combating Threat Finance

Recognizing financing as the lifeblood of illicit networks, the United States and its international allies have developed multidimensional strategies and measures to combat threat financing. International cooperation, institutional capacity-building, public-private partnerships, and intelligence and law enforcement operations are key components of these counterthreat financing strategies.

The Use of Financial Intelligence

At the Treasury Department, the Office of Terrorism and Financial Intelligence marshals the department's intelligence and enforcement functions to protect the financial system against illicit use and combat rogue nations, terrorist facilitators, WMD proliferators, money launderers, drug kingpins, and other national security threats.[36] After the 9/11 attacks, the Treasury Department established the Terrorist Finance Tracking Program (TFTP) to identify, track, and pursue terrorists such as al Qaeda and their networks. The Treasury Department is uniquely positioned to track terrorist money flows and assist in broader U.S. Government efforts to uncover terrorist cells and map terrorist networks at home and around the world.[37]

Since the start of the program, the TFTP has provided thousands of valuable leads to U.S. Government agencies and other governments that have aided in the prevention or investigation of many of the most visible and violent terrorist attacks and attempted attacks of the past decade.

As part of its vital national security mission, the Treasury Department issues subpoenas to the Society for Worldwide Interbank Financial Telecommunication (SWIFT), a Belgium-based company with U.S. offices that operates a worldwide messaging system used to transmit financial transaction information and seek information on suspected international terrorists and their networks. Under the terms of the subpoenas, the U.S. Government may only review information as part of specific terrorism investigations. SWIFT information greatly enhances our ability to map out terrorist networks, often filling in missing links in an investigative chain. The U.S. Government acts on this information and, for counterterrorism purposes only, shares leads generated by the TFTP with relevant governments' counterterrorism authorities in order to target and disrupt the activities of terrorists and their supporters. By following the money trail, the TFTP has allowed the United States and its allies to identify and locate operatives and their financiers, chart terrorist networks, and help keep money out of their hands.[38]

A successful case of following the money trail and leveraging financial intelligence uncovered the terror-crime nexus between Lebanese Hizballah and drug-trafficking rings. In February 2011, the United States accused one of Lebanon's famously secretive banks, Beirut Bank, of laundering money for an international cocaine ring with ties to Hizballah, the Shi'ite militant group. In the wake of the bank's exposure and arranged sale, its ledgers offer evidence of an intricate global money-laundering apparatus that, with the bank as its hub, appeared to let Hizballah move huge sums into the legitimate financial system despite sanctions aimed at cutting off its economic lifeblood.[39]

For the United States, taking down the bank was part of a long-running strategy of deploying financial weapons to fight terrorism. The account of the serpentine 6-year inquiry and what has since been revealed is based on interviews with government, law enforcement, and banking officials across three continents as well as intelligence reports and police and corporate records.[40] From the Treasury Department's perspective, the case is a victory in the battle against terrorism financing. Lebanon's Central Bank showed that it was willing to shut down the Lebanese Canadian Bank and sell it to a "responsible owner," said Daniel L. Glaser, assistant Treasury secretary for terrorism financing. An important avenue to Hizballah has been blocked.[41]

The Afghan Threat Finance Cell

This use of financial intelligence reaches across the interagency and across borders. As an example, threat financing is one of the focus areas in U.S. and coalition counterterrorism and counterinsurgency operations in Afghanistan. It is believed that the drug trade, ancillary crimes, and extortion, as well as foreign donations, are funding the insurgency. Tracking illicit funding has been challenging as the Taliban and its affiliates have diversified their fundraising activities and modes of moving money.[42] Richard Barrett, Director of the United Nations

Taliban and al Qaeda monitoring team, explained that "The Taliban and related groups should be seen as opportunistic when it comes to finances; they are involved on the local level in anything that makes money."[43]

One of the more innovative interdisciplinary approaches to "following the money trail" to combat illicit financing has been the establishment by the Departments of Defense and Treasury of threat finance cells in Iraq and Afghanistan. In 2009, the United States stood up the Afghan Threat Finance Cell (ATFC), a multiagency organization with specialists from the Departments of Defense, Justice, and Treasury, Central Intelligence Agency (CIA), FBI, and DEA to gather intelligence on how the insurgency finances its operations and how to disrupt its funding.[44] More specifically, the ATFC seeks to identify and disrupt financial networks related to terrorism, the Taliban, narcotics trafficking, and corruption. Led by the DEA, with Treasury serving as co-deputy, nearly 60 ATFC personnel are embedded with military commands across Afghanistan to improve the targeting of the insurgents' financial structure. Specially vetted Afghan authorities have also partnered with the ATFC on raids of hawalas suspected of illicit financial activities, including insurgent finance, narcotics trafficking, and corruption. This cooperation has resulted in the collection and analysis of tens of thousands of financial documents.[45]

Since Afghanistan does not have a traditional banking system or a well-developed financial system, the ATFC has been scrutinizing the hawala networks in Afghanistan and greater South Asia and investigating donations from abroad and international transfers. The ATFC has constructively complemented the fight against the Taliban and related insurgents on the financial front. The threat finance cell has also uncovered cases of graft and corruption and contributed to increased transparency and oversight of foreign assistance to Afghanistan. For example, on February 18, 2011, the Department of the Treasury designated the New Ansari Money Exchange as a major money-laundering vehicle for Afghan narcotics-trafficking organizations, along with 15 affiliated individuals and entities under the Foreign Narcotics Kingpin Designation Act (Kingpin Act). The sanctions were issued after an investigation by Treasury Department officials, the DEA, and the ATFC traced money movements from Afghanistan to Dubai. As DEA Administrator Michele Leonhart remarked,

> *The New Ansari network is yet another example of money launderers exploiting legitimate financial systems to launder their ill-gotten gains, including illicit drug proceeds, as part of their criminal enterprise; cash is the ultimate commodity for these criminal networks, and these proceeds often fuel insurgent activity and corruption, while undermining the authority of developing governments.*

Stuart Levey, Under Secretary for Terrorism and Financial Intelligence, noted that "Today's designation is another important step in our ongoing efforts to target money laundering and narcotics trafficking activity in Afghanistan. We will continue to work to expose the funding and support mechanisms for such illicit activity, and to take actions to safeguard the U.S. and Afghan financial sectors from abuse."[46] This case is illustrative of the successful interagency and international efforts to combat illicit financing in Afghanistan to curb support for the insurgency. While they may not put the insurgents out of business, these efforts have played a key role in degrading their networks and disrupting their activities and have provided AFTC members with exploitable vulnerabilities for future pursuit.

Law Enforcement Investigations

In its efforts to challenge drug-trafficking organizations (DTOs) and other illicit networks, the DEA considers illicit finance as a critical enabler of the drug trade. It has identified major money-laundering threats relating to movement of drug proceeds to Mexico as follows:

+ Bulk currency smuggling to include the transportation organizations that service the Mexican DTOs

+ Mexican currency exchange houses, referred to as Casas de Cambio and Centros Cambiario

+ The remission of drug proceeds through U.S.-based money remitters.

DEA has formulated a strategy that includes intelligence-based enforcement, as well as domestic and international collaborative efforts, to target the movement of bulk currency and ultimately attack the command and control targets in the United States and Mexico. By working closely with Federal, state, and local law enforcement counterparts in the United States and Mexico, DEA exploits the intelligence from bulk currency interdictions to identify, target, and ultimately prosecute the command and control targets on either side of the border.

To carry out the strategy, DEA, through the Office of Financial Operations, has established a number of national initiatives that target bulk currency smuggling and the remittance of drug money through U.S. wire remitters. In October 2004, DEA instituted the Bulk Currency Initiative, an information-sharing vehicle through which state and local counterparts can share the information they obtain from making a currency seizure, whether along the Nation's highways or in operation at an airport. Information obtained by DEA in this manner can often be tied to other investigations throughout the world. The currency seizure itself then becomes an overt act in the drug conspiracy investigation and can help to identify other coconspirators within a DTO who were previously unknown. This initiative resulted in an increase in DEA's currency seizures for FY2005 of over $80 million, from $259 million to $339.6 million, an increase of 31 percent.

To coordinate multijurisdictional bulk currency investigations, DEA's Special Operations Division (SOD) instituted the Money Trail Initiative. In the first year and a half, the initiative was responsible for the dismantlement of six national organizations involved in the transportation of bulk currency drug proceeds from various points in the United States to Mexico. As of July 2006, this initiative resulted in the arrest of 418 defendants and the seizure of $65.4 million in U.S. currency, $14.5 million in assets, 59.6 metric tons of marijuana, 9.7 metric tons of cocaine, 126.7 kilograms of methamphetamine, 9 kilograms of heroin, 249 vehicles, and 77 weapons. One of these SOD investigations, Operation *Choque*, resulted in the identification and arrest of Mexican cartel leaders Oscar, Miguel, and Luis Arriola-Marquez and the dismantlement of their organization, which, based on ledgers seized by Mexican authorities, was responsible for smuggling at least 14,000 kilos of cocaine into the United States and smuggling $240 million back. Based on information supplied by DEA, Mexican authorities have seized over $18 million of the Arriola-Marquez organization's assets.[47]

More recently, "DEA has been working with Mexican authorities to gather and use information about these criminal organizations to counter the threats they pose to both of our countries," according to DEA spokesman Lawrence Payne. "As a result of this cooperation DEA has seized illicit transnational criminal organization money all around the world."[48] In FY2010, DEA maintained 21 money-laundering investigative groups to support its Financial Attack Strategy. Through several national initiatives focused on targeting the bulk cash derived from drug proceeds, DEA seized $736.7 million in FY2010. Further, DEA denied total revenue of nearly $3 billion from drug-trafficking and money-laundering organizations through asset and drug seizures that fiscal year.[49]

Following the bulk cash smuggling trail has been an effective tool in combating transnational organized crime and other illicit actors complementing intelligence and law enforcement operations.

International Cooperation Through the Financial Action Task Force

One of the most effective efforts to combat money laundering and terrorist financing has been shepherded by the Financial Action Task Force (FATF) over the past 30 years. The task force was established by the G-7 Summit in 1989 and charged with the mission to develop and promote policies to combat money laundering at first and then terrorist financing in the wake of the 9/11 attacks. FATF is therefore a "policy-making body" that works to generate the political will to bring about national legislative and regulatory reforms in these areas.[50] Its primary role is to set standards for global anti-money-laundering efforts and for combating the financing of terrorism and to ensure the effective implementation of these standards in all jurisdictions. Enhanced global compliance with the standards reduces the money-laundering/financing-of-terrorism risks to the international financial system and increases transparency and effective international cooperation.[51]

The FATF monitors members' progress in implementing necessary measures, reviews money-laundering and terrorist-financing techniques and countermeasures, and promotes the adoption and implementation of appropriate measures globally. In performing these activities, the FATF collaborates with other international bodies involved in combating money laundering and the financing of terrorism. FATF is best known for its "40 Recommendations on Money Laundering" and the "Nine Special Recommendations on Terrorist Financing."[52]

The success of the FATF can be attributed to the Non-Cooperating Countries and Territories (NCCT) Initiative from 2000–2006, aimed at reducing the vulnerability of the financial system to money laundering by ensuring that all financial centers adopt and implement measures for the prevention, detection, and punishment of money laundering according to internationally recognized standards. Countries that were examined and failed to meet international standards were placed on the NCCT list and deemed jurisdictions of money-laundering concern.[53] This "name and shame" process underscored the threat of money laundering and terrorist financing and encouraged countries to respect and uphold FATF's international standards. FATF's NCCT sanctions impede a country's ability to do business in the international financial system and discourage investors from engaging in a country of money-laundering concern. This was the case for Indonesia and Burma, which had to reinforce their anti-money-laundering measures. To get delisted, countries must take remedial steps to address their deficiencies ranging from amending anti-money-laundering legislation to reinforcing their bank regulatory regime.

The FATF "Nine Special Recommendations on Terrorist Financing" are:[54]

+ ratification and implementation of UN instruments

+ criminalizing the financing of terrorism and associated money laundering

+ freezing and confiscating terrorist assets

+ reporting suspicious transactions related to terrorism

+ international cooperation

+ alternative remittance

+ wire transfers

+ nonprofit organizations

+ cash couriers.

Public-Private Sector Partnerships

In addition to interagency cooperation, law enforcement, military, intelligence, and other government authorities benefit greatly from collaboration with the private sector. By facilitating information-sharing among these groups, these relationships provide both public and private sector participants with knowledge that limits the access illicit networks have to legitimate financial systems. The private sector benefits from receiving timely and relevant information that protects its internal business interests from threats posed by terrorists and criminal groups. This may include information about bad actors, identified trends, red flags to detect criminal acts, and tips on when to contact regulatory or law enforcement authorities regarding dubious customers or transactions. They also have the enhanced opportunity to participate in the protection of their industries, the free-enterprise ecosystem, and the Nation as a whole—a patriotic and rational choice that most organizations and their leaders strive to embrace.

Conversely, businesses, nonprofit organizations, and academia share information and perspectives with public sector entities that are vital to the protection of national security and critical infrastructure. This assistance can come through mandatory or voluntary regulatory relationships including the reporting of suspicious activities surrounding business transactions.

When fully pursued by government agencies, it also includes knowledge about industry practices and trends that can aid in the detection of illicit networks, the identification of vulnerabilities, and the prediction of future attempts to circumvent the systems' protections. Further, the private sector shares best and worst practices in managing business operations, including human capital, finance and accounting, and program management, which helps public sector entities to improve the management of their own resources and to better understand the parallel operational structures used by the illicit networks. Where applicable, the private sector can impart an understanding of, and sense of preparedness regarding, promising technologies and services on the horizon that have the potential to be exploited by illicit networks. Finally, these groups are forums where interested parties can forge trust and partnerships prior to a crisis—a resource- and time-conserving opportunity that can speed reaction to new threats.

In recognition of these advantages, the U.S. Government–sponsored Financial Intelligence and Information Sharing (FIIS) Working Group regularly brings together executives from financial institutions, intelligence agencies, academic institutions, law enforcement agencies, and the military. Members of the forum informally discuss relevant topics, including protection of critical financial infrastructure, prevention of fraud, and obstruction of terror finance and money laundering. FIIS meetings and the relationships formed therein facilitate information flow from government to industry and vice versa, and help bridge the cultural gaps often found between the private and public sectors. The FIIS is an outgrowth of the Financial Industry Committee of the Analyst/Private Sector Pilot Project, a project of the Office of the Director of National Intelligence's Office of Private Sector Partnerships. The 6-month pilot project united intelligence analysts with members of industry to facilitate better information-sharing and protection of this facet of American critical infrastructure.[55]

U.S. Special Operations Command's (USSOCOM) Counter Threat Finance (CTF) Working Group is another successful example of partnership both among agencies and between the public and private sectors. In September 2001, USSOCOM was delegated the responsibility to synchronize U.S. agencies focused on operations within the bounds of the "War on Terror." To achieve this goal, several interagency working groups were founded to bring together action officers and policy leaders from U.S. agencies engaged in a select group of activities. In 2005, USSOCOM recognized that Department of Defense (DOD) support aimed at attacking the financial networks of terrorists and associated criminal support networks represented a tangible opportunity to diminish the adversaries' power and influence, and incorporated CTF as one of these working groups. Meeting twice annually, the DOD CTF Working Group creates personal connections and coordinates the planning and resource capabilities resident within DOD that are lacking within other agencies and departments. Finding that expertise from the financial and technology communities would benefit the discussion, this particular segment of the synchronization mission took what was an unusual step and declassified many of their meetings to allow experts from industry, think tanks, universities, nonprofit organizations, and other relevant sectors to join the conversation.[56]

On August 19, 2010, DOD Directive 5205.14, "DoD Counter Threat Finance Policy,"[57] formally designated the commander of USSOCOM as the Defense Department Lead Component for synchronizing counterthreat finance activities across DOD in support of the broader whole-of-government effort to tackle illicit funding activities. Furthermore, DOD

Directive 5205.14 charged the Defense Intelligence Agency (DIA) with providing a critical intelligence component—the Joint Threat Finance Intelligence (JTFI) Office—dedicated to the DOD CTF mission. In October 2011, the JTFI office within DIA joined USSO-COM's CTF effort. The command continues to proactively pursue private-sector partners to contribute to additional facets of its mission of providing direct support to a global whole-of-government effort. Overall, these private-sector individuals and the organizations they represent have supported DOD efforts to identify immediate threats and aid the working group members in planning for imminent vulnerabilities from foreseeable technological trends in the financial community. Over the last 4 years alone, the USSOCOM CTF Working Group has supported over 300 Executive Orders and UN Security Council Resolution designations, provided critical leads that led to over 50 CTF-related arrests and warrants, and been instrumental in the seizure of over $500 million. Since no sector and no community is immune from threat finance, public-private sector partnerships that detect new methods of money laundering and threat finance and devise effective and timely countermeasures are a critical component of strategies to combat the financing of illicit networks.

Conclusion

Financing is the lifeblood of any organization, licit or illicit. While globalization and financial innovation have provided consumers unprecedented conveniences to finance their daily activities, illicit actors have capitalized on these very developments to expand their criminal enterprises. Illicit networks through their terrorist, insurgent, and criminal activities undermine democratic institutions, governments, and international markets and present national security threats to nations around the world. For all of these activities, financing is the engine of growth and is therefore a critical enabler of illicit networks.

To finance themselves, illicit actors engage in a broad spectrum of activities to raise, move, and disguise money that are difficult to detect and disrupt. They include cash couriers, bulk cash smuggling, alternative remittance systems, charitable organizations, trade-based money laundering, and mobile and virtual payments. These modes of finance are ever evolving, and illicit actors are the first to take advantage of gaps in government regulation and oversight of these new methods. Since illicit networks use their capital to keep score and wield influence to undermine, corrupt, and co-opt state actors, attacking their financing is a key component of any strategy to fight illicit networks.

To combat threat finance, governments have developed complex interagency and international counterterrorism and countercrime strategies that struggle to keep up with all the financial innovation and illicit actors' creativity—in other words, networks to fight the networks. These strategies include law enforcement and intelligence operations that "follow the money trail" to pursue and prosecute illicit actors. Since September 11, 2001, financial intelligence has become a fundamental component of counterterrorism and countercrime efforts to detect, disrupt, and deter illicit networks, as evidenced through the Treasury Department's Terrorist Finance Tracking Program against al Qaeda, Hizballah, and their affiliates. Further afield, the mission of the Afghan Threat Finance Cell against the Taliban and the insurgency in Afghanistan has directly contributed to U.S. and coalition counterinsurgency and counterterrorism operations. International initiatives such as those undertaken by the Financial Action Task Force have raised awareness of the detrimental effects of illicit financing on international financial markets and promoted increased regulation and oversight to safeguard the international financial system. Since illicit actors threaten all sectors of society, the private sector must be aware of these threats and engage in combating transnational organized crime and illicit networks. Through public-private sector partnerships, we can glean a better understanding of financial innovation's advantages and disadvantages and of how illicit actors might be financing themselves. While illicit finance will never be fully eliminated, it is a key enabler of illicit networks that must be met head-on through comprehensive and collaborative strategies that engage a whole-of-nation approach.

Notes

[1] *Strategy to Combat Transnational Organized Crime: Addressing Converging Threats to National Security* (Washington, DC: The White House, July 2011), available at <www.whitehouse.gov/administration/eop/nsc/transnational-crime/threat>.

[2] Ibid.

[3] Treasury Department Web site, available at <www.treasury.gov/resource-center/terrorist-illicit-finance/Pages/Money-Laundering.aspx>.

[4] Financial Action Task Force Web site, available at <www.fatf.org>.

[5] Jean-François Thony, "Money Laundering and Terrorism Financing: An Overview," *Current Developments in Monetary and Financial Law* 3 (Washington, DC: International Monetary Fund Publications, 2002), available at <www.imf.org/external/np/leg/sem/2002/cdmfl/eng/thony.pdf>.

[6] International Monetary Fund, Fact Sheet, "The IMF and the Fight Against Money Laundering and the Financing of Terrorism," August 31, 2011, available at <www.imf.org/external/np/exr/facts/aml.htm>.

[7] Global Relief Foundation v. Paul H. O'Neil et al., 207 F. Supp. 2^d 779, U.S. District Court, Northern District of Illinois, Eastern Division, June 11, 2002.

[8] U.S. v. Enaam Arnaout, Case No. 02CR892, U.S. District Court, Northern District of Illinois, Eastern Division, April 2002.

[9] "How Much Does a Million Dollars Weigh?" *Fivecentnickel.com*, available at <www.fivecentnickel.com/2006/02/07/how-much-does-a-million-dollars-weigh/>.

[10] Financial transparency reporting requires Currency and Monetary Instrument Reports, which obligate the filer to declare if he or she is transporting $10,000 or more in cash or monetary instruments across the border.

[11] Rohan Gunaratna, "The Evolution of Al Qaeda," in *Countering the Financing of Terrorism*, ed. Thomas J. Biersteker and Sue E. Eckert, 58–59 (New York: Routledge, 2008).

[12] National Drug Intelligence Center (NDIC), *National Drug Threat Assessment Report 2011* (Washington, DC: Department of Justice, NDIC, 2011), 40, available at <www.justice.gov/ndic/pubs44/44849/44849p.pdf>.

[13] According to the Department of Justice, Al Barakaat operated a hybrid *hawala* in which its informal system interconnected with the formal banking system. Because Al Barakaat also used financial institutions, law enforcement was able to trace the transactions to Somalia by analyzing Suspicious Activity Reports generated by the banks pursuant to their obligations under the 1970 Bank Secrecy Act [Pub. L. No. 91-508, 84 Stat. 1114 (1970) (codified as amended in 12 U.S.C. §§ 1829(b), 1951–1959 (2000); 31 U.S.C. §§ 5311–5330 (2000)].

[14] Douglas Farah, "A Bit More on Dawood Ibrahim and Why He Matters," December 11, 2008, available at <www.douglasfarah.com/article/429/a-bit-more-on-dawood-ibrahim-and-why-he-matters.com>.

[15] NDIC, *Report* (2011), 40.

[16] The Financial Action Task Force offers a useful demonstration of this process and related case studies, available at <www.fatf-gafi.org/dataoecd/60/25/37038272.pdf>.

[17] Mercator Advisory Group, "Seventh Annual Prepaid Card Forecast," available at <www.mercatoradvisorygroup.com/index.php?doc=news&action=view_item&type=2&id=571>.

[18] Per 2009 statistics, the unbanked make up 7.7 percent, or 9 million U.S households. This amounts to 17 million American adults. See "2009 FDIC Survey of Unbanked and Underbanked Households," (December 2009), available at <www.fdic.gov/householdsurvey/2009/index.html>.

[19] Prepaid or "Stored Value" cards are not defined as "monetary instruments" under U.S. law and thus they cannot be seized under requirements to declare currency upon crossing the U.S. border.

[20] "Philadelphia/Camden High Intensity Drug Trafficking Area Drug Market Analysis 2009," NDIC, February 2009, available at <www.justice.gov/ndic/pubs32/32787/abuse.htm#foot3>.

[21] "Hackers Indicted in Widespread ATM Heist," The Wall Street Journal, November 11, 2009, available at <http://online.wsj.com/article/SB125786711092441245.html>.

[22] Specifically to address these issues, and as ordered by Congress via the "Credit Card Accountability, Responsibility, and Disclosure (CARD) Act of 2009," a new rule by the Financial Crimes Enforcement Network (FinCEN), an agency of the Treasury Department, will change how cards are issued, reloaded, and reported in the United States; this rule also renamed stored-value cards or prepaid cards and related instruments as "prepaid access products," available at <www.gpo.gov/fdsys/pkg/FR-2011-07-29/pdf/2011-19116.pdf>.

[23] "FinCEN Issues Prepaid Access Final Rule Balancing the Needs of Law Enforcement and Industry," July 26, 2011, available at <www.fincen.gov/news_room/nr/html/20110726b.html>.

[24] "Financial Crimes Enforcement Network, Amendment to the Bank Secrecy Act Regulations—Definition of 'Monetary Instrument,'" October 11, 2011, available at <www.fincen.gov/statutes_regs/frn/pdf/Prepaid_at_the_Border_NPRM.pdf>.

[25] Aaron Smith, "Smartphone Adoption and Usage," Pew Research Center, July 11, 2011, available at <http://pewinternet.org/Reports/2011/Smartphones.aspx>.

[26] Charles Golvin, "2009: Year of The Smartphone—Kinda," Forrester, January 4, 2010, available at <http://blogs.forrester.com/consumer_product_strategy/2010/01/2009-year-of-the-smartphone-kinda.html>.

[27] Juniper Research Limited, "mPay, mShop, mTransfer!" available at <http://juniperresearch.com/whitepapers/mPay_mShop_mTransfer>.

[28] Juniper Networks Global Threat Center (GTC), "Malicious Mobile Threats Report 2010/2011," May 10, 2011.

[29] Christopher Carboni, "Thoughts on Malware for Mobile Devices—Part 2," Internet Storm Center Diary, SANS Technology Institute, July 12, 2010, available at <http://isc.sans.edu/diary.html?storyid=9160&rss>.

[30] Moneyslex virtual exchange, Web site, available at <http://moneyslex.com/lindendollar2.php>.

[31] Bitcoin Web site, available at <www.weusecoins.com>.

[32] Paysafecard Web site, available at <www.paysafecard.com/us/us-paysafecard/>.

[33] Department of Justice, Press Release, April 27, 2007, available at <www.justice.gov/criminal/cybercrime/egoldIndict.htm>.

[34] Stephanie Condon, "Judge spares E-Gold directors jail time," Cnet.com, November 20, 2008, available at <http://news.cnet.com/8301-13578_3-10104677-38.html>.

[35] Ibid.

[36] Treasury Department, "Terrorist Finance Tracking Program," Fact Sheet, August 2, 2010, available at <www.treasury.gov/resource-center/terrorist-illicit-finance/Terrorist-Finance-Tracking/Documents/TFTP%20Fact%20Sheet%20revised%20-%20%288-8-11%29.pdf>.

[37] Ibid.

[38] Ibid.

[39] Jo Becker, "Beirut Bank Seen as a Hub of Hezbollah's Financing," The New York Times, December 13, 2011.

[40] Ibid.

[41] Ibid.

[42] Craig Whitlock, "Diverse Sources Fund Insurgency in Afghanistan," The Washington Post, September 29, 2009.

[43] Catherine Collins with Ashraf Ali, "Financing the Taliban: Tracing the Dollars Behind the Insurgencies in Afghanistan and Pakistan," New America Foundation, April 19, 2010, available at <www.newamerica.net/publications/policy/financing_the_taliban>.

[44] Ibid.

[45] Treasury Department, "Fact Sheet: Combating the Financing of Terrorism, Disrupting Terrorism at Its Core," available at <www.treasury.gov/press-center/press-releases/Pages/tg1291.aspx>.

[46] Treasury Department, "Treasury Designates New Ansari Money Exchange," Press Release, available at <www.treasury.gov/press-center/press-releases/Pages/tg1071.aspx>; and "U.S. Sanctions Target Afghan Money Laundering," Associated Press, February 18, 2011, available at <www.foxnews.com/world/2011/02/18/sanctions-money-exchange-outfit-afghanistan/>.

[47] Drug Enforcement Administration (DEA) Web site, available at <www.justice.gov/dea/index.shtml>.

[48] Terry Frieden, "DEA defends money-laundering operations with Mexicans," *CNN.com*, January 9, 2012, available at <http://articles.cnn.com/2012-01-09/us/us_dea-money-laundering_1_dea-agents-drug-trafficking-laundering?_s=PM:US>.

[49] Office of National Drug Control Policy, "International Money Laundering and Asset Forfeiture," available at <www.whitehouse.gov/ondcp/international-money-laundering-and-asset-forfeiture>.

[50] Financial Action Task Force (FATF) Web site, available at <www.fatf.org>.

[51] FATF, "High-Risk and Non-Cooperative Jurisdictions," available at <www.fatf-gafi.org/document/52/0,3746,en_32250379_32236992_48468340_1_1_1_1,00.html>.

[52] FATF Web site.

[53] FATF, "High-Risk and Non-Cooperative Jurisdictions."

[54] Ibid.

[55] In addition to the Financial Industry Committee, the Analyst/Private Sector Pilot Project also included committees focused on issues such as nuclear, cyber, and ground transportation security.

[56] Data derived from interviews at U.S. Special Operations Command, Tampa, Florida.

[57] Department of Defense, "DoD Counter Threat Finance Policy," DOD Directive 5205.14 (Washington, DC: DOD, August 19, 2010, incorporating Change 1, November 16, 2012).

Chapter 8
Money Laundering into Real Estate

Louise Shelley

Money laundering into real estate (MLRE) did not end with the movement of organized crime investment into Las Vegas even though many real estate professionals would like to think otherwise.[1] Rather, MLRE is an enduring but insufficiently recognized international problem. Corrupt leaders, organized crime groups, and terrorist organizations channel large quantities of illicitly obtained funds into real estate daily as a way to disguise the criminal origin of their proceeds and to integrate them into the formal economy.[2] These illicit funds are invested into residential and commercial real estate as well as into farmlands and tourist properties, often allowing the criminals and corrupt politicians to enjoy the profits of their criminal activities.[3] Thus money launderers are able to conceal the revenues from illicit activities while acquiring valuable and appreciating investments from which they benefit, thereby doubling their gains. MLRE has a long history because it has typically been a relatively safe and prestigious investment.

MLRE by criminal groups and corrupt officials has been identified on every continent but has failed to command the attention it deserves as a criminal activity. Part of the explanation may be that this form of money laundering can have significant benefits for the recipient country by contributing to construction, adding jobs, and bringing investment into the economy.

Terrorist groups have laundered money into real estate in the Middle East as well as other regions.[4] Both commercial and residential real estate are conducive to money laundering because of the easy mixing of legitimate and illegitimate funds. The consequences of money laundered into real estate are distinct from the sums that remain in cash or cash equivalents. Purchase of expensive homes by crime figures and corrupt officials shows very visually that crime really does pay. Economic bubbles can be exacerbated by MLRE and ordinary citizens can be priced out of markets distorted by money launderers.[5] This was very evident in Japan in the late 1980s. Japan was then the world's second largest economy, and "ambitious and capable *yakuza* [traditional Japanese crime organizations] members were themselves involved heavily in real-estate and stock-market speculation."[6] Estimates of one-third to one-half of the bad debt that strangled economic growth was tied to the *yakuza*.[7] Environmental destruction and overconstruction are characteristics of real estate construction undertaken to launder money, as will be discussed.[8] Moreover, MLRE has often been more beneficial to criminals than other forms of investment in the legitimate economy, as property can provide a base for other criminal operations and/or a steady parallel income stream.

The absence or deficiencies of legislation targeting real estate money laundering make it an attractive arena for corrupt officials and criminals. Corrupt officials are more interested in enjoying and preserving capital while organized criminals more frequently launder money to hide capital, generate revenue streams, and facilitate their illicit activities. Therefore, they often choose different types of property in which to launder their investments.

The failure to recognize and acknowledge the extent of the problem has resulted in little if any pressure on the real estate community to regulate itself in this arena, unlike banking and other financial sectors. Moreover, the absence of will among many real estate professionals in many parts of the world to conduct due diligence regarding their clients, and the absence of sanctions for complicity in money laundering in real estate, have exacerbated the problem. Furthermore, no sanctions have been imposed on other professionals allied with the real estate business, such as notaries and mortgage brokers, who have facilitated this laundering.

This chapter analyzes the global problem of MLRE. Even though there are certain locales particularly known for this problem such as the Mediterranean coast of Spain and France, Sicily, Dubai, and the high-end resort areas in Mexico, MLRE occurs in diverse locales in both the developed and developing worlds.

Although there are no definitive global studies, the problem of MLRE appears to have increased in frequency since restrictions were tightened on other financial transactions post-September 11, 2001. This has been possible because in some countries and regions lacking adequate controls, cash can be used to purchase real estate.

Defining the Problem

Money-laundering experts define three phases of laundering—placement, layering, and integration. *Placement* involves the introduction of dirty money into the system. *Layering* occurs when the money is already in the system and the audit trail is deliberately obscured. *Integration* occurs when the money is already functioning within the system. Real estate can be used at all stages of the laundering process. Money laundering in real estate can occur in the placement stage, where the launderer places the illegally obtained funds into real estate construction or into a house or a commercial real estate purchase. Transactions in the layering stage are intended to obscure any financial (traceable) links between the funds and their original criminal sources. In this stage, laundering typically occurs by moving funds in and out of offshore bank accounts. Overseas, the money may be used for real estate investments or may assume the form of a foreign bank loan to buy a house, when the loan is in reality the purchaser's own money parked overseas. Finally, the goal of integration is to create a "history" showing that funds were acquired legally. In the integration phase, the criminal places money in the real estate sector and is not interested in trading in real estate but in investing.[9]

Many aspects of money laundering have been thoroughly investigated. A significant body of literature examines laundering through banks, shell companies, offshore vehicles, and more unusual instruments such as art and coins. Surprisingly, there has been very little research done on the real estate market as a vehicle for money launderers. The few research studies of the problem are limited primarily to Denmark and Italy. Indeed, there is little systematic research examining this problem anywhere outside Europe.

At the same time, a number of significant prosecutions reveal that the profits of organized crime are often invested in real estate. That is particularly true of investigations into human trafficking, whose proceeds have been traced by European law enforcement agencies to villas in Romania and to hotels, bars, and restaurants in Turkey and the Netherlands.[10] A major prosecution of Russian organized crime in Germany revealed many expensive real estate purchases in the Stuttgart area by members of a well-known post-Soviet crime group.[11]

This analysis is based on interviews with top law enforcement personnel from Europol, the United States, Latin America, Asia, and Australia. It also draws on the expertise of members of the Global Agenda Council on Organized Crime of the World Economic Forum (whose members are leading international scholars and practitioners on money laundering and transnational crime). Reports, legal documents, and the limited scholarly literature have been used to the extent possible.

The slim research on money laundering and real estate means it is not possible to know the extent to which different facilitators are involved in real estate money laundering. Available cases, primarily from Canada and Europe, reveal that some lawyers, real estate agents and brokers, notaries, property management and rental firms, and managers of holiday parks either knowingly or unknowingly facilitate MLRE.[12] But without more comprehensive research, it is impossible to determine which are the most important facilitators of MLRE. Canadian investigation revealed that real estate professionals are among the financial professionals most likely to encounter criminal proceeds (see figure). But there were not a large number of Canadian cases for analysis. Moreover, comparable data is not available from elsewhere.[13]

Professionals Who Came into Contact with the Proceeds of Crime

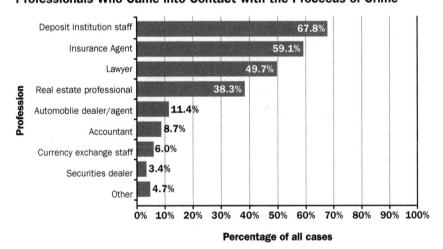

Source: Margaret Beare and Stephen Schneider, *Money Laundering in Canada: Chasing Dirty and Dangerous Dollars* (Toronto: University of Toronto Press, 2007), 135.

In countries with less strict regulations over financial transactions, particularly in the developing world, profits from crime may be placed directly into investments in restaurants, hotels, and other commercial properties.

Politically Exposed Persons (PEPS) represent a unique case and set of money-laundering patterns. According to the Wolfsberg Principles addressing money laundering, which were adopted by 11 global banks, PEPS are described as

> *persons who perform important public functions for a state. The definition used by regulators or in guidance is usually very general and leaves room for interpretation. For example the Swiss Federal Banking Commission in its guidelines on money laundering uses the term "person occupying an important public function," the U.S. interagency guidance uses "senior foreign political figure" and the BIS paper Customer due diligence for banks says "potentates."*

> *The term should be understood to include persons whose current or former position can attract publicity beyond the borders of the country concerned and whose financial circumstances may be the subject of additional public interest. In specific cases, local factors in the country concerned, such as the political and social environment, should be considered when deciding whether a person falls within the definition.*[14]

A review of examples by the author as well as some of the analyses done by Global Witness[15] suggest that MLRE involving Politically Exposed Persons more frequently flows through banks than funds derived directly from criminal activity. Banks may be hesitant to reject the accounts of some Politically Exposed Persons when the political leaders of Western powers interact with those same political leaders.[16]

The Multiple Consequences of MLRE

Money laundering into real estate differs from other forms of laundering activity. Especially in the residential real estate market, such activity can be a sign of "having made it." This contrasts with other types of laundering which have a more utilitarian function—simply hiding capital. Therefore, this form of laundering has at least a dual purpose—cleaning the money and enjoying the proceeds of crime and corruption. But as will be discussed, it can serve other purposes by providing a base on which to mount operations, thereby compounding the harms of the initial criminal activity.

Criminals show their success by buying valuable real estate—beautiful estates and villas as well as elegant and/or popular bars and nightclubs.[17] Criminological research as well as popular songs in the criminal culture, especially among Latin American crime groups, shows that visible signs of wealth serve as important tools for recruiting members into the criminal enterprise. Not surprisingly, this is especially true of impressionable youth. Real estate is one of these signs of success. Money laundering into ostentatious homes is a dramatic force for destabilization as it shows that leaders can enjoy the fruits of corruption. Popular anger and revulsion at the wealth and opulent homes of corrupt figures including political leaders can be a source of major discontent within the population, particularly in poor countries. Revelations of the sumptuous property of Arab leaders, for example, helped fuel popular resentment before the Arab Spring.

An activist in Gabon with assistance from Transparency International and Sherpa successfully initiated legal action to freeze 73 properties purchased by the families of African dictators from Gabon and two other African countries.[18] The French Supreme Court froze these properties in an unprecedented legal action against the French assets of PEPs.

Among the many other financial consequences of money laundering into real estate is the exacerbation of economic bubbles. For example, this was seen in Japan before the long-term recession. *Yakuza* members and organizations moved large amounts of their assets into commercial real estate, driving up prices. This bubble could not be sustained and many commercial loans were not repaid to the banks as Yakuza members had key roles in the directorates of the banks.[19] The same phenomenon was observed more recently in Dubai. Little scrutiny was given to the source of assets used to buy apartments, villas, and offices in the rapidly developing property market in Dubai. As investigations have shown, money from corrupt officials poured out of Afghanistan into Dubai, and the recent arrest of post-Soviet drug kingpin Kamchy Kolbaev in United Arab Emirates (UAE) is illustrative of the criminal capital that has been invested in the economy there. The illicit capital fueling the real estate boom may have contributed to the instability of the markets and the sharp decline of property values.[20]

MLRE can increase prices, making it impossible for local individuals to acquire real estate. The threefold growth of property prices in Mombasa, Kenya, within 5 years made it increasingly difficult for middle-income families to buy homes.[21] In the mid-2000s, according to the *Economist*, South Africa had the largest price increase in housing of any major market analyzed.[22] The President of the Financial Action Task Force stated in 2005 that he suspected money laundering was one of the reasons for the sharp increase in property prices there.

Money laundering can result in overconstruction of expensive housing and hotels. Properties[23] stay vacant for years because individuals cannot afford to buy or stay in them. The vacant hotel phenomenon can be seen in Central America and several Soviet successor states. The housing boom in Turkey, facilitated in part by the laundering of drug and human trafficking proceeds, has resulted in overconstruction and empty housing in some coastal areas.

MLRE can also result in severe violations of environmental regulations and destruction of protected habitats. This has been a widespread phenomenon in southern Italy. Already in the mid-1990s, citizen groups were established to combat the destructive environmental impact of housing in areas too close to the sea or other fragile terrains.[24] As a result of this citizen mobilization, some illegally constructed housing has been torn down, especially in periods of government crackdowns on the Mafia.

Money laundering into construction companies can drive out legitimate investors who cannot obtain bank credits or compete against money launderers with so much disposable cash. These cash-rich investors move money not only into the property, but into the companies that will construct new developments, thereby increasing their laundering options.

These practices do not always involve high-end properties. Money laundering into lower-end property may lead to decay and increased crime in urban areas.[25] In some cases, properties are bought to launder the money and then left vacant. This has been observed in such diverse environments as rural Ohio and central Tokyo.[26] Such decay may be allowed so the criminal investors can subsequently buy neighboring properties at depressed costs, thereby increasing their territorial influence.[27]

MLRE may yield profits or other tangible benefits allowing launderers to sustain their activities while simultaneously providing stability and security for their investments. This has been seen in Mexico, major cities in the United States, and Italy.[28] Therefore, although many forms of laundering cost launderers 10 to 20 percent of the sums they seek to clean, this rule does not always apply in the real estate sector. An advantage of laundering into real estate is often that there is not even a cost of laundering, but rather ongoing financial benefits derived from property-based revenue streams. Moreover, money laundering into apartments can be used to facilitate the work of drug traffickers who can use these venues for production of drugs and as warehouses for drug distribution.[29] These buildings can also be used to house illegal workers who are sometimes exploited in the criminal businesses of apartment owners.[30] Or as in one Arizona case, they may be used as safe houses for migrant smuggling.

Money laundering into hotel construction and tourism businesses may facilitate other illicit activity of organized crime groups such as human trafficking. This has been seen in tourist areas in Spain and Turkey where drug money has moved into the hotel sector.[31] In Southeast Asia, Japanese organized crime investment in the hotel sector has facilitated sex tourism.

Money laundering into farms and ranches can distort agricultural prices. In Colombia drug traffickers chose not to invest in agricultural lands that demanded constant cultivation. Instead, they invested in cattle ranches, resulting in depressed beef prices as their desire was not to engage in profitable cattle farming but merely to launder their money.[32]

Diverse Environments in Which MLRE Occurs

MLRE is a transnational phenomenon spanning the developing and developed worlds. Money moves from the developed world into properties in the developing world and vice versa. The mobility of capital makes it difficult for financial professionals to know their customers.[33] Moreover, money moves among different regions within the developed and the developing worlds.

The extent to which criminals and corrupt officials launder money into real estate depends on the "cultures" of organized crime groups and the effectiveness of the regulatory and confiscation policies of particular countries. In cultures where investments traditionally occur in property, there is a greater likelihood that money laundering will occur in all forms of real estate both in their home region and abroad.[34] Therefore, investment patterns of criminals reflect those of legitimate investors.

In the case of corrupt government officials, especially those from poor countries and conflict regions, expensive properties are often acquired in developed countries or havens such as the Gulf states.

With transnational crime groups, laundering can occur in their countries of origin as well as in areas where they plan on further or future operations. The former has been seen by British officials investigating welfare fraud and child trafficking.[35] Romanian criminals profiting from human trafficking crimes in the United Kingdom built enormous villas with the proceeds of their crimes. But this phenomenon was not confined to the UK.[36]

Yakuza have become global investors in real estate. They have invested in golf courses and land in Hawaii and in hotels, resorts, and other commercial properties throughout Southeast Asia. This property facilitates their activities in the area of sex tourism.[37]

MLRE occurs most easily in cash-based economies where there is no regulated price market and no specific property acquisition requirements. This is the case throughout Latin America and in many parts of Asia and Africa. These features make countries in these regions especially vulnerable to money-laundering abuse in both the commercial and residential real estate markets.

In many real estate markets in the developing world, laundering opportunities are enhanced by uncertainty in the actual market value of property and the long-term tradition of speculation in real estate markets. Real estate offers many comparative advantages relative to financial instruments as it allows launderers to dispose of considerable sums of money through construction, yet also derive a subsequent income flow, particularly in markets where housing is in high demand.[38] Real estate also produces steady returns of clean money from tourists staying at hotels and resorts.

In diverse regions of the world, real estate such as apartments, garages, and warehouses can also be used to conduct criminal activities.[39] As one European analyst concluded, "To sum up, the real estate sector is by its very nature complex and prone to criminal abuse."[40]

Earlier Patterns of Money Laundering

MLRE has evolved over the decades as criminals adapt to changes in regulations and market conditions and become more skilled global investors. Earlier identified examples show that most investment in real estate by criminal organizations has been conducted closer to home. Criminals tended to choose environments that they were familiar with.

In the 1970s, Colombian crime groups were accumulating significant amounts of capital. A major investment strategy was the purchase of large tracts of rural property as well as valuable commercial real estate in Bogota. This influx of capital as a result of laundering led to a rapid escalation of commercial real estate prices.[41] The phenomenon of laundering money into property extended for decades; the attorney general's office of Colombia estimates that $10 billion was laundered into real estate between 1997 and 2004 and only a fraction of this was confiscated.[42]

The experience of the Colombian groups in using property as a major vehicle to dispose of drug proceeds was subsequently replicated by Mexican drug cartels. Mexican cartels moved money and drugs into and through Cancun and other resorts. This provided them a steady revenue stream.[43]

The utility of property as an investment vehicle has also been observed in Asia and Europe. Japanese yakuza moved into multiple real estate markets forcing rapidly escalating purchasing prices. Individuals who refused to sell their homes to yakuza-associated real estate investors faced severe harassment. Banks could not collect on the large loans made to yakuza property developers, leading to large uncollectable loans for the banking sector that have been calculated as high as 40 percent of nonperforming loans, contributing to Japan's so-called lost decade.[44]

Organized crime groups acquired agricultural lands in Sicily, villas on the Sicilian coast, and real estate in Naples. From the 1980s to 2000, real estate was the major asset seized by the Italian state from Italian organized crime groups. Post-1993, money laundering was no longer concentrated in the major urban areas of the south but spread to secondary and tertiary cities in that region.[45]

In the Netherlands, proceeds from extensive trade in drugs as well as human trafficking have provided significant revenues for criminals. As early as 1996, the Dutch national organized crime report determined that through their illegally acquired property, organized crime groups had acquired control over Amsterdam's red light district. Amsterdam had by this time become a major center for international organized crime.[46]

The United States has a long tradition of organized crime laundering money into real estate. Organized crime groups have owned significant property in such cities as Las Vegas and New York since the 1930s. Organized crime money lies at the very heart of Las Vegas's historical development. Indeed, pension funds diverted by organized crime from the Teamsters Union helped to establish the city.[47]

Recently Identified Patterns of MLRE (post-2000)

With greater scrutiny over banks and other financial instruments since the financial controls enacted after September 11, 2001, real estate has become an especially important investment vehicle for organized crime. In some parts of the world such as the Middle East, Russia, and Africa, cash can be moved directly into real estate. Suitcases of cash are used to pay for transactions. In more developed markets such as the United States and Europe, money laundered into real estate appears more often to be a secondary transaction after it has already entered into banks or another part of the financial system.

In Europe, with its long-term traditions of maintaining assets in residential and commercial real estate, property remains a favored investment by crime figures. It may have become more popular after 9/11 because of increased scrutiny of bank financial records.[48] In this respect, the investment patterns of those who participate in the illicit economy are no different from those in the legitimate economy. An analysis of over 50 cases of money laundering in the Netherlands revealed that in 30 to 40 percent of the identified cases, laundering had occurred by way of immovable property.[49] An Italian study of all property confiscated from organized crime revealed that laundering into real estate was no longer confined to the south but had spread throughout Italy including the north.[50] This is a consequence of the movement of organized crime groups out of their traditional strongholds in the south into the more lucrative and attractive markets of northern Italy. As previously mentioned, the illicit assets of corrupt African dictators have been found in France and the revelations of the Arab Spring have provided evidence of Libyan, Turkish, and Egyptian leaders and their families transferring their countries' looted assets and the proceeds of their corrupt activities into European real estate markets.

The United States continues to serve as an investment destination for the funds of foreign corrupt leaders. Money stolen by corrupt foreign leaders is often destined for high-end real estate that can be enjoyed by the launderers and their families. Former Ukrainian Prime Minister Pavel Lazarenko bought a $7 million home in California and was charged and prosecuted for money laundering in 2004.[51] A recent Senate investigation examined four cases of MLRE by Politically Exposed Persons from Africa. These were high-level officials such as Teodoro Obiang, a son of the president of Equitorial Guinea, and their families.[52]

In Africa as well, there is now significant evidence of laundering by organized crime groups into local property markets. In recent years, the population growth rates of several African countries have been among the highest in the world. Illicit money from neighboring countries flows into countries with high growth rates such as Kenya. This phenomenon is also identified in southern parts of western Africa where growth rates are not as high as in Kenya. Ransom money associated with the piracy attacks off the coast of Somalia has also contributed to the recent real estate boom in Nairobi. New cocaine transport routes through West Africa have also contributed to a real estate boom in that region. Expensive villas have been built with drug proceeds in Guinea-Bissau, for example, a poverty stricken country with no legitimate source of funds for such construction.[53] Money has flowed from other parts of Africa and Asia into real estate in South Africa as well because of its well-developed infrastructure.[54]

There are certain locales that, because of their large-scale real estate development and failure to scrutinize the source of funds, have become magnets for laundering money. Two examples, as previously mentioned, are the southern coast of Turkey and Dubai in UAE. Large-scale development has occurred on the Mediterranean and Aegean coasts of Turkey because real estate purchases can be made in cash, soaking up tens of millions of dollars that flow back from criminal enterprises in Western Europe.[55] The proceeds of corruption and the drug trade in Afghanistan are laundered in real estate, particularly in commercial and residential real estate in UAE, notably Dubai.[56] Money from organized crime in the post-Soviet area is also laundered there. As previously mentioned, Central Asian drug kingpin Kamchy Kolbaev was arrested in his villa in Dubai. The profits from a very large case of fraud against the Russian government committed by law enforcement officials following a corporate raid against the Hermitage Company resulted in profits in the tens if not hundreds of millions of dollars for the perpetrators.[57] The Russian lawyer for the Hermitage firm who opposed the fraud died in prison after being convicted of trumped-up charges. Some of the stolen money has been traced to expensive properties owned by the perpetrators in Dubai.[58]

Latin American drug organizations continue to launder money into real estate both in their home states and abroad. Members of Colombian and Mexican drug cartels continue to launder money in their own countries but also place it in Central and South America. Favorite locales are Argentina and Venezuela.[59] Money laundering into Uruguayan real estate has also become a major problem. A recent study reports that 90 percent of money laundering in Uruguay occurs through real estate.[60] Panama has been singled out for money laundering in the real estate sector,[61] and Honduras has unoccupied hotels, a phenomenon also visible in Argentina and other Latin American countries.[62]

In the United States, as previously mentioned, drug cartels are using their profits to buy property, particularly in border states. Canada has also identified drug traffickers using property across Canada as a vehicle for money laundering.[63]

Distinct patterns have been noted since 2008 and the start of the financial crisis. The sharp decline of real estate prices has been exploited by criminals for both their operational benefit and to purchase real estate at sharp discounts. Banking regulation encourages criminals to keep their assets in cash, enabling them to quickly take advantage of low prices. Criminals, therefore, have been major beneficiaries of the crisis because many of them held cash and have not been negatively affected by stock market declines. This has occurred in the United States and in some depressed markets in Southern and Northern Europe.[64] In Italy, the Mafia is now estimated to control 7 percent of the economy and extorts and works through construction businesses and urban properties, particularly in Rome.[65]

U.S. authorities have estimated that approximately $19 billion in cash returns to Mexico annually as a consequence of drug sales in the United States. In one well publicized case, over $200 million in dollar bills was found in a single property in Mexico City.[66] Mexican drug organizations operating in both the United States and Mexico have benefited from this cash liquidity. They have been ready to buy from motivated property sellers at reduced and often distressed prices.

Under these circumstances, it is hardly surprising that investigations by the office of the Attorney General of Arizona have discovered that Mexican crime groups are buying depressed properties in Arizona and are then using the purchased buildings as safe houses for their human smuggling operations. In this way, they are taking advantage of low cost real estate that can then be used to benefit their criminal activity. This is more ambitious than most money laundering, which merely seeks to move money obtained in the illegitimate economy into the legitimate system. Real estate can disguise the illicit proceeds of crime but can also simultaneously embed transnational crime within the community.

How Is Money Moved into Real Estate Investment?

There is insufficient research and analysis to fully reveal how money is laundered into real estate. Knowledge of how real estate markets work in different parts of the world can provide some insight. Distinct patterns appear to exist in both the developing and developed worlds. In the more highly regulated developed countries, there is little likelihood that a full real estate purchase will be exclusively in cash, requiring a physical transfer of bills. In many parts of the developing world, on the other hand, cash payments are normal.

In developed countries, financial institutions have to be the initial point of the laundering process for most real estate transactions because suitcases of cash are automatically suspect and should trigger reports of suspicious financial transactions. Yet in some cases, cash payments outside the regular banking system may be provided for part of the purchase price. For example, 10 percent of the purchase price may be paid in cash outside the formal transaction thereby permitting the purchaser to launder smaller amounts of money even in the more regulated markets.[67]

False documents play a key role in the laundering process in Europe. Europol reports that false documents are used to conduct mortgage fraud in the purchase of premises.[68]

The reason the yakuza have been able to launder such large quantities in Japanese real estate is that they are key actors in the banking sector, owning shares of banks and sitting on the boards of directors of institutions. They have been able to move their money through the banks they control and obtain loans for future construction, demonstrating that MLRE is far from limited to developing countries and indeed occurs on a large scale even in the world's currently third largest economy. Even though the yakuza have been hurt financially in the past by the decline of real estate, they continue to be major players in valuable commercial real estate markets in Tokyo and elsewhere in Japan.[69] Many multinational companies buy, partner, and sell to companies controlled or dominated by Japanese organized crime as they seek a stake in the Japanese economy.

In parts of the developing world where cash easily enters real estate markets, there are a variety of ways in which money is transferred to the desired destination. These include wire transfers,[70] the use of cash couriers and underground banking procedures such as the *hawala* system (particularly present in Africa and the Middle East), and alternative remittance systems directly or through third parties. In the Sneep case, involving human traffickers operating in the Netherlands, couriers carried cash to Antalya, Turkey, where it was retained and used to make real estate purchases in the region's booming economy.[71] In the Dubai markets discussed earlier, large shipments of cash from Afghanistan were used to buy villas. Yet in the Dubai markets, money was not necessarily laundered in cash. As in the Hermitage case, the illegally obtained money was first deposited in a Swiss bank and transferred to Dubai.[72]

In many parts of the world, the individual buying the property with money derived from criminal or corrupt activity seeks to disguise his/her identity.[73] In both developed and developing country markets, efforts are made to separate the identity of the purchaser from the individual on the property register. This can be done in several ways. Lawyers and others in positions of trust can purchase property as front people for the real owners. Especially in the case of corrupt officials, properties are bought under false names or by relatives or close associates. A ready sign of laundering in many geographic contexts is that a company is established and then used as a vehicle to launder money into real estate. Once the purchase is made, the company is liquidated.

The absence of transparency in ownership documents and the ability of front companies to buy property and hide actual owners make trafficking into real estate a very viable option for laundering significant sums.

Conclusions

This chapter has focused primarily on the mechanisms by which large amounts of illicitly obtained money are invested in both residential and commercial real estate markets globally. The ease with which criminal organizations do this on such a large scale, as is apparent with yakuza laundering into Japan's highly developed and valuable real estate market, indicate that money laundering through real estate transactions should be an important concern if one is trying to improve the transparency of international financial markets and investment vehicles.

MLRE is also connected to core themes of the global political economy including disparity of income, access to decent housing, and problems of quality of urban life in densely populated areas. The ability of corrupt leaders to steal so much from their countries and to launder their money into valuable properties overseas contributes to stunted economic development in many countries and to destabilizing income disparities. The price bubbles created by laundering that have been seen in parts of Africa as well in other markets deny ordinary citizens access to adequate housing. In cities, and especially in large cities in the developing world where organized crime groups are major investors, criminal economic control of territory through ownership of residential and commercial property undermines and determines the quality of life in the community.

The global financial crisis that began in 2008 was caused, in part, by problems in the real estate market and inflated valuations. Fraud and organized crime played a role in these market manipulations in some regions of the world. But as cash-rich investors in the extended recession, they have the possibility to position themselves for growth and to facilitate their operations in the future.

As the global financial community works to establish regulation to prevent such future financial crises, the impact and role of organized crime and corrupt elites in real estate markets should be carefully examined. Market reforms must include greater oversight of the sources of funds entering into real estate markets just as they do for other financial markets. There must be greater accountability and reporting requirements for real estate professionals who facilitate MLRE. Market analyses must also address susceptibility to money laundering. Greater transparency is required in land registries, privatizations, and property transfers. Law enforcement must follow the money trails into both commercial and residential real estate and seize properties as they do other laundered assets. They must give higher priority to countering the placement of money in real estate and the ongoing revenue streams going to the organized criminals who own residential and commercial real estate.

In Italy, properties that have been confiscated from organized crime are used for the social good. Farms once owned by the Mafia have been used to provide work for individuals with special needs. Villas have been used as social centers for children and for other functions in the community. In contrast to countries that just sell the confiscated properties linked to organized crime, the Sicilian experience has another purpose: to ensure that "crime does not pay" and properties that once served people who harmed others now serve the public good.[74] This concept could be replicated in other communities that seek to wrest authority away from organized criminal organizations.

The decade since 9/11 has seen much attention placed on regulation of banks and global financial markets. As the collapse of the real estate bubble worldwide has shown, there must be much greater attention paid to real estate markets and to funds used for real estate investment. Little has been done to institutionalize the recommendations of the Financial Action Task Force (the chief international body addressing money laundering) to ensure that real estate markets are properly regulated and guarded against money laundering. Harmonized measures have not been put in place to ensure that real estate professionals, lawyers, notaries, and accountants are accountable and understand their responsibility to "know their customers." There is an absence of reporting worldwide on financial irregularities and laundering in real estate markets. Not enough has been done to ensure that real estate professionals make countering money laundering a higher priority. This is a difficult time to enhance requirements when sales are down and customers are hard to find, yet it is needed if the global financial community is to prevent future crises.

Failing to take these steps has consequences for society that transcend the present financial crisis. They include loss of homes, displacement of people, and a more precarious life in communities in which organized crime groups are major investors. As the events of the Arab

Spring are assessed, it is evident that MLRE by corrupt leaders was a galvanizing force for the anger citizens felt at the waste, corruption, and self-indulgence of their leadership.

As we examine the forces for financial and political stability in the first decades of the 21st century, it is clear that abuse, corruption, and money laundering into domestic and commercial real estate have played an important role in destabilizing markets and societies. The abuses of property markets deserve greater attention and must be more fully integrated into efforts to regulate markets and establish transparency and accountability.

Acknowledgment

This chapter has been developed from a report prepared by the author for the Organized Crime Global Agenda Council of the World Economic Forum. The author appreciates the support of the council members who provided sources to conduct this research and also highlighted important trends they observed in their own regions.

Notes

[1] Based on interviews with specialists in real estate at the Global Agenda Council meeting of the World Economic Forum, Fall 2010.

[2] Mary Beth Sheridan, "Oxon Hill Development Has Ties to Terror," *The Washington Post*, April 19, 2004, available at <www.washingtonpost.com/ac2/wp-dyn?pagename=article&contentId=A22576-2004Apr18&notFound=true>.

[3] Financial Crimes Enforcement Network, "Suspected Money Laundering in the Residential Real Estate Industry: An Assessment Based upon Suspicious Activity Report Filing Analysis," Washington, DC, April 2008.

[4] Money laundering into real estate by terrorist groups has been observed in the Middle East. See Joe Becker, "Lebanese Bank Seen as a Hub of Hezbollah's Financing," *The New York Times*, December 13, 2011, available at <www.nytimes.com/2011/12/14/world/middleeast/beirut-bank-seen-as-a-hub-of-hezbollahs-financing.html?_r=1&scp=2&sq=hizbollah%20and%20money%20laundering&st=cse>.

[5] Illustrative of this is the real estate sector in Dubai, which received money from Afghanistan and elsewhere that contributed to inflated prices. Dexter Filkins, "Depositors Panic over Bank Crisis in Afghanistan," *The New York Times*, December 2, 2010, available at <www.nytimes.com/2010/09/03/world/asia/03kabul.html?scp=5&sq=afghan%20laundering%20into%20Dubai&st=cse>. It also had one of the steepest corrections in its real estate with drops of over 40 percent when the bubble burst. For a discussion of the aftermath, see Liz Alderman, "After Crisis, Dubai Keeps Building, But Soberly," *The New York Times*, September 29, 2010, available at <www.nytimes.com/2010/09/30/business/global/30dubai.html?ref=mohammedbinrashidalmaktoum>.

[6] Peter B.E. Hill, *The Japanese Mafia: Yakuza, Law, and the State* (Oxford: Oxford University Press, 2003), 185.

[7] Ibid., 186–187; also discussed by Grant Newsham, "Yakuza Influence on the Japanese Financial Sector: Security and Solvency in Japan," April 25, 2011, at Terrorism, Transnational Crime and Corruption Center, George Mason University, Fairfax, VA, available at <http://traccc.gmu.edu/events/previously-hosted-events/>.

[8] See the Web site of the Italian environmental organization *Legambiente* (League for the Environment) that has been closely following the illegal construction of the Mafia since the 1990s. A whole section of its site is devoted to this topic. See <www.legambiente.it/temi/ecomafia>; also see Shasta Darlington, "Italian 'Eco-Mafia' Booms, Outstrips Economy," Reuters, June 1, 2005, available at <www.enn.com/top_stories/article/1675>.

[9] Brigitte Unger et al., *Detecting Criminal Investment in the Dutch Real Estate Sector* (Dutch Ministry of Finance, Justice and Interior Affairs, January 2010), 202–203; Brigitte Unger and Joras Ferweda, *Money Laundering in the Real Estate Sector, Suspicious Properties* (Cheltenham, UK, and Northampton, MA: Edward Elgar, 2011).

[10] Reports at the Organization for Security and Cooperation in Europe (OSCE) and United Nations Office on Drugs and Crime (UNODC) Expert Seminar on Leveraging Anti-Money Laundering Regimes to Combat Human Trafficking, October 3–4, 2011, Vienna; Louise I. Shelley, *Human Trafficking: A Global Perspective* (Cambridge: Cambridge University Press, 2010).

[11] Stuttgart Public Prosecutor's office, statement of charges against members of the Izmailovskaya organized crime group, March 15, 2007.

[12] Margaret Beare and Stephen Schneider, *Money Laundering in Canada: Chasing Dirty and Dangerous Dollars* (Toronto: University of Toronto Press, 2007); Unger et al.; and Europol, *Organized Crime Threat Assessment* (OCTA) 2011, The Hague, the Netherlands, May 4, 2011, available at <www.europol.europa.eu/content/press/europol-organised-crime-threat-assessment-2011-429>.

[13] Beare and Schneider.

[14] Web site available at <www.wolfsberg-principles.com/faq-persons.html>.

[15] Global Witness, *Undue Diligence: How Banks Do Business with Corrupt Regimes*, March 11, 2009, available at <www.globalwitness.org/library/undue-diligence-how-banks-do-business-corrupt-regimes>.

[16] Interviews with European money-laundering specialists, Fall 2011.

[17] Unger et al., 51.

[18] Basel Institute on Governance, available at <www.assetrecovery.org/kc/node/4418aa8c-12f8-11de-b174-1f0207abe99a.9>.

[19] Hill, 177–247.

[20] J.D. and Aman Agarwal, "Keynote Address on Money Laundering: The Real Estate Bubble," available at <www.mse.ac.in/.../20070629-MLREB%20(AFFI%20%20BSE).pdf>; "Kyrgyz Crime Czar Arrested in UAE," July 22, 2011, available at <www.universalnewswires.com/centralasia/viewstory.aspx?id=4497>.

[21] Denis Gathanju, "Laundered ransoms cause Kenyan property boom," *Baird Maritime*, February 25, 2010, available at <www.bairdmaritime.com/index.php?option=com_content&view=%20article&id=5693:piracy-cash-lands-in-kenya-property-market-in-nairobi-skyrockets-&catid=113:ports-and-shipping&Itemid=208>.

[22] Charles Goredema, ed., "Money Laundering Experiences," ISS Monograph no. 124, June 2006, available at <www.issafrica.org/pgcontent.php?UID=1690>, 21.

[23] Financial Action Task Force/Groupe d'action Financière, *Money Laundering & Terrorist Financing through the Real Estate Sector*, 2007, 27, identifies hotels and luxury resorts as prime targets for money launderers.

[24] The *Legambiente* (League for the Environment) and its work discussed in note 9 illustrate this.

[25] Unger et al., results based on analysis by the mayor of Maastricht, 49.

[26] Katrin Anacker, "Big Flipping Schemes in Small Cities? The Case of Mansfield, Ohio," *Housing and Society* 36, no. 1 (2009), 5–28; interview with vice president of major investment firm in Tokyo responsible for monitoring investments.

[27] Interview conducted by World Economic Forum members.

[28] On Mexico, see Jana Schroeder, "Illegal finance animates Mexico's narco-cartels. So why isn't it being attacked?" available at <www.borderlandbeat.com/2011/04/mexicos-drug-money-addiction.html>; Ernesto Savona, "Exploring OC Portfolio Investments for Rethinking Confiscation Policies," presentation at Pan-European High Level Conference on Asset Recovery Offices, Budapest, March 7–8, 2011, on the Italian experience; and Howard Abadinsky, *Organized Crime*, 7th ed. (Belmont, CA: Wadsworth/Thomson Learning, 2003), 235–236, on the United States.

[29] Unger et al., 174.

[30] Ibid., 177.

[31] Dutch criminal investigations have reported that the proceeds of the drug trade are returning to Turkey for hotel and real estate construction in Istanbul and on the coasts.

[32] Francisco Thoumi, *The Political Economy & Illegal Drugs in Colombia* (Boulder, CO: Lynne Rienner, 1995).

[33] This is illustrated by the problem in some of the smaller Gulf states that were once closed communities. As they open up their real estate projects to international investors, they do not have the capacity or the skills to vet their international customers. Interview with money-laundering specialist who served in a Gulf state, January 2012.

[34] Savona and discussions with financial crime investigators.

[35] Report of Jennifer MacLeod, Europol representative at the OSCE and UNODC Expert Seminar on Leveraging Anti-Money Laundering Regimes to Combat Human Trafficking, October 3–4, 2011, Vienna.

[36] Financial Action Task Force, *"Risks Arising from Trafficking of Human Beings and Smuggling of Migrants,"* July 2011, available at <www.fatf-gafi.org/document/41/0,3746,en_32250379_32237202_48412265_1_1_1,00. html>.

[37] Shared Hope International, *Demand: A Comparative Examination of Sex Tourism and Trafficking in Jamaica, Japan, the Netherlands and the United States,* July 2007, 113–141, available at <www.sharedhope.org/Resources/ demand.aspx>.

[38] Unger et al., 14; also, yakuza are deeply involved in construction through supply of labor and protection. See Hill, 96.

[39] Ibid., 13; H. Nelen, "Real estate and serious forms of crime," *International Journal of Social Economics* 35, no. 10 (2008), 751–761.

[40] Unger et al., 26.

[41] Thoumi, 60, 237–238.

[42] International Crisis Group, *War and Drugs in Colombia,* Latin America Report no. 11, January 27, 2005, 26.

[43] "Cancun Mayor Gregorio Sanchez Faces Drug Charges," available at <www.bbc.co.uk/news/10209580>.

[44] Hill, 185–190; David Kaplan and Alec Dubro, *Yakuza: Japan's Criminal Underworld,* expanded edition (Berkeley and Los Angeles: University of California Press, 2003), 196–220.

[45] Savona, slides 11–41.

[46] Unger et al.; Unger and Ferweda, 2011.

[47] Abadinsky, 235–236.

[48] See Savona on spread of the phenomenon.

[49] Unger et al., 14.

[50] Savona.

[51] Pavel Lazarenko money-laundering case; to see other reports on money laundering into real estate by corrupt officials, see Basel Institute of Governance, available at <www.assetrecovery.org/kc/node/1698185c-4768-11dd-a453-b75b81bfd63e.html>.

[52] U.S. Permanent Subcommittee on Investigations, "Keeping Foreign Corruption Out of the United States: Four Case Histories," February 4, 2010, available at <http://hsgac.senate.gov/public/index.cfm?FuseAction=Hearings.Hearing&Hearing_ID=dd873712-eb12-4ff7-ae1a-cbbc99b19b52>.

[53] Ashley Neese Bybee, *Narco State or Failed State? Narcotics and Politics in Guinea-Bissau,* dissertation, School of Public Policy, George Mason University, Fairfax, VA, September 2011.

[54] Intergovernmental Action Group Against Money Laundering in West Africa, 2008, available at <www.giaba. org/index.php?type=c&id=37&mod=2&men=1>.

[55] Illustrative is the Sneep case in the Netherlands, a multiyear investigation of a large-scale human trafficking ring in which cash and other payments were transferred back to Antalya for investment in nightclubs, hotels, and other businesses. The form of transfer and the destination were ascertained by wiretaps. Presentation by Inge Schepers, National Prosecutor for Human Trafficking, Netherlands; the Prosecutor at the OSCE and UNODC Expert Seminar on Leveraging Anti-Money Laundering Regimes to Combat Human Trafficking, October 3–4, 2011, Vienna, and subsequent discussion of the case with Ms. Schepers.

[56] Brad Reagan, "Afghan Central Bank Calls on UAE for Help in Freezing Assets," September 15, 2010, available at <www.thenational.ae/business/afghan-central-bank-calls-on-uae-for-help-in-freezing-assets>.

[57] Carl Mortished, "Bill Browder: At Home after Mean Streets of Moscow," *The Sunday Times,* August 4, 2008, available at <http://business.timesonline.co.uk/tol/business/industry_sectors/banking_and_finance/ article4453893.ece>.

[58] "Hermitage Reveals Russian Police Fraud," available at <www.youtube.com/watch?v=ok6ljV-WfRw>. Also, conversation with a lawyer who helped trace the money, October 2011.

[59] Web site available at <www.wantedsa.com/index.php?option=com_content&view...id...www.cfatf-gafic.org/.../Venezuela_3rd_Round_MER_(Final)_English.pdf;en.mercopress.com/.../argentina-under-pressure-to-approve-anti-money-laundering-legislation;www.gatewaytosouthamerica-newsblog.com/.../new-argentine-rules-against-money-laundering/>.

[60] Web sites available at <www.antilavadodedinero.com/wap/new_det.php?id=2304>; and <www.knowyourcountry.com/uruguay1111.html>.

[61] Web site available at <primapanama.blogs.com/_panama.../how-serious-a-problem-is-money-laundering-in-panama.html>.

[62] The author has seen these unoccupied hotels. They were pointed out by local specialists on corruption.

[63] Agarwal and Agarwal, 24.

[64] Interview with a member of the staff of the President of the NATO Army Armaments Group (NAAG), Fall 2011; Frances D'Emilio, "Italy's Mafia Thrives in Global Financial Meltdown," Associated Press, April 25, 2009, available at <www.breitbart.com/article.php?id=D97PHLVO0>; Walter Kego, Erik Leijonmarck, and Alexandru Molcean, eds., *Organized Crime and the Financial Crisis: Recent Trends in the Baltic Sea Region* (Stockholm: Institute for Security and Development Policy, 2011), 36, as well as a more general discussion of the problem.

[65] Nick Pisa, "Mafia is now Italy's 'biggest bank'—and they're squeezing the life out of small business (quite literally)," *Daily Mail*, January 12, 2012, available at <www.dailymail.co.uk/news/article-2085209/Mafia-Italys-No-1-bank-profits-100bn-year.html>.

[66] Illustrative of this is the house found in Mexico with $205 million in cash. See Hector Tobar and Carlos Martinez, "Mexico meth raid yields $205 million in U.S. cash," *The Los Angeles Times*, March 17, 2007, available at http://www.latimes.com/la-fg-meth17mar17,0,709967.story>.

[67] Interview with individual selling real estate in Europe.

[68] Europol, OCTA, 2011, 13.

[69] Hill; Kaplan and Dubro and expertise provided by members of the organized crime group of the World Economic Forum.

[70] Report on Romania in OSCE and UNODC, Expert Seminar on Leveraging Anti-Money Laundering Regimes to Combat Human Trafficking, October 3–4, 2011, Vienna; Shelley.

[71] In markets such as Turkey, the actual sale price of the house is never reflected on the deed of sale, allowing ample opportunities for laundering money.

[72] Discussions with a Swiss banking expert, October 2011.

[73] In some countries where corrupt officials know there is no capacity to investigate their activity or to recover assets overseas, they make no effort to disguise their identity in making purchases overseas. The author, around 2000, did training with the anticorruption commission in Honduras. Through working with the online property registrations for the city of Miami, it was possible to locate the undeclared homes of many of the country's top politicians. There was no attempt to disguise these purchases by themselves or their family members.

[74] In 2000, the author toured both rural agricultural and urban properties in Sicily that have been confiscated and saw how they have been turned into institutions that serve the community.

Part III.
The Attack on Sovereignty

Chapter 9
The Criminal State

Michael Miklaucic and Moisés Naím

Corruption in public administration dates from time immemorial. And though agencies such as Transparency International rate some states as more or less corrupt than others, no state is above, beyond, or immune to public corruption. One need only review the serial revelations of corruption within the governments of the District of Columbia, Spain, or Hong Kong in recent years to see that it can and does happen everywhere. However, as this book argues throughout, the recent proliferating interaction among criminal, terrorist, and insurgent networks and the exponentially greater magnitude of their commerce made possible by the processes of globalization have moved the overall threat posed by state collusion with transnational illicit networks from the status of international nuisance to a substantial threat to the contemporary international order.

While corruption and criminalization are outcomes of human agency and choice, the most threatening aspect of the current situation is not the morally fallen individual. The corruption and criminality we describe in this book is, rather, the presence of intensively networked international organizations engaged in a full range of illicit activities including trafficking in all manner of contraband and counterfeit. This book describes the emergence of such networks from the shadows of the criminal world into the world of national and international security.

This chapter examines a frequently mentioned but infrequently elaborated new and disturbing development: the emergence of the criminal state. We begin with a prefatory discussion of some of the impediments to clear analysis and understanding of converging criminal networks, as well as the unique obstacles to full comprehension that criminal states pose. These have been iterated in previous works of one of the authors,[1] but they bear reiteration to serve as the backdrop for the further discussion. We argue that state criminalization exists along a spectrum characterized by varying degrees of corruption and various functions of corruption within the state. At one end of the spectrum is "criminal penetration," which occurs when an illicit network, be it criminal, terrorist, or insurgent, is able to place "one of its own" into the state structure. That agent may carry out formal functions on behalf of the state but also carries out actions in support of an illicit network or criminal enterprise. "Criminal infiltration" occurs when the infection has spread throughout the state apparatus within the given country, and the linkages to external illicit networks proliferate. When this penetration becomes pervasive and affects critical nodes of decisionmaking, it becomes an advanced pathology of dysfunctional governance characterized by the presence of criminals and criminal agents in positions of state authority where they facilitate and implement criminal activities for the illicit networks for which they ultimately work, thus undermining the pillars of government legitimacy and functionality. We describe "Criminal capture" as the condition of dysfunctional governance in which criminal agents are so sufficiently prominent in positions of state authority that their criminal actions cannot effectively be restrained by the state. At some point, it may become part of state or substate institutional doctrine to engage in illicit activity. At this level, the state is not only subverted by illicit elements but is directly complicit in those activities and endorses those activities. Pervasive institutionalization defines the fully "Criminal sovereign," which functions effectively as a criminal enterprise with the apparatus of the state itself engaged directly in criminal activity as a matter of policy. Such a state exists outside the realm of the family of states, which is governed if only nominally by the international law regime and associated norms. Such a state is a threat to all other states, to the international rule-based system of states, and to the contemporary structure of global order. To animate this spectrum, we discuss examples of the various stages along the spectrum.

Conceptual Challenges

Networks of criminals, smugglers, counterfeiters, insurgents, and terrorists are not a new phenomenon. Indeed, such networks have operated on the criminal fringe of communities in all societies and at all times. These vocations are surely among the oldest professions. Yet our understanding of the operations and structures of the organizations that dominate these trades is woefully inadequate. In this field, we are at a very primitive level of conceptual development. Indeed, we are only beginning to know what we need to understand. There is a crippling shortage of reliable data regarding the operations and structures of illicit networks. When data are available, we lack tested and proven tools to interpret them. This does not refer to technically advanced computer models, but to the absence of the fundamental conceptual models required for a disciplined analytic approach. Even concepts as basic as "licit" and "illicit" remain contested. A basic and widely accepted taxonomy of the objects of study—illicit networks and organizations—is lacking.

This conceptual underdevelopment has many causes, but professional, bureaucratic, and international fragmentation are important sources of confusion and poor analysis when studying transnational illicit networks. Sociologists perceive the phenomena from the vantage point of intellectual frameworks and models emerging from their discipline, emphasizing the dynamics of collective human behavior. Criminologists are prone to view transnational criminal networks as an extension of individual criminality, best addressed within a law enforcement framework. Anthropologists, political scientists, and international relations specialists all perceive the phenomenon through the lenses of their disciplines; none owns it, yet none can dismiss it. This leads inevitably to conceptual confusion, competing models, and interdisciplinary competition for the right to claim the correct or best analysis.

As the various intellectual disciplines approach transnational illicit networks from their niched perspectives, so do the different professions engaged in the fight against them. If you speak with lawyers, you will discover that they see a world that the people in finance do not see, or see in a completely different way. If you are speaking with finance professionals, you will find that they perceive the problems in a way completely unlike those in law enforcement. Those officials responsible for containing the international trade in counterfeits see yet a different world—they worry about different things and use different tools than people who are worried about international licit trade.

The many agencies involved in countering transnational criminal networks in each country have their own organizational cultures, methods, mandates, authorities, and idiosyncrasies. In the United States, the effort to counter illicit networks involves the State Department, Department of Defense, Department of Homeland Security, Treasury Department, and Department of Justice, as well as national intelligence organizations such as the Central Intelligence Agency, Federal Bureau of Investigation, and the Defense Intelligence Agency. Also involved are specialized agencies such as the Drug Enforcement Administration, the Bureau of Alcohol, Tobacco, Firearms, and Explosives, and U.S. Customs and Border Protection. Law enforcement organizations typically understand their challenge as building cases against individuals, whereas military organizations see the challenge in terms of a battle campaign. Information sharing—or more specifically, intelligence sharing—among civilian and military, intelligence, and operational agencies can be very strained. Problems of effective collaboration—exacerbated by differing interests, cultures, mandates, authorities, jurisdictions, and methods—complicate and compromise effective action against illicit networks.

Finally, the conceptual seams created by different perceptions of illicit networks and illicit commerce within different countries confound multilateral and even bilateral efforts to address the problem. In some countries, narcotic trafficking is seen as a supply problem—others see it as a demand problem. Some countries view counterfeiting as a violation of international law and norms—others view it as a jobs program. Borders are heaven—they are nirvana for traffickers and for the illicit networks in which they function. National borders are what create the price differentials that drive the immense profits of illicit commerce. They also provide the shields for criminals to hide behind, guarding them from law enforcement agencies and governments seeking to disrupt their activities. Of course, there are coordinating mechanisms and alliances between agencies from different governments. But for any government agency to work in an alien jurisdiction is exceedingly challenging. To operate outside our national habitat, we require a full array of equipment and infrastructure. For any government official working in another country, there are many difficulties. Even ambassadors depend on a panoply of regulations, conventions, and institutional arrangements to operate. So while for governments, borders are very confining, for traffickers borders are what justify their existence, protect them, make their life possible, and make their business profitable.

Our understanding of transnational illicit networks, and of the challenges they represent, is clouded by three critical delusions. The first is that crime is crime, and it has been with us throughout human history, and that in this regard, "There is nothing new under the sun." The Bible chronicles many cases of crime, black markets, and smuggling—all is part of the human experience. Unfortunately, this is a delusion—the wrong way to think about it. The velocity and magnitude of illicit commerce today are unprecedented. Estimates of the global black market range from 2 to 25 percent of global product. The mobility of illicit commodities shipped in vast quantities via standardized shipping container units, and thus hidden, across the world is unprecedented. And the interconnections between illicit networks defying historical geographical, cultural, and motivational barriers are unprecedented. Ultimately, the clear and present danger constituted by these threats is something quite new under the sun.

A second debilitating delusion is that illicit networks and their activities are all about crime and criminals. We argue it is no longer merely about crime; it is beyond crime. Today it is about criminalized states and the future configuration of power within the state system. If you approach this challenge as merely a problem of crime and criminals, the traditional way of dealing with it requires the typical institutions: law enforcement, courts, and jails. Today's challenge is about incentives and reinforcing the value of service in the public interest and the integrity of public administration. These and other normative values must be incubated and fortified in schools, churches, community organizations, and in the media through the disciplined application of incentives and disincentives. The delusion that you can deal with these issues only through traditional ways of crime fighting is very dangerous.

The third delusion is that illicit behavior is an aberration and the people involved in this business are deviants. The notion of essential deviant character has its origins in the work of Cesare Lombroso's theory of "anthropological criminology."[2] Essentially, the argument is that you are a criminal if you are deviant. In many countries, normalcy is defined by involvement in what we are here calling illicit networks and illicit commerce. Approximately 8 to 10 percent of China's gross domestic product (GDP) is associated with the manufacture and export of counterfeit goods. This endeavor involves a lot of people, a lot of families, and a lot of economic activities related to something we call crime. For them, it is their daily life. Sixty percent of Afghanistan's GDP is associated with poppy production, cultivation, and distribution. How many workers does this require? China and Afghanistan are two prominent country examples, but, in fact, every day millions of people all over the world wake up, go to work, work all day, and bring bread to their families, all thanks to doing something we label a criminal activity. Moreover, the counterfeit commodities and contraband they produce are consumed by ordinary individuals throughout the world, sometimes unknowingly, but very often in full knowledge of their illicit origin—for instance, when a packet of counterfeit cigarettes is purchased or a counterfeit car part is installed. Are all of these consumers deviants as well? Are the multitudes of gardeners, farm and restaurant workers, and nannies working illegally all deviants?

Ambiguity pervades this field. Three specific ambiguities that make analysis particularly difficult are the blurring between legal and illegal politics, between the legal and illegal private sector, and between legal and illegal philanthropy. The transnational criminal organizations that animate global illicit networks are businesses. For any highly regulated big business, it is absolutely normal to dedicate part of its revenues to influencing the regulators. The financial, pharmaceutical, defense, and food industries—indeed, all industry groups—engage in such activity. Public institutions including universities, hospitals, and media organizations do the same. In some countries, this is called corruption; in others, it is called lobbying. In the United States, the practice of using some of your business revenue to influence the very government officials who make decisions affecting your bottom line is pervasive. If a company that we respect and whose products we routinely use (for example, Ford, Chevron, Microsoft, Google, Citigroup, or Harvard University) does this, why would we expect criminal enterprises not to do it as well? Because these companies are heavily regulated and their survival depends on governmental rules, it would be unreasonable to expect them not to try to influence those rules. The same applies to criminal enterprises, but to a much larger and dangerous extent. This represents the gray area between criminal enterprises and politics.

A fundamental rule of doing business and investing is diversification. If you have substantial assets and much of your revenue comes from a highly risky but high-yield activity, it is normal to take some of your accumulated profits and put them in a less risky activity. That is why many large criminal enterprises are investing in licit business activities. They invest in banks and in shipping, real estate, and trucking companies to support and expand their illicit activities. In other cases, they are just happy owners of lucrative and licit enterprises. When the ill-gotten gains of illicit commerce are recycled through normal and legal means into the financial system and generate additional revenue, is the knock-on activity licit or illicit? The answer to this important question differs country by country. This ambiguity between licit and illicit enterprise is very difficult to disentangle.

A third ambiguity is in the field of philanthropy. It is normal for large corporations to have significant philanthropic activities. They fund sports clubs, museums, hospitals, and schools with a broad range of concerns. If you look at the list of the owners of soccer clubs in the Balkans, you will see an amazing overlap between the ownership of these sports institutions that are part of the pride of the nation and the "most wanted" list. Recall that Pablo Escobar, a pioneer in the area of criminal philanthropy, started a lot of these activities in the 1980s in Medellín, Colombia. Though best known as the murderous leader of the Medellín Cartel, he was well known locally as one of the main providers of social services, sports activities, and protection for local youth. Hizballah, in Lebanon, is famous for its local provision of public services normally expected from the state, including public protection, home loans, public sanitation, and social welfare. Moreover, you will find that drug traffickers provide many of the social services in Brazil. You might also investigate the list of major financial supporters of big museums in Russia or Ukraine. The ambiguity between the philanthropy that we know and admire and the philanthropy that is undertaken by these criminal networks confuses the terrain for those trying to combat illicit networks.

Governments Give Way, Introducing the Criminal State

The international criminal networks that trade in all sorts of counterfeit and contraband— drugs, humans, weapons, illegal timber, pharmaceuticals, car parts, cigarettes, etc.—have become so large, so global, so powerful, and so politically connected that they have evolved into a threat potentially greater even than that from international terrorism. Yet while terrorism looms large in the global security dialog, international criminal networks often continue to be seen dismissively as just crime, criminals, and law enforcement issues. These perceptions reinforce other characteristics of the field, for example, the profound asymmetry between those in charge of containing the growth of these illicit enterprises and the enterprises themselves. We are dealing with entrepreneurs who are at the forefront of globalization. These are groups and individuals that see international opportunities earlier than most of us, even earlier than other multinationals. They are capable of recruiting and hiring some of the best talent in the world; they can afford highly trained and talented professionals in information technology, law, finance, and banking, as well as politicians, journalists, generals, judges, and diplomats. Meanwhile, the government agencies that have to deal with them are often highly resource-constrained.

By the way, we are not talking about the U.S. Government or the governments of its wealthy allies. The greatest threat is to governments in the developing world that have very limited resources and multiple compelling and competing demands on those resources. Many of these nations, though often small and poor, are the partners upon which we must increasingly depend to support our efforts in counterterrorism, combating transnational crime, providing humanitarian and disaster relief, peacekeeping efforts, and meeting other emerging challenges. These include not-so-small countries such as Mexico, Nigeria, and Iraq. Even in the case of the United States, there is an asymmetry between the capabilities, resources, dynamism, and speed of networked illicit organizations and networks and those of the agencies tasked to combat them.

Looking back over the past generation one cannot easily find many instances of strategic success in the struggle against illicit networks. There have undoubtedly been successful campaigns against specific organizations or individuals. One can find instances in which a law enforcement agency or a government has neutralized a specific network or ring or taken down notorious individuals. People go to jail—some are executed—and networks are dismantled. But any government would be very hard-pressed to show even one market for illicit internationally traded goods or activities in their country that is smaller today than it was 20 years ago. The Medellín and Cali cartels that once terrorized Colombia have been dismantled, and the poster child drug lord Pablo Escobar was killed. Yet the enterprise was replaced by a less monopolistic (or duopolistic) arrangement and the quantity of cocaine coming from Colombia has not decreased. Indeed globally, "Long-term seizure trends show that cocaine, heroin, and morphine as well as cannabis seizures—in volume terms—almost doubled between 1998 and 2009, while seizures of ATS [amphetamine-type stimulants] more than tripled over the same period."[3] According to a much-cited 2008 Organisation for Economic Co-operation and Development report, "The share of counterfeit and pirated goods in world trade is estimated to have increased from 1.85 percent in 2000 to 1.95 percent in 2007."[4] More recent estimates project the global value of counterfeit and pirated goods could reach $1.77 trillion by 2015. The same growth trend is likely to be found in illicit trafficking in persons, weapons, natural resources (for example, timber, minerals, and wildlife), and cultural items (art and archeological items, among others).

States and governments around the world are losing the battle. Moreover, the problem has evolved beyond criminal networks. It has evolved beyond even the convergence, globalization, and diversification of illicit networks described and discussed throughout this book. What we must worry about today and tomorrow is the emergence of criminal states; governments that have succumbed to this scourge and been taken over by criminals, and in fact governments that have become themselves criminal enterprises. Our collective lack of understanding of how criminal enterprises become governments, and how some governments have taken over criminal enterprises not to dismantle them but to use them for their financial, political, and military advantage, represents a dangerous gap. Criminals are becoming politicians and government officials. And top government officials and political leaders are doubling as the heads of vast and often international criminal enterprises.

The term *criminal state* has been used to characterize countries as diverse as North Korea, Russia, and Guinea-Bissau. What exactly do we mean by "criminal state"? In the simplest terms, a *criminal* is one who has committed (or has been convicted of committing) a crime. By extension, a *criminal state* is a state that has committed a crime. But we mean more than that; by *criminal state* we generally imply systematic behavior and systemic characteristics. We certainly mean more than mere complicity in illegal acts by corrupt individuals encumbering official positions, though that is perhaps where we should begin. We are implying an ongoing pattern of engagement in activities that contravene international law. *Criminal state* implies not only an ongoing pattern of criminal behavior by individuals, but by a large number of officials, or even by whole organs of the state apparatus, and the deployment of governmental assets and resources in support of criminal undertakings. In some cases—which we will describe in this chapter—organs of the state will even have formal assigned responsibility for engaging in criminal activities.

Rachel Locke suggests that criminal networks interact with the state in three ways: 1) corruption describes a relationship in which a state permits or condones the actions of illicit networks; 2) infiltration is a symbiotic relationship in which agents of illicit organizations or networks are positioned within the government itself; and 3) competition is the situation in which the state attempts to enforce its primacy over illicit organizations and networks.[5] In only the second characterization—infiltration—is the state itself directly complicit in the illicit activity. This typology, though useful in other contexts, is not effective in disaggregating the complex of phenomena described broadly as criminal states. It ignores the situation in which the state itself is the criminal agent, as opposed to just a few individuals at high levels.

Michael Hanlon defines *criminal states* as "organizations that fulfill the role of the state, even though they are not recognized as an official governing authority."[6] This definition seems inherently problematic though, as he specifically distinguishes the organizations he is describing (revolutionary movements) from the existing state apparatus, arguing that the former has usurped certain narrow roles of the latter. But the autonomous state itself remains in this construct—perhaps not effective, but not criminal either. This line of analysis is followed by Cockayne and Lupel, who distinguish between predatory groups that compete with the state in open conflict, parasitic groups that extract rents from populations and authority structures, and symbiotic groups that "coexist with existing authority structures, either through overlaps of membership or through other clandestine arrangements of reciprocity and joint venture arrangements."[7] In this analysis the symbiotic groups would come closest to our understanding of the criminal state. If symbiosis is understood as a process, its advanced stages, in which the state and the criminal organization cannot be distinguished, constitute a level of criminalization of the state itself.

Bayart, Ellis, and Hibou provide a more comprehensive definition of *criminal state*:

> *The criminalization of politics and of the state may be regarded as the routinization, at the very heart of political and governmental institutions and circuits, of practices whose criminal nature is patent, whether as defined by the law of the country in question,*

or as defined by the norms of international law and international organizations or as so viewed by the international community, and most particularly that constituted by aid donors.[8]

This is closer to the approach we take, as it pins criminal complicity not merely on actors external to the formal state apparatus, but on the state itself. The authors further identify six indicators of state criminalization that collectively provide what might be thought of as a comprehensive state criminalization profile:

+ the use for private purposes of the legitimate organs of state violence by those in authority, and the function of such violence as an instrument in the service of their strategies of accumulation of wealth

+ the existence of a hidden, collective structure of power which surrounds and even controls the official occupant of the most senior political office, and which benefits from the privatization of the legitimate means of coercion, or is able with impunity to have recourse to a private and illegitimate apparatus of violence, notably in the form or organized gangs

+ the participation by this collective and semiclandestine power structure in economic activities considered to be criminal in international law, or which are so classified by international organizations or in terms of moral codes that enjoy wide international currency

+ the insertion of such economic activities in international networks of crime

+ an osmosis between a historically constituted culture that is specific to the conduct of such activities in any given society and the transnational cultural repertoires that serve as vehicles for processes of globalization

+ the macroeconomic and macropolitical importance, as distinct from the occasional or marginal role, of such practices on the part of power holders and of these activities of accumulation in the overall architecture of a given society.[9]

However, by their own admission very few African states—and by implication other states—fulfill the criteria they propose as characterizing the criminal state. This attempt at a comprehensive definition is perhaps overreaching in that by its comprehensiveness it renders itself inapplicable to many of the states we might wish to cover in a discourse on criminal states. Though nuanced and useful in creating an intelligible narrative, it assumes a binary distinction between the criminal and the noncriminal state. Perhaps the discussion need not necessarily be focused on labeling a state *criminal* or *noncriminal*, but rather on elaborating on the degree of criminalization in the state.

Some authors have focused on the morphology of criminal states. Bunker and Bunker distinguish four criminal state forms "derived from jihadi insurgency, state failure-lawless zones, criminal takeover, and oligarchic regimes."[10] An innovative attribute of this framework is its inclusion of criminal states animated by religious or ideological motivations, such as Afghanistan's Taliban or Lebanon's Hizballah. These differ from the typical assumption of the venal motivations of criminals and criminal states. According to this pair of scholars, "Such a state is anathema to the modern state-form and criminal in its orientation because, ultimately, it was to merge with other conquered territories to form the new Caliphate envisioned by al-Qaeda."[11] Less persuasive is their description of criminal states emerging from state failure and lawless zones. It can be argued that what they are describing here is a condition arising from the absence of a state, rather than an entity engaged in criminal activity.

What we propose below is an exploratory effort to disaggregate further the basket of states referred to as criminal states, based not on their motivations or morphology but on their degree of criminalization. We have argued elsewhere that "increasingly, fighting transnational crime must mean more than curbing the traffic of counterfeit goods, drugs, weapons, and people; it must also involve preventing and reversing the criminalization of governments."[12] Likewise, to preserve the sovereignty of legitimate states, we must prevent and reverse this corrosive process. But to meet these challenges, we must distinguish between states with corrupt officials, states that operate large-scale criminal enterprises, and states that are themselves large-scale criminal enterprises. The typology developed below is intended to assist in this effort at differentiation. Four characterizations are proposed to describe differing levels of criminal network entanglement with the state. To be clear, the dynamics of transition from one level of criminal consolidation to another must be the focus of future research. We do not assume that these levels represent a linear progression in which a state progresses from one level to the next with inevitability. We do not yet know if that is true. We merely know that there are different degrees of state criminality and that effectively preventing or reversing the further decline into criminalization will require a full array of tools and approaches, as well as a high degree of sensitivity to individual case and context-specificity.

Criminal Penetration

As noted above, corruption and illicit networks exist in all states alongside the normal complement of governmental and civil society structures and networks. Their presence may be more or less extensive, more or less corrosive, and more or less disruptive, but they are always there. In some states there may be a terrorist or insurgent presence alongside traditional criminal networks and organizations. Though terrorists, insurgents, and criminals do not inevitably interact, it is well known that their mutual hesitation to interact has declined and such interactions are becoming more common. At a certain point, what we might call the normal—or an acceptable level of—consolidation among illicit networks within a state is exceeded, and a process of erosion or atrophy of the state itself may begin and the legitimacy of the state, already often tenuous, becomes even further compromised in the eyes of its citizens. The presence or expansion of criminality becomes an international security threat when the impact of criminal, insurgent, or terrorist activity is felt in other states. And such cross-border effects inevitably become dramatically more likely when groups on either side of a border establish persistent connectivity and interaction.

While disturbing, this is not the essence of the criminal state phenomenon. It is when the governing apparatus of the state itself is co-opted and even transformed—thus giving access to public resources and other assets to non- or even antigovernment interests—that we encounter the criminal state. A customs agent in the pay of contraband traffickers may compromise an inspection. A tax official may be paid to overlook money-laundering activity. A judge may be on retainer to protect the interests of criminals or other agents acting against the public interest. Such penetration pits the assets and resources of the state against the public interest, a central characteristic of a criminal state.

South Africa

In 2006, the heads of 152 nations met in Brazil for the 75th general assembly of the International Criminal Police Organization (Interpol), one of the few institutions we have to deal globally with what is inherently a global problem. The opening ceremony was addressed by South African Mr. Jackie Selebi, at the time the National Commissioner of the South African Police Service and President of Interpol. He said that the Interpol mission is to ensure and promote the widest possible mutual assistance among all criminal authorities globally, and that we need to find systems to make sure that our borders and border controls are on a firm footing. This is a very noble cause. Ironically, at the very moment he was speaking the Commissioner had another job—shielding drug traffickers from police scrutiny in South Africa. Today Mr. Selebi is in jail in South Africa, convicted of profiting personally from the activities of international drug trafficker Glenn Agliotti. He was effectively an embedded agent of a drug-trafficking organization.

Is South Africa a criminal state? It is a state struggling with profound economic problems and a painful and not yet completed political transition from its apartheid past, but it would be unfair to label South Africa a criminal state. In the Failed State Index for 2012, South Africa ranks 115 out of 178 states in terms of vulnerability to state failure[13]—a reasonable ranking given South Africa's history, economy, and demographic profile. Unquestionably the major organs of the South African state—the legislature, the judiciary, the executive, the army, and the police—have instances of corruption, but none appears to engage in a pervasive and persistent pattern of violation of the law or of the public interest. South Africa has retained its Freedom House designation as a "free" country since the mid-1990s.[14] Transparency International ranked South Africa 64th of 183 countries surveyed in 2011 for perceived corruption—not a glowing recommendation, yet not abysmal either.[15] However, with corrupt individuals like Selebi, criminal penetration has begun. And the proof is that he was caught and imprisoned. The same is true in Mexico, where the state still has the will and capability to attack powerful cartels even though the government has been penetrated.

Criminal Infiltration

Russia

The Russia that emerged from the ashes of the USSR in 1991 was already permeated by corruption and deeply rooted organized crime—a legacy of the "400 year history of Russia's peculiar administrative bureaucracy . . . especially shaped into its current form during the seven decades of Soviet hegemony that ended in 1991."[16] Organized crime remains pervasive throughout Russia today, but its nature has evolved. "Now, you have a bunch of law enforcement people who are essentially organized criminals with unlimited power to ruin lives, take property and do whatever they like and that's far worse than I have ever seen in Russia before. Russia is essentially a criminal state now."[17] While this characterization of Russia as a criminal state may be an exaggeration by an individual involved in a legal dispute, criminality has metastasized throughout the Russian state. A study under the auspices of the U.S. National Institute of Justice (NIJ) states, "What is unique in this situation is the degree to which criminal groups have infiltrated the political process. The Uralmash and central groups in particular have succeeded in electing political candidates favorably disposed towards them, in opposing other candidates, and even in electing their own members to office."[18] This is not only true in the Urals, but has parallels in St. Petersburg, Moscow, the Far East, and all other regions. What is distinctive here is that illicit organizations are "electing their own members to office." Once illicit networks are able to infiltrate "their own members" into positions of governmental responsibility they can subvert the legitimate intent of the state from the inside for ulterior—criminal—purposes.

While Russian state institutions continue to engage in traditional governmental activities—making and enforcing laws, adjudicating disputes, providing services admittedly often without competence or efficiency—the government often practices either neglect or complicity with illicit network activity throughout the country. It is ranked by Transparency International as 143 out of 183 in terms of the perception of public corruption.[19] The NIJ report goes on to say, "These abuses are not idiosyncratic to particular persons who happen to be in power at a particular time; they are endemic to the system of power."[20] This situation corresponds to Locke's "corruption," or "infiltration."[21]

Russia would appear to represent a more pervasive criminalization of the state than South Africa, where an individual or several individuals were abusing state power for illicit purposes, yet the legitimacy of the state itself has not been fatally compromised. Where Russia differs from South Africa is in the degree to which it has become delegitimized within its own population. What does this mean? How do we know when a state has been delegitimized? As noted, the public perception of corruption in Russia is significantly greater than in South Africa, Russia ranking 143rd out of 183 countries, compared to 64th for South Africa (the lower score corresponding to a public perception of less corruption). Inevitably such a judgment must be subjective—that is, it is the opinion of the analyst; however, there is some work toward measuring state legitimacy through the use of substitutive variables. Gilley measures state legitimacy in terms of public perceptions of state legality and moral justification, as well as public "acts of consent" that reflect recognition of state authority and compliance with its requirements. By Gilley's methodology, South Africa ranks 19th out of 72 countries in legitimacy, following immediately behind Australia, Spain, and Britain. Russia, on the other hand, ranks 72nd out of 72 countries, falling just behind Armenia and Pakistan.[22]

Kenya

The case of Kenya is illustrative if ambiguous. Kenya is by some measures a successful African state, ranked "partly free" by Freedom House, and demonstrating acceptable economic performance that led the World Bank to say, "with growth at 5 percent, Kenya's per capita income now exceeds US$800 for the first time, and the country is firmly on the path to Middle Income status."[23] Yet despite this reported respectable political and economic performance, Kenya exhibits the characteristics of advanced criminal infiltration, with a very low level of legitimacy as perceived by Kenyans. According to the Fund for Peace, it ranks 16th most prone to state failure out of 180 states, with a very poor score for "legitimacy," one of the 12 indicators utilized by Fund for Peace in constructing the "Failed States Index."[24] Reporting on a 1-year study of transnational organized crime, Gastrow concludes, "Criminal networks have penetrated the political class and there are growing concerns about their ability to fund elections and to exercise influence in parliament and in procurement processes."[25] Gastrow describes "a slow degeneration [of state institutions] until it becomes clear that they are no longer functional."[26]

Ambiguity arises, however, with respect to the degree to which Kenyan authorities are able to resist the gradual process of criminalization and execute effective countermeasures against embedded criminal interests. Gastrow argues that "impunity is rife for political and government elites and hardly any prosecutions for transnational organized crime have taken place."[27] Furthermore, "Kenya's police no longer have the capacity to effectively to investigate and prosecute powerful suspected organized crime figures."[28] Yet ultimately, Gastrow equivocates, "Even though Kenya is experiencing high levels of corruption and organized crime, it would be wrong to conclude that it has reached the stage where a central function of its political and government institutions is to routinely engage in criminal practices."[29] It therefore fails to meet the "routinization standard" set out by Bayart, Ellis, and Hibou,[30] leading Gastrow to state, "The research findings do not justify Kenya being labeled as a criminalized state, but its foundations are under attack."[31]

Criminal Capture

A significant threshold is crossed when the commercial interests of corrupt individuals or specific organizations give way to institutional imperatives that support illicit activity. In the case of South Africa, clearly Selebi was acting in his own self-interest and was under no duress from the state or its organizations to do so. Others have likely been doing the same, but not to the degree that the fundamental legitimacy of the state is compromised. In Russia, organized crime and transnational illicit networks are prolific and have infiltrated many state organs. Indeed, the Kremlin itself houses many top-level "coordinators." However, though the legitimacy of the Russian state—already questionable for a variety of reasons—may be further compromised in the view of many or even most Russians, the state maintains independence and the ability to pursue effective countermeasures against illicit networks. Russia suffers from, but is not yet hostage to, the illicit networks it tolerates or engages. This is a threshold point—while the Russian state does not systematically or effectively impede the activities of illicit networks, when there is any doubt of who is in charge, the state can and does enforce its will. Indeed, Russian President Putin has shown the capacity and persistent willingness to demonstrate that no power in Russia can stand up to the power of the presidency. The state thus keeps control of those illicit networks it "owns."

It is this latter capacity, to take effective countermeasures against embedded illicit networks, that suggests Kenya is not yet a captured state. Gastrow acknowledges that "despite

Kenya's huge problems with issues of corruption, crime and governance, there is a critical mass of senior government officials in place that is determined to get to the root of these problems in order to solve them.... These officials and the departments they work in have real potential to make Kenya succeed in its fight against corruption and organized crime."[32] The ability to fight back to cure itself demonstrates state resilience and resistance to capture.

The captured state has lost the capacity to take effective action against embedded illicit networks. Hellman and Kaufmann define state capture as, "the efforts of firms to shape the laws, policies, and regulations of the state to their own advantage by providing illicit private gains to public officials."[33] The data underlying their path-breaking work was drawn from the Business Environment and Enterprise Performance Survey, conducted under the joint auspices of the World Bank and the European Bank for Reconstruction and Development (EBRD), and consists of interviews with owners and managers of hundreds of firms in the EBRD area. In cases of state capture, the state is caught in a vicious cycle of misgovernance from which it cannot extract itself and in which it suffers damage it is unable to repair. "Institutional reforms necessary to improve governance are undermined by collusion between powerful firms and state officials who reap substantial private gains from the continuation of weak governance."[34]

While pioneering, their focus on interactions between ordinary firms and state officials limits the utility of their analysis by excluding the complex and even more insidious interactions between the state and entities outside the ordinary market economy, such as organized crime groups. Kupferschmidt expands Hellman and Kaufmann's definition by developing a matrix of 13 cases of "strategic uses of illicit political finance used toward state capture."[35] In this expanded conceptualization, the agents of state capture include warlords, organized crime, drug cartels, guerrillas, rebel groups, oligarchs, and rogue governmental elements, as well as corporations. The matrix includes countries as diverse as Afghanistan, Colombia, Guinea-Bissau, Italy, and the United States.

Venezuela

The Kupferschmidt matrix does not purport to be a comprehensive register of state capture but rather offers an illustrative sample. It does not include the interesting case of Venezuela. General Henry Rangel Silva currently serves as the Venezuelan Minister of Defense. In 2008, the U.S. Treasury ordered the freezing of Silva's U.S. accounts due to his involvement in assisting the Colombian terrorist organization FARC with their narcotrafficking enterprise. According to the United Nations Office on Drugs and Crime, Venezuela now accounts for more than half of the cocaine shipments to Europe, often via Africa. The cocaine produced in Andean countries goes to Venezuela and from there to Europe, mostly via Africa. In the past decade, Venezuela has become the heart of a vast clandestine network that moves cocaine in great quantities around the world. Maps of flights of private planes going from the Caribbean to Africa show that in 2000 there was almost no traffic, while today there is a dense network of flights. This evidence strongly supports the case that Venezuela is well along in becoming a criminal state.

The question is whether Venezuela has the capacity to heal itself or has been irreversibly captured. Obviously, a change of regime could have a significant impact on this question.

However, under the current regime of Hugo Chávez, the answer appears to be no. Farah argues that a criminal state is one

> *where the senior leadership is aware of and involved—either actively or through passive acquiescence—on behalf of the state in transitional criminal enterprises, where TOC* [transnational organized crime] *is used as an instrument of statecraft, and where levels of state power are incorporated into the operational structure of one or more TOC groups.*[36]

Arguably, this definition fuses two distinct phenomena; the participation either passive or active of senior officials in criminal activity, and the convergence of state and criminal organizations as crime is used as a tool of statecraft and state organs become functional parts of criminal operations. Venezuela clearly meets Farah's first criterion: senior officials are involved both actively and passively in a wide range of criminal activities. At this point in time it is no longer clear to what degree President Chávez remains in control. He undoubtedly continues steadfast in pursuit of his Bolivarian project, which though we may find it repulsive and misguided is not inherently criminal. However, Venezuela may have passed the point where Chávez's will is sufficient to clean up the country, or at least effectively counter the criminal activity within its borders.

Guinea-Bissau

Guinea-Bissau, on the other hand, may have crossed over the threshold. This small West African state is deeply mired in narcotics trafficking, serving as a major transit point between source regions in Latin America and demand centers in Europe and the United States. It is among the least developed countries in the world, ranking 176th out of 187 countries in the "2011 Human Development Index" published by the UN Development Programme.[37] Under the Drug Kingpin Act, Washington has imposed financial sanctions on several members of the ruling elite, including air force head Ibraima Papa Camara and former navy chief Jose Americo Bubo Na Tchuto.[38] Guinea-Bissau ranks 154th out of 172 in the Transparency International "Corruption Perceptions Index" for 2011.[39] Though not ranked in the Gilley state legitimacy index, the succession of military coups in recent years and the absence of state capacity in virtually all sectors should cast doubt on any pretense to legitimacy. If we accept the above, we can comfortably place Guinea-Bissau in the ranks of criminally infested states.

Does Guinea-Bissau have the ability to fight back? The evidence suggests not. Symptomatic is a 2006 case in which two Latin Americans were detained while in possession of 670 kg of cocaine. The army secured their release and the disposition of the cocaine went unknown. In 2007 Guinea-Bissau police were able to seize 635 kg of cocaine, but "traffickers escaped with the remainder of the consignment believed to total around 2.5 mega ton of cocaine [which had been flown into a military airstrip] because police did not have the manpower or vehicles to give chase."[40] Government spokespersons routinely dismiss allegations of official involvement while acknowledging the state is woefully unequipped to meet the challenge.

Criminal Sovereignty

The conceptualizations of state capture above might be characterized as agent-centric. This level of criminalization goes beyond individual or even a multiplicity of agents; it is the dynamic of institutionalization. This is to say that the very institutions of the state are subject to subversion and perverse evolution, becoming in the process criminal enterprises. Thompson explains:

> [L]ike all forms of corruption, the institutional kind involves the improper use of public office for private purposes. But unlike individual corruption, it encompasses conduct that under certain conditions is a necessary or even desirable part of institutional duties. . . . What makes the conduct improper is institutional in the sense that it violates principles that promote the distinctive purposes of the institution. It is still individuals who are the agents of institutional corruption and individuals who should be held accountable for it, but their actions implicate the institution in a way that the actions of the agents of individual corruption do not.[41]

In her robust description of state capture, Grzymala-Busse builds her definition around four strategies; clientelism, predation, redistribution without democratic competition, and capture of state assets under competitive conditions:

> All these strategies involve the formation of distinct state institutions and capacities. State seizure does not simply corrode the state. Although extractive rulers seek to maximize their discretion by weakening regulation and oversight, they also construct rules and durable practices of redistribution, budgeting, and authority. It is not simply the case that "clientelism thrives when government institutions are weak" (Manzetti & Wilson, 2007, p. 955), but rather, that specific institutions are built to serve the extractive goals of rulers, sometimes with unintended consequences (Tilly, 1992).[42]

This builds on Hellman, Kaufmann, Kupferschmidt, and Thompson by further clarifying the malignant evolution of the state institutions themselves within the process of advanced criminalization and previews a sound characterization of the criminal state. In the criminal state, individuals are no longer able to function effectively without bending to or even supporting the new, illicit organizational mission. Indeed, to resist complicity in the criminal mission generally entails significant personal risk and cost or even loss of life.

Charles Taylor's Liberia

Convicted war criminal Charles Taylor was president of Liberia from 1997 to 2003 and controlled most of the territory of the small West African state from the early 1990s on. During that period, he effectively turned the apparatus of the state into a criminal enterprise for the private enrichment of himself and his cronies through extraction of national resources. He also exploited political instability in neighboring countries to extend his reach into their resource bases. Farah writes, "What set Taylor apart was his ability to connect the state apparatus with different criminal and terrorist groups in order to form a single enterprise—an enterprise that would benefit its members tremendously, but threaten the stability of West Africa and eventually the world."[43]

During his administration, Taylor incorporated the security services, state administration agencies, natural resource management, and the economic and financial organs of the state into a criminal enterprise. This is not a case of corrupt officials presiding over criminal enterprises that operate outside the reach of the state, but the transformation of the state into a criminal enterprise with no other purpose than its criminal agenda. Under Taylor, the customs and immigration services and passport agency supported the international transit of known criminals in league with Taylor and his accomplices. They were also critical in making Liberia a meeting place and safe haven for terrorists, criminals, insurgents, and the like. The shipping and aircraft registries were used to provide cover for criminal air and seafaring fleets transporting all manner of contraband. In the words of the late Liberian journalist Tom Kamara, Taylor created "a criminal state with the capacity of extending its wings around West Africa and beyond. Illegal passports issuance, money laundering, the proliferation of dubious banks, and drug trafficking are finding an accommodating environment in Liberia."[44] It is the fusion of the state with the illicit networks comprised of insurgent groups (such as the Revolutionary United Front), transnational terrorist organizations (such as al Qaeda), and transnational criminal organizations that distinguish the criminal state. Taylor's Liberia became "a functioning criminal enterprise while continuing to enjoy—and use for illegal purposes—many of the international benefits of statehood (e.g., the ability to issue recognized diplomatic passports, maintain shipping and airplane registries, control border entry and exit points, and collect taxes."[45]

North Korea

Perhaps the most extreme example of a criminal state is the People's Democratic Republic of Korea (DPRK). The DPRK government is reportedly directly involved in a wide range of transnational criminal activities sanctioned at the very highest level of the state, and implicating officials at all levels. These activities include illicit drug production and trafficking, foreign currency counterfeiting, trafficking in counterfeit cigarettes and pharmaceuticals, insurance fraud, and trafficking in humans, among others. Estimates of the illicit revenues from these activities range from $500 million to $1 billion per year.[46] What is distinctive about North Korea is the degree to which the state apparatus has been converted to criminal enterprise. This conversion involves the DPRK diplomatic service, the security services, many organs of public administration, the public manufacturing and agricultural sectors, state-owned trading companies, and unique state organs established explicitly for the purpose of criminal activity.

The criminalization of the North Korean state has taken place over a long period. As early as 1976, 17 North Korean diplomats were expelled from Norway, Sweden, Denmark, and Finland for trafficking in black market cigarettes, liquor, and hashish. According to the Danish Prime Minister, their actions were not for personal gain, implying that the actions were sanctioned by their masters in Pyongyang.[47] With the discontinuation of substantial subsidies from the defunct Soviet Union in 1991, followed by deteriorating economic conditions and the famine of 1994–1998, and perhaps influenced by the leadership succession from Kim Il Sung to Kim Jong Il in 1994, state criminalization appears to have accelerated throughout the 1990s. Perl, writing in 2004, stated, "Since 1976, North Korea has been linked to over

50 verifiable incidents involving drug seizures in at least 20 countries. A significant number of these cases involved arrest or detention of North Korean diplomats or officials."[48] Indeed, overseas missions and embassies are essentially used as autonomous profit centers responsible for covering their own costs in country and sending excess revenues back to Pyongyang.[49]

An extraordinary office known as Bureau 39 (or Office 39) was established in 1974 to generate foreign currency by whatever means necessary on behalf of the regime. "To commit its criminal acts, Office 39 uses the powers and resources of a nation-state—merchant and military vessels, diplomatic and embassy posts, as well as state run companies and collective farms."[50] Thus, criminal activity is not ancillary to the work of Bureau 39 or the corruption of specific officials; it is rather the primary function of the organization. Bureau 39 runs numerous front companies with offices throughout the world and "is involved in drug production and distribution, smuggling, money laundering, currency counterfeiting and product piracy and illegal gold trafficking."[51] This is the signature attribute of a criminal state. As Kan, Bechtol, and Collins write, "North Korea directs its national instruments of power to *commit* crimes in other states."[52] Bureau 39 is the hub of this extensive criminal enterprise, managing the complicit organizations, activities and processes, and revenue.

In North Korea, we do not have a case of criminals penetrating and infiltrating the state. It is not the case in the DPRK that criminals have become sovereigns and attained the prerogatives of sovereignty. Rather it is the state itself —the sovereign—that is the criminal. Criminal enterprise is its purpose and there can be no personal success for a North Korean official who defies this institutional imperative. Again quoting Kan, Bechtol, and Collins:

> North Korea practices a form of "criminal sovereignty" that is unique in the contemporary international security arena. North Korea uses state sovereignty to protect itself from external interference in its domestic affairs while dedicating a portion of its government and the KWP [Korean Workers' Party] to carrying out illicit international activities in defiance of international law and the domestic laws of numerous other nations.[53]

This completes the spectrum ranging from criminal penetration, common to most if not all states, to the level of criminal infestation leading to loss of legitimacy, criminal capture in which the state is helpless to repair or reform itself, to criminal sovereignty; the state as criminal. Is this level of criminal sovereignty unique to North Korea? It is doubtful. Charles Taylor's Liberia, Saddam's Iraq, and others past and present may have been criminalized to this extent, but deeper examination is necessary. Ironically, in the case of North Korea the combination of endemic economic failure and comprehensive international sanctions drives it farther into the realm of transnational crime and illicit networks. Its networks for trafficking contraband and counterfeit items, money laundering, etc., are extensive. "These networks are adaptive, resilient, covert, and widespread; they could be put to use for the transfer of nuclear material from North Korea to other dangerous regimes or to violent nonstate actors."[54] The existence of these networks combined with the DPRK's track record of operating in flagrant violation of the international rule-based system, including its 2003 withdrawal from the Nuclear Non-Proliferation Treaty, justify grave trepidation.

An Existential Threat

What are existential threats? Are they limited to those that result in mushroom clouds? Throughout the Cold War years, the Soviet Union, with its terrifying military machine, was clearly perceived as a clear and present threat to U.S. national and international security. But also perceived as posing an existential threat was the international spread of the Marxist/ Communist ideology. The Soviet threat led to substantial institutional innovation within the United States and the West, such as the creation of the North Atlantic Treaty Organization. It also resulted in various tools the U.S. Government created largely to fight the ideological battle against the existential threat posed by international communism, such as the U.S. Agency for International Development, U.S. Information Agency, Radio Free Europe, and Radio Liberty. Though these latter efforts were not notably effective, they indicated the magnitude of the threat perceived.

Spanish prosecutor Jose Grinda Gonzalez was responsible for investigating organized crime particularly coming from Russia, Eastern Europe, Ukraine, Belarus, Chechnya, and elsewhere. In Grinda's view, in the nations he calls mafia states the interlocking networks of criminals and political leaders is vast, deep, and permanent. One cannot differentiate between the activities of the government and organized criminal groups; they are one in the same. He reports examples of going after some criminals and criminal groups and discovering that he was not only dealing with them; they were able to mobilize their nations' diplomats, military, governments, judiciary, and so on. "Here I am, a prosecutor, a magistrate in Spain, but on the other side of me I face the best lawyers that can be hired in Spain and I face nation-states that are deploying diplomats, the military, intelligence and spies." This invokes the point made above concerning asymmetry and underscores the challenges we face in containing, let alone reversing, the decay within the global rule-based system of states.

Does the emergence of criminal states pose an existential threat to U.S. national interests or to international security or is this just a marginal issue? We believe unequivocally that this threat is at the least very substantial, and has the potential to become existential if the trends it generates remained effectively unchallenged. Although our analytic, policy, development, and military communities are still at a very primitive stage of understanding and intellectual development, the evidence of risk is pervasive and frightening. While not measurable in megatons or body counts, the rapid erosion of a fundamental principle of modern governance—the dedication of the state to the public interest—is an insidious threat to the state system as we know it. The proliferation of criminal states would represent a profound disruption in the contemporary system for the preservation of global order.

We need to be selective and prioritize the greatest threats, perhaps even letting go for the time being those that might be characterized as nuisance risks. If we have scarce resources, what should we prioritize? Should we spend time, money, and other resources pursuing those who produce and sell counterfeit children's books or dedicate those resources to pursuing people who are trafficking in children? Should we put the same effort into going after people who are stealing Microsoft software and plagiarizing and copying programs and apps as we put into the effort to stop people from selling centrifuges for nuclear enrichment on the black market?

Fighting these illicit networks is no longer about drugs, counterfeits, weapons, terrorists, or insurgents. It is about defending the integrity of the system of viable sovereign states and the fundamental structure of global order. The challenge is preventing and reversing the criminalization of states and governments. In meeting the emerging challenges of the 21st century, we cannot be everywhere, do everything, prevent every wrong, and protect every right. There is absolutely no hope in fighting everything, from the counterfeiting of Harry Potter books and Prada bags, to trafficking of humans, narcotics, and every other kind of contraband, and to countering terrorism, insurgency, and the many other destructive activities of illicit networks around the world. We must be selective. Concentrating on preserving the legitimacy and sovereignty of governments and of the rule-based structure of global order must be the priority. Many states have already suffered criminal infestation and are well on their way to criminal state capture or even criminal sovereignty. These are the same governments on which the United States and other countries must depend as partners in the struggle against international terrorism, transnational organized crime, and other emerging international security challenges. These often-feeble states must deal with the international financial system and its regulatory system. They will have to deal with fighting poverty and with developing alliances to deal with the long litany of problems that are no longer amenable to solutions by any country acting alone. The need for global collaboration has never been greater, nor has the risk to our potential partners as they lose ground in the struggle to maintain sovereignty free of criminalization.

Notes

[1] Moisés Naím, "Mafia States: Organized Crime Takes Office," *Foreign Affairs* 91, no. 3 (May/June 2012); see also, Moisés Naím, *Illicit: How Smugglers, Traffickers, and Copycats Are Hijacking the Global Economy* (New York: Doubleday, 2006).

[2] Cesare Lombroso (1835–1909) was an Italian criminologist who founded the "Italian positivist school of criminology." His key argument was that criminal nature is inherited and represents a regression from normal human development.

[3] United Nations Office on Drugs and Crime (UNODC), "2011 World Drug Report" (New York: UNODC), 22.

[4] Organisation for Economic Co-operation and Development (OECD), *The Economic Impact of Counterfeiting and Piracy 2008* (Paris: OECD), available at <www.oecd.org/sti/counterfeiting>.

[5] Rachel Locke, *Organized Crime, Conflict and Fragility: A New Approach* (New York: International Peace Institute, July 2012).

[6] Michael Hanlon, "The Cooption of Revolutionary Movements into Criminal States," January 25, 2009, available at <http://bss.sfsu.edu/economics/newsevents/pacdev/Papers/Hanlon.pdf>.

[7] James Cockayne and Adam Lupel, "Introduction: Rethinking the Relationship Between Peace Operations and Organized Crime," *International Peacekeeping* 16, no. 1 (February 2009), 4–19.

[8] Jean-François Bayart, Stephen Ellis, and Béatrice Hibou, *The Criminalization of the State in Africa* (Bloomington: Indiana University Press, 1999).

[9] Ibid.

[10] Robert J. Bunker and Pamela L. Bunker, "Defining Criminal States," *Global Crime* 7, nos. 3–4 (August–November, 2006).

[11] Ibid.

[12] Naím, "Mafia States."

[13] Fund for Peace, "Failed States Index 2012," available at <www.fundforpeace.org>.

[14] South Africa has maintained a Freedom House rating of at least 2/2 since 1994–1995. See Freedom House Web site, available at <www.freedomhouse.org/report-types/freedom-world>.

[15] Transparency International, "Corruption Perceptions Index 2011," available at <http://cpi.transparency.org/cpi2011/results/>.

[16] James O. Finckenauer and Yuri A. Voronin, *The Threat of Russian Organized Crime* (Washington, DC: National Institute of Justice, June 2001).

[17] Statement of William Browder, chief executive officer of Hermitage Capital Management, BBC News, November 23, 2009.

[18] Finckenauer and Voronin.

[19] "Corruption Perceptions Index 2011."

[20] Ibid.; see also, Mark Galeotti, "The Criminalization of Russian State Security," in *Criminal States and Criminal Soldiers,* ed. Robert J. Bunker (London: Routledge, 2008).

[21] Rachel Locke.

[22] Bruce Gilley, "The Meaning and Measure of State Legitimacy: Results for 72 Countries," *European Journal of Political Research* 45 (2006), 499–525.

[23] The World Bank, "Kenya Economic Update," 6th ed., June 2012.

[24] "Failed States Index 2012."

[25] Peter Gastrow, *Termites at Work: A Report on Transnational Organized Crime and State Erosion in Kenya* (New York: International Peace Institute, 2011).

[26] Ibid.

[27] Ibid.

[28] Ibid.

[29] Ibid.

[30] Bayart, Ellis, and Hibou.

[31] Gastrow.

[32] Ibid.

[33] Joel Hellman and Daniel Kaufmann, "Confronting the Challenge of State Capture in Transition Economies," *Finance and Development* 38, no. 3 (International Monetary Fund, September 2001).

[34] Ibid.

[35] David Kupferschmidt, *Illicit Political Finance and State Capture* (Stockholm: International Institute for Democracy and Electoral Assistance, August 21, 2009).

[36] Douglas Farah, *Transnational Organized Crime, Terrorism, and Criminalized States: An Emerging Tier-One National Security Priority* (Carlisle, PA: Strategic Studies Institute, July 2012), available at <www.strategicstudies-institute.army.mil/pubs/download.cfm?q=1117>.

[37] United Nations Development Program (UNDP), "2011 Human Development Index," *Human Development Report 2011: Sustainability and Equity, A Better Future for All* (New York: UNDP, November 2, 2011).

[38] BBC News, April 9, 2010, available at <http://news.bbc.co.uk/2/hi/africa/8610924.stm>.

[39] "Corruption Perceptions Index 2011."

[40] UNODC, *Cocaine Trafficking in Western Africa* (New York: UNODC, October 2007).

[41] Dennis F. Thompson, *Ethics in Congress: From Individual to Institutional Corruption* (Washington, DC: Brookings Institution Press, 1995).

[42] Anna Grzymala-Busse, "Beyond Clientelism: Incumbent State Capture and State Formation," *Comparative Political Studies* 41, nos. 4–5 (April/May 2008), 638–673.

[43] Douglas Farah, *A Volatile Mix: Non-State Actors and Criminal States* (Washington, DC: Center for American Progress, October 2004).

[44] Tom Saah Kamara, "Liberia: The Emergence of a Criminal State," *ThePerspective.org*, July 2000, available at <www.theperspective.org/criminalstate.html>.

[45] Farah, *A Volatile Mix.*

[46] Liana Sun Wyler and Dick K. Nanto, *North Korean Crime-for-Profit Activities*, RL33885 (Washington, DC: Congressional Research Service, August 25, 2008), available at <www.fas.org/sgp/crs/row/RL33885.pdf>.

[47] *Ludington Daily News*, November 10, 1976.

[48] Raphael Perl, "State Crime: The North Korean Drug Trade," *Global Crime* 6, no. 1 (February 2004).

[49] David L. Asher, "The North Korean Criminal State, Its Ties to Organized Crime, and the Possibility of WMD Proliferation," Policy Forum Online 05-92A (Berkeley: Nautilus Institute, November 15, 2005).

[50] Paul Rexton Kan, Bruce E. Bechtol, Jr., and Robert M. Collins, *Criminal Sovereignty: Understanding North Korea's Illicit International Activities* (Carlisle, PA: Strategic Studies Institute, March 2010).

[51] Mark Galeotti, "Criminalization of the DPRK," *Jane's Intelligence Review* 13 (March 2001), 10–11.

[52] Kan, Bechtol, Jr., and Collins.

[53] Ibid.

[54] Ibid.

Chapter 10
How Illicit Networks Impact Sovereignty

John P. Sullivan

This chapter examines the *problématique* of illicit networks and transnational crime's impact on sovereignty.[1] Transnational criminal organizations (TCOs) and gangs challenge states and sovereignty in a variety of ways. These include eroding state solvency through corruption, subtle co-option of state officials and institutions, direct assault on state functions, and in the worst case, state capture or failure under the threat of criminal challengers. Rarely do criminal enterprises totally supplant states; rather they change the nature of state functioning. This chapter looks at how gangs and TCOs influence state change at local, state, federal, and transnational levels. It specifically examines networked TCOs and transnational gangs in Mexico and Latin America to gauge the current and emerging threats to state functions, and the reconfiguration of power within states.

State reconfiguration, including co-option, and the rise of criminal enclaves result from the growth and proliferation of lawless or other-governed zones—including failed communities or failed zones—through corruption and the application of force by private nonstate armies. As part of this exploration, the concept of "criminal insurgency" will be examined in the context of a "battle for the parallel state" (dual sovereignty) and the potential rise of "narco-states" and "narco-networks." The chapter illuminates the logic structure of criminal state-challengers as they attempt to establish "neo-feudal" governance structures.

We begin with an examination of deviant globalization and dark side actors using examples from Mexico and Latin America. The discussion then looks at *narcocultura* and state-making potentials. Finally, we examine the impact on the state and the contours of state reconfiguration, and the potential—or actuality—of gangs and criminal cartels to emerge as new warmaking and potentially networked state-making entities. Within this construct, criminal insurgency is defined as a means of removing the criminal enterprise from the control of the state, enabling it to pursue its goals to dominate the illicit economy. Here the illicit economy and globalization converge, conferring advantages to criminal enterprises.[2] Despite this economic agenda, a political agenda also emerges. As a result, gangs and TCOs become political actors influencing the reconfiguration of states.

Essentially, the chapter is concerned with the potential for illicit networks to reconfigure states. Such reconfiguration could include erosion of state capacity (or the exploitation of a state capacity gap), corrupting and co-opting state organs (government, the police, the judiciary) in all or part of the state through the development of criminal enclaves, or, at the extreme edge, state failure. State reconfiguration appears to be a more common outcome than abject state capture or state failure.[3] While failure and reconfiguration may appear to be the same phenomenon, they have distinct features. State capture (StC) involves criminals subverting and seizing control of key political functions at the central or national level (politicians, judges, police, etc.) through corruption. Co-opted state reconfiguration (CStR) involves the systematic alteration of governance to benefit the criminal enterprise.[4]

Co-opted state reconfiguration is a distinct, advanced form of state capture. CStR involves the participation of lawful and unlawful groups seeking economic, criminal, judicial, and political benefits together with a quest for social legitimacy. Coercion, political alliances (complementing or replacing bribery), and impacts on all branches and levels of government are core features of this dynamic. As Garay and Salcedo-Albarán argue, "The co-opted reconfiguration' concept accepts that co-optation can be carried out in any direction. In a CStR situation, it is therefore possible to find scenarios in which legal agents—candidates or officials—are co-opting illegal agents— paramilitary or subversive groups—and vice versa."[5] In a CStR process, state institutions are manipulated and even reconfigured from inside. When officials are being captured and manipulated from outside, it is reproduced as an StC situation.[6] TCOs as deviant social networks exploit both dynamics.

Deviant Social Networks and Deviant Globalization

Social networks are important elements of contemporary social and political processes. Certainly this is not new; social networks have been around since man began assembling in political groups at all levels of society. Yet the information age is bringing important changes to the nature of networks. Members of networks can communicate across vast distances in real time, changing the pace and shape of their individual and collective influence. The importance of information age networks is a key to understanding emerging conflict.

David Ronfeldt has argued that societies have moved through four distinct (albeit overlapping) phases of organization: tribes, institutions, markets, and networks.[7] Ronfeldt and Arquilla discussed the conflict and security dimensions of networks and netwar in their landmark collection *Networks and Netwars*.[8] Netwar is essentially an emergent form of low-intensity conflict, crime, and activism waged by social networked actors, including TCOs, terrorists, and gangsters. Manuel Castells has also outlined the rise of the networked, information society in his landmark trilogy *The Information Age: Economy, Society, and Culture*.[9] Specifically, Castells envisioned the emergence of powerful global criminal networks as one facet of the shift to a new state/sovereignty structure where the state no longer controlled all aspects of the economy and society. Networks currently take two shapes: positive networks that inform civil society and dark side or negative networks that exploit society. These dark side actors are essentially "criminal netwarriors."

Transnational gangs and cartels operating as netwarriors are a threat to the sovereignty of nations. "When states fail to deliver public services and security, criminals fill the vacuum."[10] This situation leads to a "time of anomalies and transitions," according to Juan Carlos Garzón.

Complex criminal networks where multiple criminal factions interact by "cooperating and competing for the control of illicit markets are impacting democratic environments and transforming themselves into a real force that could end up determining the destiny of institutions and communities."[11] A consequence of TCO action is diminished state capacity. TCOs are thus challenging the sovereignty and solvency of states

Dark Side Actors

Transnational crime, gangs, terrorism, and insurgency are threats influencing the current and future conflict environment.[12] These separate—and increasingly linked or networked—threats result in a diffuse security environment that blurs distinctions between crime and war. A consequence of this convergence is the rise of new political and economic actors including gangs and TCOs that alter the internal and external security dynamics of states and the relationship between states and their citizens.

Global cities are home to transnational criminals and global gangs.[13] Specifically, "Third Generation gangs" (3 GEN) are developing increased complexity and impact. 3 GEN gangs differ from First Generation gangs, which are essentially turf organizations that engage in opportunistic crimes, and the more market-focused Second Generation gangs that sometimes operate on a national level. 3 GEN gangs are internationalized, networked, and complicated structures that evolve mercenary or political aims.[14] In the Americas, *maras* frequently fit the third generation definition operating at the high ends of three continua: internationalization, sophistication, and politicization (see figure[15]). Maras are essentially gangs; they are separate from cartels (which are more evolved enterprises that generally seek to dominate the illicit economy), although both interact. Maras frequently interact with, and are allied with, cartels to serve as transnational subcontractors or mercenaries for cartel organizations.

Generations of Gangs

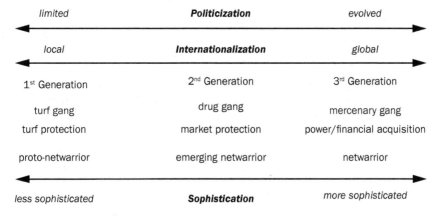

Source: John P. Sullivan, "Third Generation Street Gangs: Turf, Cartels, and Net Warriors," *Transnational Organized Crime* 3, no. 3, Autumn 1997.

The maras frequently mentioned in this context, including Mara Salvatrucha (MS-13) and Eighteenth Street (M-18), both originally from Los Angeles, now operate throughout North and Central America. Some discount the complex, networked nature of these gangs, but sensitive investigations have shown intricate, networked command and control relationships operating along a line of communication known as the "Bloody Triangle," which refers to Los Angeles, San Salvador, and Northern Virginia, where jailhouse coordinators in all three locations collaborate to authorize "green lights" for gang assassinations.[16] In addition to domination of criminal turf, maras can exhibit evolved political and/or mercenary aims. These dimensions are usually present within specific cliques (nodes or subelements) of the mara network rather than an enterprise-wide objective.

Cartels and Gangs: Criminal Insurgencies in Mexico and Central America

Mexico's drug war has killed an estimated 40,000 persons since 2006 when President Felipe Calderón declared war on the cartels.[17] "Mexico is imploding in a series of interlocking 'criminal insurgencies' culminating in a virtual civil war."[18] The wars are primarily the result of battles among drug cartels. The aim is to dominate criminal enterprise and control the "plazas" or drug corridors, as well as battles with the state—both police and military—in order to operate freely. As noted in one paper:

> The cartels are joined by a variety of gangs in the quest to dominate the global criminal opportunity space. Third generation gangs—that is, gangs like Mara Salvatrucha (MS-13) that have transcended operating on localized turf with a simple market focus to operate across borders and challenge political structures—are both partners and foot soldiers for the dominant cartels. Gangs and cartels seek profit and are not driven by ideology. But the ungoverned, lawless zones they leave in their wake provide fertile ground for extremists and terrorists to exploit.[19]

This is not to say that cartels and gangs aren't potentially insurgent. They can and do have insurgent potential and are actively challenging the state in parts of Mexico and Latin America. The cartels, while not driven by explicit ideology, are actively seeking power and reconfiguring the state (CStR).

The current levels and scope of network interconnectivity among Mexico's "criminal net-warriors" are complex and ever shifting. Los Zetas, initially an enforcement arm of the Gulf Cartel, now operates as an independent enterprise while cross-border gangs vie for a piece of the action and alliances morph and shift as the battle of "all against all" progresses.

The shifting alliances of Mexican drug cartels seem to be leading to the rise of competing alliances or "mega cartels." The Mexican daily *Excélsior* reports that SEDENA (the Mexican national defense/intelligence organization) assesses that eight drug-trafficking organizations in Mexico are potentially uniting to form two competing blocs in order to gain control of the "narco-sphere."[20] Currently, the largest players in the drug war in Mexico are the Sinaloa Cartel and Los Zetas.[21] Other cartels and gangs are present in the conflict and the tensions of all-against-all remain, but the spoils may be favoring these two major blocs.

An example of the networked cooperation among criminal gangs is found in the recent disclosure of a joint hit squad comprised of gangsters from *Los Linces* (a *La Linea*/Juárez Cartel enforcer cell) and *Barrio Azteca* (a Texas-based street/prison gang hybrid). Daniel Borunda of the *El Paso Times* writes about the joint cell: "A hit team described as the shock troops of the Juárez drug cartel and Azteca gangsters are suspected of being involved in the recent massacre of students in Juárez, according to information from the Chihuahua state attorney general."[22] The emergence of a hit squad composed of formerly competing factions is indicative of the trend toward networked cooperation and an example of organizational adaptation.

Drug cartels threaten to turn Mexico into a "narco-state" with rising lawlessness that echoes 1990s-era Colombia. According to a recent feature in the *Arizona Republic*, Mexico's drug war "has exposed criminal networks more ingrained than most Americans could imagine: Hidden economies that employ up to one-fifth of the people in some Mexican states. Business empires that include holdings as everyday as gyms and a day-care center."[23] Essentially, the cartels have adapted, improved their armaments, and are perfecting simultaneous terrorist-style attacks resulting in pockets of ungovernability. Even President Calderón, who once angrily rebutted suggestions that Mexico was becoming a "failed state," is now describing his crackdown as a fight for territory and "the very authority of the state."[24]

In "Killing for kudos—the brutal face of Mexico's 21st-century war," Ed Vulliamy looked at barbarism in Mexico's drug war in the aftermath of the January 2010 slaughter of 16 teenagers in Juárez. That "bloodfest" was the culmination of the bloodiest month to date in the battle of the border drug cartels. Vulliamy found that "this brutality defines a war very much of its time, the first 21st-century war, because it is, in the end, about nothing."[25] In actuality, the war does have an aim; the quest for raw power with the objective of rolling back state authority. Vulliamy also emphasized the power of new media: "Mexico's war is fought through YouTube and mobile phones as well as back-room torture chambers. Cartels and killers use YouTube to threaten rivals and public officials, and boast of their killings, or set up rogue websites to broadcast their savagery."[26]

The reach of new media, combined with the power of illicit economic circuits, allows Mexico's cartels to operate throughout North America, Central America, and South America, with their tentacles reaching into Africa and Europe.

The interaction among TCOs, enforcer groups, gangs, and potentially conventional terrorists deserves scrutiny. As noted in one article:

> *Terrorists, gangs, and organized crime can exist as independent threats, but increasingly, they interact in a number of ways. Terrorists or insurgents may exploit organized crime; criminal gangs may act as middlemen in small arms, explosives or human trafficking; drugs may finance operations; and actors on both sides of the house may facilitate or conduct attacks for each other.*[27]

Mexico is currently embroiled in a protracted drug war. Mexican drug cartels and allied gangs are currently challenging states and substate polities (in Mexico, Guatemala, El Salvador, and beyond) to capitalize on lucrative illicit global economic markets. As a consequence of the exploitation of these global economic flows, the cartels are waging war on each other and state institutions to gain control of the illicit economy. Essentially, they are waging a "criminal insurgency" against the states and their institutions. In doing so, they are becoming political as well as economic actors. Criminal insurgency is a battle for power; essentially it results in a struggle for who governs.[28]

Criminal insurgencies can exist at several levels:

+ Local insurgencies in a single neighborhood or "failed community" where gangs dominate local turf and political, economic, and social life. The criminal enterprise collects taxes and exercises a near-monopoly on violence. TCOs and gangs foster a perception that they are community protectors (i.e., "social bandits"). The criminal gang is seeking to develop a criminal enclave or criminal "free-state." Since the nominal state is never fully supplanted, development of a parallel state or "dual sovereignty" is the goal.

+ A battle for the parallel state where cartels and gangs battle over their own operational space, but violence spills over to affect the public at large and the police and military forces that seek to contain the violence and curb the erosion of governmental authority in the criminal enclave or criminal enterprise.

+ Combating the state where criminal enterprises directly confront the state to secure or sustain an independent range of action. The cartels are active belligerents against the state.

+ State implosion when high-intensity criminal violence spirals out of control. This potential would be the cumulative effect of sustained, unchecked criminal violence and criminal subversion of state legitimacy through endemic corruption and co-option.[29]

Criminal insurgency is directed toward retaining freedom of action for the criminal enterprise. TCOs and gangs may directly attack the state to co-opt or corrupt its political processes. The tools or methods of criminal insurgency include:

+ symbolic and instrumental violence including attacks on journalists, police, the military, and elected and judicial officials (targeted assassinations and mass attacks)

+ exerting control over turf through violence and social cleansing, resulting in refugees and internally displaced persons

+ information operations including corpse-messaging, *narcocorridos*, *narcomantas/narcopintas* (banners and graffiti), *narcomensajes* (communiqués), *manefesacions* (demonstrations), *narcobloqueos* (blockades), and *leventons* (express kidnappings)

+ utilitarian provision of social goods

+ resource extraction

- usurping state fiscal roles (street taxation, extortion)

- co-opting and corrupting government actors

- use of marked vehicles, uniforms, and insignia to confer legitimacy

- usurping the protective security (enforcement and punishment) role of the state

- promulgation of alternative identity narratives (including narcocultura) to secure legitimacy and community support (or tolerance) including adopting the mantle of social bandit.[30]

In El Salvador, both cartels and maras are adopting the mantle of social bandit. For example, National Public Radio (NPR) News reports, "In El Salvador, there's fear that the Mexican cartels are aligning themselves with the country's ubiquitous street gangs." The two main gangs, MS-13 and M-18, are so powerful and volatile that their members get sent to separate prisons. Impoverished neighborhoods in the capital, San Salvador, are clearly divided turf, belonging to one gang or the other. The maras violently and effectively rule their turf, "controlling street-level drug sales, charging residents for security and battling to exclude their rivals."[31]

According to the NPR report:

> The maras could offer—and according to some security analysts, already are offering—the Mexican cartels access to a vast criminal network. The maras have stashes of weapons, established communications networks and ruthless foot soldiers who have no qualms about smuggling drugs or assassinating rivals—for a price. . . .

> Blue [an MS gangster] talks of the MS as a social organization that protects the "civilians" in the neighborhood. They help get water lines connected. They're refurbishing the community hall. To him, it's normal that residents have to pay rent to the gang for these services.

Essentially, in El Salvador gang leaders are stating that they are social workers and that their gangs are providing social goods. While reporting for his three-part series on drug trafficking in Central America, NPR's Jason Beaubien spoke at length with "Blue," the second in command of the Mara Salvatrucha gang in El Salvador. Beaubien reported that

> gang members "really believe that they are doing good in the community. They believe that their gang structure . . . replaces what the state isn't giving"—security, water, a community hall. If Mexican cartels move in to work with the gangs in El Salvador . . . the power and money from the Mexicans combined with the organizational structure of the gangs would create "a terrible, terrible combination."[32]

Diane Davis provides insight into the dynamics of the situation. According to Davis, "Mexico's cartels constitute 'irregular armed forces'—well-organized, flexible urban gangs that make money smuggling drugs and other goods—buttressed by Mexico's socioeconomic problems."[33] The cartels, Davis contends, are different from rebel groups. They do not seek to remove the whole government, but instead to usurp some of its functions. In doing so, they use violence to protect their "clandestine networks of capital accumulation."[34] This leads some analysts (including Davis) to perceive that Mexico's drug wars involve physically dispersed, evolving organizations that could be viewed more as self-sustaining networks than antistate insurgents.[35] That conception is challenged here. The cartels are not antistate insurgents in the conventional sense. They are criminal insurgents who challenge the state in a quest for economic and raw political power. They are not seeking to replace the state, but rather to remove the state's authority over their operations. As a consequence, they are engaged in state reconfiguration (CStR) and become "accidental insurgents" with the potential to become new state-making entities.

Narcocultura, Social Bandits, and State-making

Davis also observed: "[The] random and targeted violence increasingly perpetrated by 'irregular' armed forces poses a direct challenge to state legitimacy and national sovereignty."[36] According to her analysis, cartels and gangs are "transnational non-state armed actors who use violence to accumulate capital and secure economic dominion, and whose activities reveal alternative networks of commitment, power, authority, and even self-governance."[37]

This situation has clear neofeudal dimensions. Consider Los Zetas in light of feudalism. Alfredo Corchado, a journalist covering Mexico's drug wars, points out indicators of cartel (especially Zeta) erosion of state institutions. These include territorial control and neofeudalism. Corchado explained, "Beset by violence and corruption, Guatemala teeters on the edge of being a failed state. In recent years, Guatemala has proved to be especially vulnerable to the Zetas, who rule over communities across the country like tiny fiefdoms."[38] Concrete examples of neofeudal characteristics include maintaining private armies, collecting taxes, extracting resources, and dominating local political and community organizations, as well as paying tribute to other criminal enterprises and government actors to retain freedom of movement within their sphere of influence.

Corchado observes that by leveraging the proceeds from billions of dollars in drug profits from U.S. sales, Mexican organized crime groups have taken control in parts of Guatemala, forming alliances with local criminal groups and undermining that state's fragile democracy. In Mexico, the Zetas now control chunks of territory in the Yucatan peninsula, northwestern Durango state, and the northern states of Tamaulipas, Nuevo Leon, and Coahuila (all bordering Texas).[39]

The result is "other governed spaces," "neo-feudal zones," and "criminal enclaves." In a report entitled "Drug cartels taking over government roles in parts of Mexico," Corchado explored cartel intrusion into sovereignty. He found:

The "police" for the Zetas paramilitary cartel are so numerous here—upward of 3,000, according to one estimate—that they far outnumber the official force, and their appearance further sets them apart. The omnipresent cartel spotters are one aspect of what experts describe as the emergence of virtual parallel governments in places like Nuevo Laredo and Ciudad Juarez—criminal groups that levy taxes, gather intelligence, muzzle the media, run businesses and impose a version of order that serves their criminal goals.[40]

As a consequence, "entire regions of Mexico are effectively controlled by nonstate actors, i.e., multipurpose criminal organizations," according to Howard Campbell, an anthropologist and expert on drug cartels at the University of Texas at El Paso. "These criminal groups have morphed from being strictly drug cartels into a kind of alternative society and economy. . . . They are the dominant forces of coercion, tax the population, steal from or control utilities such as gasoline, sell their own products and are the ultimate decision-makers in the territories they control."[41]

Narcocultura

Narcocultura is a form of social-environmental modification. Specifically, alternative belief systems are exploited by gangs and cartels to alter primary loyalties and erode the authority and legitimacy of the state. Narcocultura is instrumental in reshaping sovereignty. By creating alternative identity narratives, illicit networks project their power and offer themselves as an alternative to traditional state structures. Myths, folk songs (*narcocorridos*), imagery, and identity symbols (including patrons, heroes, and saints) are used to bolster the legitimacy and influence of the TCOs and gangs.

Guillermoprieta defines *narcocultura* as:

The production of symbols, rituals and artifacts—slang, religious cults, music, consumer goods—that allow people involved in the drug trade to recognize themselves as part of a community, to establish a hierarchy in which the acts they are required to perform acquire positive value and to absorb the terror inherent in their line of work.[42]

According to Bunker and Bunker, social environmental modification is an element of nonstate warfare. "This warfare—manifesting itself in 'criminal insurgencies' derived from groups of gang, cartel, and mercenary networks—promotes new forms of state organization drawn from criminally based social and political norms and behaviors."[43] Key elements of social/environmental modification by the aforementioned authors include alternative worship or veneration of "narco-saints," symbolic violence (including beheadings and corpse messaging—attaching a message to a corpse), and the use of narcocorridos (epic folk songs) and social media to spread messages and confer legitimacy on a cartel.

The mantle of "social bandit" confers legitimacy on the gangs' or cartels' actions within the community they by necessity dominate.[44] *La Familia, Los Cabelleros Templarios*, the Sinaloa Cartel, and even Los Zetas have embraced this role through skillful use of information operations and utilitarian provision of social goods to delegitimize their adversaries (rival gangs and the state alike). This is a potent example of the assertion authority and the use of alternative identity narratives as a means of stealing the mantle of sovereignty from the state.

The control of territorial space—ranging from "failed communities" to "failed regions"—is a critical element of the erosion of state capacity. This includes the exploitation of weak governance and areas (known as "lawless zones," "ungoverned spaces," "other governed spaces," or "zones of impunity") where state challengers have created parallel or dual sovereignty, or "criminal enclaves," in a neofeudal political arrangement. The use of instrumental violence, corruption, information operations (including attacks on journalists, alternative identity narratives including narcocultura, and assuming the mantle of social bandit), street taxation, and provision of social goods in a utilitarian fashion are among the tools employed by criminal actors to secure their freedom of movement and erode the authority of the state.

Impact on the State

Specific variables/indicators that are germane to understanding the impact of TCOs and networked criminal enclaves on the state include: violence both among cartels and directed at the state, corruption, degree of transparency, cartel/gang reach, effectiveness of governance/policing, community stability, effectiveness of economic regulation, and the degree of territorial control (loss or gain by the state vs. cartels).

The impact of transnational criminal enterprises on state capacity, control of territory, and legitimacy is critical. All of these activities occur across time. Some changes are slow-moving, while some are rapid in their expression. Key factors in the pace of change include:

+ social/environmental modification (such as the use of social networking media—Facebook and Twitter—propaganda/information operations, e.g., narcomantas and narcocorridos) to further a criminal gang's perceived social legitimacy

+ connections (or network connectivity) between and among criminal enterprises (i.e., nodal analysis and social network analysis)

+ impact of illicit economic circuits (including connections among criminal actors) on the legitimacy of borders in global cities and border zones, as well as criminal penetration and reach

+ usurpation of state fiscal roles (taxes, tariffs) by criminal enterprises through street taxation, protection rackets, and other diversion of public goods or funds

+ force including the use of instrumental and symbolic criminal violence (armed attacks, terrorist campaigns, "corpse messaging," kidnapping, attacks on police, attacks on journalists and public officials, and the development and employment of private armies) challenging the state's monopoly on legitimate force.

These factors are evident in three crucial spatial settings: networked diasporas, border zones, and criminal enclaves (themselves often the result of the first two).

Networked Diasporas

Terrorists, organized criminal enterprises, and transnational gangs increasingly operate across borders as global nonstate actors. This transnational dimension of crime and extremism makes coordination of foreign and domestic intelligence and law enforcement operations essential. Frequently, transnational criminal and terrorist actors find sanctuary in overseas diaspora communities. While this is not new (the Mafia or *Costa Nostra* in the United States is a classic case of Sicilian immigrants banding together and forming the nucleus of criminal gangs and enterprises), Russian *mafiya* gangs, Chinese triads, and now Mexican cartels and others operate globally. The information age allows these groups to coordinate and synchronize operations and make alliances in new ways. This includes the use of new media to transmit messages; announce alliances, attacks, and operations (including beheadings and corpse messaging or the use of symbolic brutality); broadcast threats and challenges; solicit recruits and alternative identity narratives (i.e., *narcocultura*); and transmit operational directives across multiple areas of operations in real or chosen time.

Networked diasporas are one consequence of the information age and require attention. Diaspora communities can provide extremists with a permissive environment favoring the emergence of extremist cells. Radical enclaves may arise within diaspora communities and serve as catalysts for broader radicalization. When linked to lawless zones and other radical enclaves through social networks and Internet media, a powerful "networked diaspora" results.[45]

Border Zones

Border zones are frequently areas where dark side actors and criminal netwarriors can gain a foothold. As such, they serve as incubators of conflict. Criminal gangs, insurgents, and terrorists exploit weak state presence to forge a parallel state and pursue their criminal enterprises. Their efforts are frequently sustained by fear, violence, and brutality.[46] For example, in Mexico, the northern frontier with the United States and southern border with Guatemala are contested zones. These zones have become the center of gravity for Mexico's drug cartels. These drug-smuggling corridors and plazas translink the borders. The bloody competition over these plazas and the spread of criminal reach and power throughout the state and across frontiers are significant security concerns.

Border zones in Latin America, for example, are at risk of territorial capture by armed groups and narcotrafficking networks. As Ivan Briscoe observed, government authority in these zones has been "hollowed out" and replaced by evolving global-local networks of criminal authority.[47] As noted by Sullivan and Elkus:

> *Border zones, such as Guatemala's Petén province, and sparsely policed areas like the Atlantic Coast of Nicaragua are incubators of instability and ideal venues for refueling, repackaging product, and warehousing drug stockpiles. While this was first realized in*

> *Guatemala and Honduras, El Salvador, Panama and Costa Rica, indeed all Central*
> *America, are currently at risk of being caught in the "cross-border" spillover of Mexico's*
> *drug wars. Controlling these border zones is key to transnational gangs and cartels.*
> *Los Zetas, for example, not only train in sparsely populated border areas, they seek to*
> *sustain military control of the frontier and adjoining terrain on both sides of Mexico's*
> *southern border between Chiapas and Guatemala.*[48]

The informal economies that have emerged in Latin American border zones demonstrate the transition of states that is emerging from the twin engines of globalization and the information age. The shift of government authority from the state (or in cases where the state has always been weak, the rise of criminal governance) to dark side criminal actors/criminal netwarriors is a consequence of globalization impacting loose frontier economies that serve as in-between zones for illicit goods within a common regional network. This exploitation of regional economic circuits, albeit illicit, illustrates the transition into a reconfiguration of power within the state, where traditional informal networks link with globalized forms of illicit commerce to create a new base of power.

Criminal Enclaves

In Mexico and parts of Central America, cartels and gangs have gained control over specific *plazas* ranging from a few city blocks to entire states or subnational regions. Exploiting weak state capacity in urban slums or rural border zones[49] either from the aftermath of civil war (Central America) or during the transition from one party rule (Mexico), criminal mafias of various stripes have exploited the vacuum of power. In Mexico, for example, the drug-trafficking organizations were traditionally moderated by the ruling party, the *Partido Revolucionario Institucional* (PRI). The end of the PRI monopoly on power allowed the cartels to seek new business and political arrangements. Cartels, now free from the influence of the PRI, could strike independent arrangements with local political actors. This freedom converged with the increasing globalization of crime. As a result, organized crime could then establish boundaries for the authorities, not the other way around.[50]

Drug cartels and criminal gangs are challenging the legitimacy and solvency of the state at the local, state, and national levels in Mexico and Central America. As Max Manwaring stipulated, these state challengers are applying the "Sullivan-Bunker Cocktail," where nonstate actors challenge the de facto sovereignty of nations.[51] In Manwaring's interpretation, gangs and irregular networked attackers can challenge nation-states by using complicity, intimidation, and corruption to subtly co-opt and control individual bureaucrats and gain effective control over a given enclave.

Essentially, the cartels and their networked 3 GEN gang affiliates exploit weak zones of governance, expanding their criminal turf into effective areas of control. They start by corrupting weak officials, co-opting the institutions of government and civil society through violence and bribes. They attack police, military forces, judges, mayors, and journalists to leverage their sway, communicate their primacy through information operations, and cultivate alternative social memes adapting environmental and social conditions toward their goals. Then they conduct social cleansing, killing those who get in their way and forcing others out of their area of operations; they are then able to effectively collect taxes, extract wealth and resources such as the diversion of oil and gas from Petroleos Mexicanos (PEMEX, Mexico's state-owned petroleum company), and basically control the territory.

This situation allowed a range of networked, local, and transnational criminal enterprises—gangs and cartels—to seize new criminal, economic, social, and political opportunities. Parallel or "dual sovereignty" over large swaths of the state was the result.[52] Provision of social goods (often wearing the mantle of social bandits) is one manifestation of increasing cartel power. Often this provision of social goods is purely utilitarian. The cartels seek to appease the populace to gain their complicity in fending off the state's enforcement imperative.[53] As a consequence, the cartels are exerting real territorial control:

> Mexico's periphery has become a lawless wasteland controlled largely by the drug cartels, but the disorder is rapidly spreading into the interior. In a cruel parody of the "ink-blot" strategy employed by counterinsurgents in Iraq, ungoverned spaces controlled by insurgents multiply as the territorial fabric of the Mexican state continues to dissolve.[54]

This territorial control varies in scope from a few blocks or *colonias* to entire regions. The cartels and gangs need to provide social goods to sustain their impunity, consolidate their power, and ultimately expand their reach through displacement of the state or political accommodation—whichever comes first. Accordingly, they apply a "reverse inkblot" strategy to alter states.

Leveraging the power gained by dominating the plazas and criminal enclaves, these criminal networks have the opportunity to expand their domains by exerting dual sovereignty or actual political control over their corrupt vassals to forge narco-states. In either case, their expanding reach challenges nations and polities at all levels, potentially ushering in new forms of stratified (or dual/parallel) sovereignty. These may very well become network-states.

The "Network-State"

These new power configurations may result in new forms of sovereignty or new state forms—that is, new ways of organizing state power. Philip Bobbitt argues that this will be the "market state,"[55] Ronfeldt envisions a "nexus-state" ruled by "cyberocracy,"[56] and others, such as Sullivan and Elkus, make the case for the network-state, arguing, "States are not so much declining, failing and yielding as transforming their very nature. The network is the right metaphor to grasping the new state's complexity."[57] Network-states would share greater portions of power with transnational organizations, civil society actors, other states, and substate organs. Power would not be concentrated only in single geographic states, but certain state functions would transcend the boundaries of single states and state confederations. This is already occurring, as in the case of the suprastate European Union and multilateral institutions such as the World Trade Organization.

The capture, control, or disruption of strategic nodes in the global system and the intersections between them can have cascading effects. For instance, Saskia Sassen expands on the notion of border zones to posit a growing "frontier zone," a zone of difference where identities, allegiances, and organizational forms exist in a state of constant flux. This state of flux is the consequence of far-reaching economic trends resulting in a structural "hollowing" of many state functions while bolstering the state's executive branch and its emphasis on internal security.[58] This hollowing out of state function is accompanied by an extranational stratification of state function. According to Sassen, a "much overlooked feature of the current period is the multiplication of a broad range of partial, often highly specialized, global assemblages of bits of territory, authority, and rights."[59] These fractal bits are increasingly contested, with states and (in the case discussed here) criminal enterprises seeking their own "market" shares.

The potential for a state centralizing power is an insight found in Ronfeldt's exploration of the "nexus-state." Centralized information controls can become a systematic apparatus of control to dominate or manipulate the populace. Essentially, the state becomes the decentralized arbiter of network protocols that define the nodal interactions among a complicated set of networks, actors, and relationships. The state continues to exist, but control over key functions is transferred to cities, corporations, and issue-specific transnational organizations. Here is the crux of the battle between states and nonstate criminal netwarriors, between the light side and dark side networks. Which set of organizational entities will dominate the shift to new state forms?

Conclusion

State change and shifts in sovereignty are a potential consequence of the erosion of state authority, legitimacy, and capacity. Outcomes of such shifts could include failed states, the capture of state authority by transnational criminals, and the emergence of new state forms. Insurgencies, high-intensity crime, and criminal insurgencies that challenge state legitimacy and inhibit governance are a key national and global security issue.

"State failure" is one potential outcome of insolvent governance (the inability of weak states and corrupt state officials to sustain legitimacy and effectively govern) and extreme instability. This issue has been a concern to the global security and intelligence communities for several years (specifically since the implosion of the Somali state). According to King and Zeng, the term "refers to the complete or partial collapse of state authority. . . . Failed states have governments with little political authority or ability to impose the rule of law. They are usually associated with widespread crime, violent conflict, or severe humanitarian crises, and they may threaten the stability of neighboring countries."[60] While the absolute failure of an entire state is the extreme outcome, erosion of state capacity, the establishment of criminal enclaves, and ultimately the overall reconfiguration of power within the state (or "substate failure") are likely outcomes.

Nonstate actors such as TCOs, gangs, warlords, private armies, terrorists, and insurgent networks on the dark side, and private military or security corporations, global corporations, civil society, nongovernmental organizations, and evolving state, substate, and suprastate institutions on the light side, demand the development of new security and intelligence structures to ensure global stability and human security.[61] These structures and mechanisms are needed to address the direct assaults on state functions. More analysis is needed to define the appropriate structures to negotiate this transition.

This is a battle for information and real power among global networks, social media groups, and nongovernmental organizations to secure political power. Illicit networks are controlling turf and capturing state functions, including the legitimate protective security roles of the state, assuming the power to enforce their will and punish those who do not comply. This is a battle that favors the agile and those with the will to use brute strength and force (cyber or otherwise). The criminal netwarriors increasingly employ barbarization and high-order violence combined with information operations to seize the initiative and embrace the mantle of social bandit to confer legitimacy on themselves and their enterprises. States must adapt and react to these changes and challenges to sovereignty in order to maintain collective security and retain effective control of their territory, borders, and populace.

Notes

[1] This chapter draws from my ongoing research into the impact of transnational organized crime on sovereignty. Specifically, it recaps three prior papers: John P. Sullivan, "Social Networks and Counternetwar Response," presented to the Panel on Security, Resilience, and Global Networks at the 51st Annual International Studies Association Convention in New Orleans, February 20, 2010; "Intelligence, Sovereignty, Criminal Insurgency, and Drug Cartels," presented to the Panel on Intelligence Indicators for State Change and Shifting Sovereignty at the 52nd Annual International Studies Association Convention, Montreal, March 18, 2011; and "From Drug Wars to Criminal Insurgency: Mexican Cartels, Criminal Enclaves, and Criminal Insurgency in Mexico and Central America: Implications for Global Security," presented to the Seminar on Netwars and Peacenets, Institute of Global Studies, Maison des Sciences de l'Homme, Paris, June 27–28, 2011.

[2] See Nils Gilman, Jesse Goldhammer, and Steven Weber, eds., *Deviant Globalization: Black Market Economy in the 21st Century* (New York: Continuum, 2011). My contribution, chapter 16, "Future Conflict: Criminal Insurgencies, Gangs and Intelligence," complements the analysis found in this chapter.

[3] See Luis Jorge Garay-Salamanca, Eduardo Salcedo-Albarán, and Isaac De León-Beltrán, *Illicit Networks Reconfiguring States: Social Network Analysis of Colombian and Mexican Cases* (Bogota: METODO, 2010).

[4] Luis Jorge Garay and Eduardo Salcedo-Albarán, "State Capture and Co-opted State Reconfiguration," in "Drug Trafficking, Corruption and State," prepublication draft, 2011, 29–31.

[5] Ibid., 31.

[6] Ibid.

[7] See David Ronfeldt, *Visions from Two Theories*, available at <http://twotheories.blogspot.com/>, for a comprehensive view of his network social theory.

[8] John Arquilla and David Ronfeldt, eds., *Networks and Netwars: The Future of Terror, Crime, and Militancy* (Santa Monica, CA: RAND, 2001).

[9] See Manuel Castells, *The Information Age: Economy, Society, and Culture: The Rise of the Network Society*, Vol. I (Hoboken, NJ: Wiley-Blackwell, 2010); *The Power of Identity*, Vol. II (Malden, MA: Blackwell, 2004); and *The End of Millennium*, Vol. III (Malden, MA: Blackwell, 2008).

[10] United Nations Office on Drugs and Crime, "The Globalization of Crime: A Transnational Organized Crime Threat Assessment," June 2010.

[11] Juan Carlos Garzón, *Mafia & Co.: The Criminal Networks in Mexico, Brazil, and Colombia* (Washington, DC: Woodrow Wilson International Center for Scholars, Latin American Program, 2008).

[12] See John P. Sullivan, "Public-Private Intelligence Models for Responding to the Privatization of Violence," paper presented to Intelligence Studies Section of the International Studies Association (ISA), *2007 ISA Annual Convention*, Chicago, February 28–March 3, 2007, as a complement to this chapter. Also see Robert J. Bunker, ed., *Non-State Threats and Future Wars* (London: Frank Cass, 2003), and Robert J. Bunker, ed., *Networks, Terrorism and Global Insurgency* (London: Routledge, 2005), for a detailed discussion of the global threat environment.

[13] See John P. Sullivan and Adam Elkus, "Global cities–global gangs," openDemocracy.net, December 2, 2009, available at <www.opendemocracy.net/opensecurity/john-p-sullivan-adam-elkus/global-cities-%E2%80%93-global-gangs>.

[14] See John P. Sullivan, "Transnational Gangs: The Impact of Third Generation Gangs in Central America," *Air and Space Power Journal, Spanish Edition*, July 2008, available at <www.airpower.au.af.mil/apjinternational/apj-s/2008/2tri08/sullivaneng.htm>.

[15] John P. Sullivan, "Third Generation Street Gangs: Turf, Cartels, and Net Warriors," *Transnational Organized Crime* 3, no. 3 (Autumn 1997). See also John P. Sullivan, "Future Conflict: Criminal Insurgencies, Gangs and Intelligence," *Small Wars Journal*, May 31, 2009, available at <http://smallwarsjournal.com/blog/2009/05/future-conflict-criminal-insur/>.

[16] Personal interview with gang intelligence investigators, identities withheld for operational security. See also Samuel Logan, *This Is for the Mara Salvatrucha* (New York: Hyperion Books, 2009), and the *This Is for the Mara Salvatrucha* Web site, available at <http://www.thisisforthemarasalvatrucha.com/index.html>.

[17] The Mexican press speculates that between 38,000 and over 40,000 persons have been killed in the conflict since 2006. In January 2011, the Mexican government pegged the toll at 34,600. No official updates have been provided since. See "Mexico Debates Drug War Death Toll Figure amid Government Silence," *Latin America News Dispatch*, June 3, 2011, available at <http://latindispatch.com/2011/06/03/mexico-debates-drug-war-death-toll-figure-amid-government-silence/>.

[18] John P. Sullivan, "Criminal Netwarriors in Mexico's Drug Wars," GroupIntel, December 22, 2008, available at <www.groupintel.com/2008/12/22/criminal-netwarriors-in-mexico's-drug-wars/>.

[19] Ibid.

[20] John P. Sullivan and Adam Elkus, "Cartel v. Cartel: Mexico's Criminal Insurgency," *Small Wars Journal*, February 1, 2010, available at <http://smallwarsjournal.com/blog/2010/02/cartel-v-cartel-mexicos-crimin/>.

[21] "2 powerful cartels dominate Mexico drug war," CBS News, October 1, 2011, available at <www.cbsnews.com/2102-501715_162-20114298.html>.

[22] Daniel Borunda, "'Lynxes,' Azteca formed hit squad: birthday party attack directed by cartel, gang," *El Paso Times*, February 11, 2010, available at <www.elpasotimes.com/news/ci_14378578>.

[23] Chris Hawley, "Drug cartels tighten grip; Mexico becoming 'narco-state,'" *Arizona Republic*, February 7, 2010, available at <www.azcentral.com/arizonarepublic/news/articles/2010/02/07/20100207mexico-lawlessness.html>.

[24] Ibid.

[25] Ed Vulliamy, "Killing for kudos—the brutal face of Mexico's 21st-century war," *The Guardian/Observer*, February 7, 2010, available at <www.guardian.co.uk/world/2010/feb/07/mexico-drug-war>.

[26] Ibid.

[27] Sullivan, "Criminal Netwarriors in Mexico's Drug Wars."

[28] See John P. Sullivan and Robert J. Bunker, "Rethinking insurgency: criminality, spirituality, and societal warfare in the Americas," *Small Wars & Insurgencies* 22, no. 5 (December 2011), 742–763.

[29] John P. Sullivan, "Post-Modern Social Banditry: Criminal Violence or Criminal Insurgency?" paper presented at conference, Drug Trafficking, Violence, and Instability in Mexico, Colombia, and the Caribbean: Implications for U.S. National Security, co-hosted by University of Pittsburgh's Matthew B. Ridgway Center for International Security Studies and Strategic Studies Institute, U.S. Army War College, in Pittsburgh, PA, October 29, 2009.

[30] Sullivan, "Intelligence, Sovereignty, Criminal Insurgency, and Drug Cartels."

[31] Jason Beaubien, "El Salvador Fears Ties Between Cartels, Street Gangs," NPR News, June 1, 2011.

[32] Mark Memmott, "In El Salvador: Gang Leaders Who Say They're Social Workers," The two-way (NPR News Blog), available at <http://m.npr.org/story/136858037>.

[33] Diane Davis is quoted in Peter Dizikes, "An altered state," PHYSORG.com, available at <http://phys.org/news190880266.html>.

[34] Ibid.

[35] Ibid.

[36] Diane E. Davis, "Irregular Armed Forces, Shifting Patterns of Commitment, and Fragmented Sovereignty in the Developing World," *Theory and Society*, MIT Open Access Article, April 19, 2010.

[37] Ibid.

[38] Alfredo Corchado, "Traffic in illegal drugs spawns violence and corruption on path north," *Bellingham Herald*, April 28, 2011.

[39] Ibid.

[40] Alfredo Corchado, "Drug cartels taking over government roles in parts of Mexico," *Vancouver Sun*, May 4, 2011.

[41] Ibid.

[42] Alma Guillermoprieto, "The Narcovirus," *Berkeley Review of Latin American Studies*, Spring 2009, 3–9, available at <www.clas.berkeley.edu/Publications/Review/...pdf/BRLAS-Spring2009-full.pdf>. Also see Alma Guillermoprieto, "Days of the Dead: The New Narcocultura," *The New Yorker*, November 10, 2008, available at <www.newyorker.com/reporting/2008/11/10/081110fa_fact_guillermoprieto>.

[43] Pamela L. Bunker and Robert J. Bunker, "The Spiritual Significance of ¿Plata O Plomo?" *Small Wars Journal*, May 27, 2010, available at <http://smallwarsjournal.com/jrnl/art/the-spiritual-significance-of-¿plata-o-plomo>.

[44] See Eric Hobsbawn, *Bandits* (New York: The New Press, 2000).

[45] See John P. Sullivan, "Policing Networked Diasporas," *Small Wars Journal*, July 9, 2007, available at <http://smallwarsjournal.com/blog/2007/07/policing-networked-diasporas/>.

[46] See John P. Sullivan and Adam Elkus, "Border zones and insecurity in the Americas," openDemocracy.net, November 24, 2009, available at <www.opendemocracy.net/opensecurity/john-p-sullivan-adam-elkus/border-zones-and-insecurity-in-americas>.

[47] See Ivan Briscoe, "Trouble on the Borders: Latin America's New Conflict Zones," Madrid: FRIDE, July 2008, available at <www.humansecuritygateway.com/documents/FRIDE_LatinAmerica_troubleontheborders.pdf>.

[48] Sullivan and Elkus, "Border zones and insecurity in the Americas."

[49] Border zones are potential incubators of conflict. Criminal gangs exploit weak state presence to forge a parallel state and prosecute their criminal enterprises sustained by fear, violence, and brutality. See Sullivan and Elkus, "Border zones and insecurity in the Americas."

[50] The Institutional Revolutionary Party (known as the PRI in Spanish) traditionally set all power boundaries in Mexican political and economic life, both legal and illicit. That changed with the implementation of a true multiparty state. The criminal mafias exploited that new power-generating opportunity. See Nik Steinberg, "The Monster and Monterrey: The Politics and Cartels of Mexico's Drug War," *The Nation*, May 25, 2011, available at <www.thenation.com>.

[51] See Max G. Manwaring, "Sovereignty Under Siege: Gangs and other Criminal Organizations in Central America and Mexico," *Air & Space Power Journal—Spanish Edition*, July 1, 2008, available at <www.airpower.au.af.mil/apjinterntional//apj-s/2008/2tri08/manwaringeng.htm>.

[52] Michoacán was an early example of emerging cartel political action. In that state, La Familia forged a parallel government generating employment, keeping order, providing social and civic goods, collecting (street) taxes, and co-opting legitimate governmental administrative and security functions. See George W. Grayson, *La Familia Drug Cartel: Implications for U.S.-Mexican Security* (Carlisle Barracks, PA: U.S. Army War College, Strategic Studies Institute, December 2010). Los Zetas started providing similar social goods in 2010–2011, leading the author to observe that they were acting as "accidental insurgents."

[53] See Shawn Teresa Flanigan, "Violent Providers: Comparing Public Service Provision by Middle Eastern Insurgent Organizations and Mexican Drug Cartels," 52nd Annual ISA Convention.

[54] John P. Sullivan and Adam Elkus, "Mexico's Criminal Insurgency," *Defense and the National Interest* and *Small Wars Journal*, November 9, 2008, available at <http://smallwarsjournal.com/blog/2008/11/mexicos-criminal-insurgency/>.

[55] Philip Bobbitt, *The Shield of Achilles: War, Peace and the Course of History* (New York: Knopf, 2002); and Philip Bobbitt, *Terror and Consent: The Wars for the Twenty-First Century* (New York: Knopf, 2008).

[56] David Ronfeldt and Danielle Varda, "The Prospects for Cyberocracy (Revisited)," Social Science Research Network, December 1, 2008, available at <http://papers.ssrn.com/sol3/papers.cfm?abstract_id=1325809>.

[57] See John P. Sullivan and Adam Elkus, "Security in the network-state," openDemocracy.net, October 6, 2009, available at <www.opendemocracy.net/article/state-change-sovereignty-and-global-security>.

[58] See Saskia Sassen, "The new executive politics: a democratic challenge," openDemocracy.net, June 28, 2009, available at <www.opendemocracy.net/article/the-new-executive-politics-a-democratic-challenge>; and Sullivan and Elkus, "Security in the network-state."

[59] Saskia Sassen, "Neither global nor national: novel assemblages of territory, authority and rights," *Ethics & Global Politics* 1, no. 1–2 (2008), 61.

[60] Gary King and Langche Zeng, "Improving Forecasts of State failure," *World Politics* 53 (July 2001), 623–658.

[61] John P. Sullivan, "Fusing Terrorism Security and Response," in *Countering Terrorism and WMD: Creating a Global Counter-terrorism Network*, ed. Peter Katona, Michael D. Intriligator, and John P. Sullivan, 272–288 (London: Routledge, 2006).

Chapter 11

Counterinsurgency, Counternarcotics, and Illicit Economies in Afghanistan: Lessons for State-Building

Vanda Felbab-Brown

Since 2001, Afghanistan has become synonymous with the narco-state and the spread of crime and illegality. In 2007 and 2008, the Afghan drug economy reached levels unprecedented since at least World War II. Although the drug economy has declined since, the decrease has largely been driven by the saturation of the global drug market and by poppy crop disease rather than the policies of the international community and the Afghan government. Although several other illicit economies thrive in Afghanistan including the smuggling of legal goods, narcotics receive by far the most attention because they generate the largest profits and the greatest international opprobrium.

Narcotics production and counternarcotics policies in Afghanistan are of critical importance not only for drug control there and worldwide, but also for security, reconstruction, and rule-of-law efforts in Afghanistan. Unfortunately, many of the counternarcotics policies adopted after 9/11 not only failed to reduce the size and scope of the illicit economy in Afghanistan but also had serious counterproductive effects on peace, state-building, and economic reconstruction.

In 2009, the Obama administration wisely decided to scale back eradication efforts in Afghanistan, courageously breaking with 30 years of counternarcotics policies that focused on ineffective forced eradication of illicit crops as a way to reduce the supply of drugs and to bankrupt belligerents. But the effectiveness of its counternarcotics policies there—interdiction focused on Taliban-linked traffickers and alternative livelihoods efforts—has been challenged by implementation difficulties and is ultimately dependent on major progress in improving the security situation and governance in Afghanistan. As of fall 2011, governance in Afghanistan had been steadily deteriorating, with corruption and ethnic tensions rising and political patronage networks becoming more exclusionary, while any security improvements following the 2010 U.S. military surge remain extremely fragile. A civil war post-2014 remains a very likely outcome, with the corollary thriving of the drug trade.

This chapter first details the evolution of U.S. counternarcotics policy in Afghanistan since 2001, situating the changes in the policy within two conceptual frameworks. Next, it describes how the Obama administration broke with the dominant counternarcotics framework in an attempt to synchronize counternarcotics policies with its counterinsurgency efforts. That section also analyzes the implementation challenges President Barack Obama's counternarcotics strategy encountered—from the side effects of its interdiction focus, to poor governance and the inability to decide whether and how to combat broader corruption in Afghanistan, to defining alternative livelihoods efforts as narrow buying support programs rather than long-term sustainable development. Next, the chapter considers the likely security and political conditions in Afghanistan after a reduction in U.S. combat forces there in 2014. Subsequently, it explores two oft-ignored but potentially problematic side effects of any future counternarcotics success in Afghanistan: what illegal economy may replace the opium poppy economy if it is reduced, and where the opium poppy economy is likely to shift. In conclusion, the chapter offers broader lessons for dealing with illicit economies in the context of counterinsurgency and state-building.

Evolution of Counternarcotics Policy in Afghanistan Since 2001

The initial objective of U.S. intervention in 2001 was to degrade al Qaeda capabilities and institute regime change in Afghanistan. Dealing with the illicit economy was not considered integral to those military objectives. Thus, until 2003, U.S. counternarcotics policy in Afghanistan was essentially laissez-faire. The military understood that it would not be able to obtain intelligence on the Taliban and al Qaeda if it tried to eradicate poppy production. Meanwhile, to provide intelligence on the Taliban and to carry out direct military operations against the Taliban and al Qaeda, it relied on key warlords who had often been deeply involved in the drug economy since the 1980s.[1]

Under the concept of "lead nations" for the international assistance mission in Afghanistan, with specific countries assigned responsibility for specific sectors, the United Kingdom (UK) was tasked in 2002 with managing the international counternarcotics effort. Sensitive to the political problems of eliminating the rural population's livelihood, the UK at first deployed a "compensated eradication" program. Thus, during the 2002–2003 poppy growing season, the UK promised to pay farmers $350 for each *jerib* (unit of area) of poppies they themselves eradicated, with $71.75 million committed for the program.[2] From the outset, the policy was plagued by problems including corruption and moral hazards, and the policy was aborted in less than a year.[3]

By 2004, increased interdiction was undertaken instead. Its goal was to target large traffickers and processing laboratories. Immediately, however, the effort was manipulated by local Afghan strongmen to eliminate drug competition and ethnic/tribal rivals. Instead of targeting top echelons of the drug economy, many of which had considerable political clout, interdiction operations were largely conducted against small, vulnerable traders who could neither sufficiently bribe nor adequately intimidate the interdiction teams and their supervisors within the Afghan government. The result was a significant vertical integration of the drug industry.[4]

The other—again undesirable—effect of the interdiction effort was that it allowed the Taliban to integrate itself back into the Afghan drug trade. Having recouped in Pakistan, the Taliban was once again needed to provide protection to traffickers targeted by interdiction.[5]

Alarmed by the spread of opium poppy cultivation, some public officials in the United States in 2004 and 2005 also started calling for a strong poppy eradication campaign including aerial spraying.[6] From 2004 to 2009, manual eradication was carried out by central Afghan units trained by DynCorp International as well as by regional governors and their forces. That immediately ignited violent strikes and social protests. Another wave of eradication took place in 2005 and achieved a reduction in poppy cultivation. Most of the reduction was due to cultivation suppression in Nangarhar Province, traditionally one of the largest producing areas, which in 2004 produced approximately a quarter of all Afghan opium. Through promises of alternative development and threats of imprisonment, production there was slashed by 90 percent.[7]

Alternative livelihoods never materialized for many. The cash-for-work programs reached only a small percentage of the population in Nangarhar, mainly those living close to cities. The overall pauperization of the population there was devastating.[8] Unable to repay debts, many farmers were forced to sell daughters as young as 3 years old as brides or abscond to Pakistan, where refugees have frequently ended up in the radical Deobandi madrasas, refilling the ranks of the Taliban.

Apart from incorporating the displaced farmers into their ranks, the Taliban also began to protect the opium fields in addition to protecting traffickers. In fact, the antagonized poppy farmers came to constitute a strong and key base of support for the Taliban, denying intelligence to the International Security Assistance Force (ISAF) while providing it to the Taliban.[9]

One of the standard myths about Afghanistan's drug economy is that the Taliban forces farmers to grow poppies. That is almost never the case. Rather, the Taliban presence both enables and necessitates poppy cultivation as insecurity may interfere with farmers' ability to get legal crops to markets. However, such interference also often comes from the Afghan National Police (ANP), which often extorts farmers along illegal checkpoints until the "taxes and tolls" exceed the possible profits to be made in a legal market.[10] Farmers can avoid such extortion regarding opium because traders often pick up opium at the farm gate. Rather than being forced by the Taliban to grow poppies (to the extent that farmers are "forced" at all), pressure comes more from the economic, security, and political constraints they face. For example, access to loans, which many Afghan households need to cope during the winter months and to buy both consumer goods and durables, is linked to opium. Similarly, many sharecroppers are only able to rent land if they dedicate a portion of their acreage to opium poppy cultivation so they can pay landowners in opium and have valued collateral; legal crops such as wheat are not considered to be collateral of any value.

Like interdiction, eradication has been plagued by massive corruption problems, with powerful elites able to bribe or coerce their way out of having their opium poppy fields destroyed. The elites can also direct eradication against their political opponents, with the poorest farmers, who are most vulnerable to the Taliban's mobilization, bearing the brunt of eradication.[11]

Moreover, the reductions in opium poppy cultivation due to eradication have not been sustained. By 2007, cultivation in Nangarhar reached nearly the same level as before the 2005 eradication campaign.[12] Since then, Governor Gul Agha Sherzai has managed to keep cultivation negligible by a combination of buy-offs of influential *maliks* (tribal elders), promises of alternative livelihoods, and threats of eradication of the poppy crops and imprisonment. While farmers close to the provincial capital of Jalalabad have often managed to cope by switching to vegetable crops, increased dairy production, and working in cash-for-work construction programs, those distant from the provincial center, such as in the districts of Achin, Khogyani, and Shinwar, have suffered great economic deprivation. As their income has crashed by as much as 80 percent, and no alternative livelihoods programs have been available to them, their political restlessness has steadily grown.[13] Those areas have seen great levels of instability; intensified tribal conflict over land, water, and access to resource handouts from the international community; rebellions of young men against the local maliks supporting eradication; physical attacks on eradication teams; intense Taliban mobilization; and increased flows of militants into and through the province from Pakistan.

By 2009, eradication had the following effects: It did not bankrupt the Taliban. In fact, the Taliban reconstituted itself in Pakistan between 2002 and 2004 without access to large profits from drugs, rebuilding its material base largely from donations from Pakistan and the Middle East and from profits from another illicit economy—the illegal traffic of licit goods between Pakistan and Afghanistan. Rather, eradication strengthened the Taliban physically by driving economic refugees into its hands. Eradication also alienated the local population from the national government as well as from local tribal elites who agreed to eradication efforts, thus creating a key opening for Taliban mobilization.[14] Crucially, eradication critically undermined the motivation of the local population to provide intelligence on the Taliban to the counterinsurgents; instead, it motivated the population to provide intelligence to the Taliban. Moreover, the local eradicators themselves were in a position to best profit from counternarcotics policies, being able to eliminate competition—commercial and political alike—and alter market concentration and prices, at least in the short term, within their region of operations.

Although alternative livelihoods programs were part of the counternarcotics package, they were clearly a secondary and inconsistent mechanism designed to partially alleviate the pain that eradication brought to many rural households. Alternative livelihoods programs were slow to reach most of the population. To the extent they were extended, it was primarily in areas that had experienced eradication, but many areas subject to bans on cultivation or eradication did not receive any livelihood assistance programs. Where alternative livelihoods programs were extended, they did not sufficiently relieve the immediate economic losses, nor did they address the structural drivers of opium poppy cultivation. A legal microcredit system, for example, was absent in most of Afghanistan. The lack of security, along with increasing insurgency in southern Afghanistan, halted many of the alternative livelihoods projects. Although some areas, such as Helmand, had been showered with aid, much of it failed to reach ordinary farmers.[15] Projects such as the Kajaki Dam, the centerpiece of the U.S. Agency for International Development (USAID) efforts in the south for much of the 2000s, while important in the long or even medium run, frequently made little difference in the immediate economic conditions of the farmers at the village level and failed to be completed because of insecurity. At the same time, economic development programs even in the more permissive environments, such as northern Afghanistan, often simply did not materialize, although bans on poppy cultivation were secured through promises of alternative livelihoods.[16]

U.S. Counternarcotics Policy in Afghanistan

Recognizing the counterproductive effects of eradication, the Obama administration broke with decades of U.S. counternarcotics policies and defunded centrally led eradication in Afghanistan.[17] Although the U.S. Government continues to provide limited funding and technical assistance to Afghan governors who decide to proceed with eradication, the two core components of the administration's counternarcotics policy have been 1) interdiction of Taliban-linked drug traffickers and 2) rural development. Scaling back eradication strongly enhanced the new counterinsurgency policy focus on providing security to the rural population. However, success in reducing instability and the size of the drug economy also depends on the actual operationalizing of the strategy.

The interdiction element has been geared primarily toward Taliban-linked traffickers. ISAF units from countries that want to participate in the interdiction program—mainly U.S. and UK forces—have concentrated on reducing flows of weapons, money, drugs, precursor agents, and improvised explosive device (IED) components, with the goal of degrading the Taliban's finances and physical resources. Although hundreds of interdiction raids have now been conducted, especially in the south, and large quantities of opium and IEDs have been seized, it is questionable whether the impact on the Taliban's resource flows has been more than local. Large-scale military operations to clear the Taliban from particular areas, such as in Marja, Helmand, have also affected the insurgents' funding capacity and resource flows in those areas. But so far, the cumulative effects of the narcotics interdiction effort to suppress financial flows—including both generalized anti-Taliban interdiction and raids solely dedicated to counternarcotics—do not appear to be affecting the Taliban at the strategic level. Taliban fund-raising policy has long been to tax any economic activity in areas where the insurgents operate—for example, sheep herding in the north, illegal logging in the east, or National Solidarity Program projects in the center. The strongest effect of focusing interdiction on Taliban-linked traffickers appears to be, at least temporarily, to disrupt Taliban logistical chains because many of its logistical operatives handle both IED materials and drugs and because most raids are dual-use. In combination with ISAF's targeting focus on mid-level commanders, prioritization of the counternarcotics-interdiction focus is probably palpably complicating the Taliban's operational capacity in the south, where both the military surge and counternarcotics efforts have been prioritized.

Whatever its benefits regarding disrupting Taliban logistical chains, the interdiction policy has had two negative side effects. First, under the dual-use interdiction policy—which combines searching for mid-level Taliban commanders and "mere" supporters with searching for drugs—opium seizures have become too prevalent. The great frequency of night raids[18] and the tendency to seize or destroy any opium found in any searched household has blurred the strategic distinction between traffickers and farmers. Since most rural Afghans do not have access to and do not trust a banking system, these families hold their life savings and assets in opium. ISAF interdiction searches that destroy any found opium, perhaps under the belief that they are destroying Taliban stockpiles, are in fact often wiping out the entire savings of an Afghan household. Thus, in areas that have been subject to intense interdiction raids such as Marja and Nad Ali, the effects of supposedly "selective" and politically sensitive interdiction have come to resemble the effects of blanket eradication. The consequences have been the same: intense alienation of the local population from ISAF forces and the central government.[19]

Second, the interdiction policy signals to Afghan power brokers that the best way to conduct the drug business is to be a member of the Karzai government, further undermining the domestic legitimacy of the Kabul government and compromising the rule of law. But tackling corruption in Afghanistan is no easy task because of the international community's continuing dependence on problematic but "useful" interlocutors, competing priorities, and the domestic political sensitivities and debts of the Karzai government.

Beyond the matter of the drug trade, ISAF's reliance on corrupt and abusive warlords for intelligence, logistics, and direct counterterrorism operations often comes at the price of ignoring governance issues. Some of the most notorious power brokers, such as Ahmed Wali Karzai (before his assassination in July 2011), Matullah Khan, and Gul Agha Shirzai, know how to get things done to facilitate international operations in Afghanistan. The internationals are often too isolated behind the Hesco gravel bags at their compounds to be aware how rapacious and discriminatory some of their key Afghan interlocutors have been, or they just choose to ignore their problematic aspects.

Especially early on, the Obama administration accorded great importance to fighting corruption by building up various civilian structures, such as the Major Crime Task Force, and ultimately similar equivalent units within ISAF, such as its anticorruption task force, Shafafiyat. But it often demanded reform with an intensity that ignored the realities and political complexities of a system in which the highest to the lowest government officials, line ministries, banking centers, and most international contracts are pervaded by corruption.[20] The Obama administration's anticorruption campaign thus secured dramatic promises from President Hamid Karzai to tackle corruption with little actual follow-up. Moreover, the lack of prioritization as to which corruption needs to be addressed first and cannot be compromised often ignores the political debts President Karzai owes and his internal entanglements and dependencies. Karzai thus often seeks to reverse such anticorruption efforts as indictments of powerful corrupt officials and the development of the anticorruption and anticrime institutions the international community is trying to stand up.[21] His efforts often succeed.

But as the Obama administration began to scale down its military presence in Afghanistan, U.S. officials started vacillating once again in their determination to take on corruption. Many in the U.S. Government have begun to argue that tackling corruption is a luxury the United States can no longer afford; instead it needs to prioritize stability. This school of thought holds that limiting the military mission mostly to remotely-delivered airborne counterterrorist strikes could permit working through the local warlords and power brokers instead of being obsessed with their criminal entanglements and discriminatory practices and the means they used to acquire their power.[22]

Meanwhile, absent a coherent policy on corruption, the Obama administration and ISAF have failed to develop mechanisms and structures to work around and marginalize the problematic power brokers and often continue to be dependent on their services. As a high-ranking ISAF official in Kandahar told me in the fall of 2010, "In the current struggle for Kandahar, our nightmare is having to take on the Taliban and Wali [the then-alive Ahmed Wali Karzai] at the same time. But we understand that he has alienated some people in Kandahar." He went on to enthusiastically describe how then-Colonel Abdul Razziq—a notorious power broker and smuggler from Spin Boldak—recently cleared Mahlajat, a troubled subdistrict of Kandahar City, of the Taliban, "something even the Soviets couldn't do."[23] ISAF has since brought Razziq to Arghandab and other areas of Kandahar to conduct military operations and he has been named police chief in the city. A former high official of the U.S. Provincial Reconstruction Teams (PRTs) in Kandahar explained the difficulties the international community has faced when trying to impose redlines on power brokers such as Razziq. There was to be no undermining of provincial and district officials and no interference with the Peace Jirga (a body established by President Karzai to set up the broad framework for reconciliation with the Taliban) or with parliamentary elections. "But very quickly they violated all of the redlines we gave them. But they are effective in getting things done. We can't go after them at the same time as we are fighting the Taliban. When the Taliban is defeated, the Afghans will take care of the power brokers themselves."

The infusion of tens of billions of dollars of foreign aid has also generated corruption.[24] Large amounts of aid money appear to have been siphoned off by clever power brokers.[25] At other times, these financial flows have strengthened existing power brokers, who can get their hands on the money and who have developed vested interests in preventing others and the population at large from accessing these financial flows.[26] Some of the contract wars in places like Kandahar have been *actual* wars with rival businessmen linked to prominent tribes and power brokers, such as the Popolzais and Ahmed Wali Karzai on the one hand and the Barakzais and Gul Agha Shirzai on the other, physically shooting each other to get access to the contract money and setting up coercive monopolies under the guise of business associations to control the rents.[27] At other times, the aid flows have given rise to new "khans," further undermining both traditional institutions and the official government the international community has struggled to stand up. Many have financially profited from the insecurity that generates demands for private security companies and militias and that prevent effective monitoring.[28]

The international community's strategy has thus oscillated between tolerating corruption for the sake of other goals—with the justification that Afghans are used to corruption anyway—or confronting it head on, but with little effectiveness. Ignoring corruption is often justified as prioritizing stability, but since corruption and the lack of rule of law are key mobilizing mechanisms for the Taliban and a source of Afghans' anger with their government, it is doubtful that stability can be achieved without addressing at least the most egregious abuse.

Yet the system is so pervasively corrupt and so deeply and intricately linked to key structures of power and networks of influence that some prioritization of anticorruption focus is required. Such prioritization could include a focus on systematic tribal discrimination, corruption and ethnic discrimination in Afghan National Security Forces, corruption that undermines fragile legal markets such as illegal road tolls, and massive fraud in the banking system. Intolerable to most Afghans is not that they are required to pay bribes but that the bribes exceed tolerable norms, such as high and unpredictable extortion along illegal ANP checkpoints that take away all the profit from legal crops the farmers want to transport to markets.

Much of the corruption that is deeply grating to local populations and is thus politically explosive derives from the fact that dominant power brokers privilege their networks and tribes while marginalizing others and preventing them from getting access to international contracts and other economic opportunities. In Kandahar, for example, most legal and illegal economic activity is divided between the Karzai and Shirzai families, with few others being able to cut in. In Uruzgan, Matullah Khan has systematically discriminated against the Ghilzai Pashtuns, blocking their access to markets, imposing higher toll taxes on them, and denying them access to national government resources.[29] Anticorruption efforts need to focus on limiting such narrow patronage networks and on making sure that entire tribes and subtribes are not isolated. The political crisis surrounding the massive theft at the Kabul Bank once again drove home to ordinary Afghans that the privileged few, such as a narrow network around Vice President Mohammad Fahim, key northern leaders, and President Karzai's family, could steal hundreds of millions of dollars from public funds as well as money belonging to poor Afghans with almost complete impunity.[30]

The corruption in the Afghan National Security Forces (ANSF) is similarly linked to narrow patronage networks. Not only do anticorruption efforts need to ensure that commanders do not steal the wages of ordinary soldiers, but they also need to target commanders who systematically discriminate against soldiers from rival ethnic groups, for example by disproportionately placing them in the most dangerous areas.[31] A failure to mitigate the ethnic factionalizing of the Afghanistan National Army could be a crucial contributor to a post-2014 civil war in Afghanistan. Finally, influential power brokers whose economic and political power is threatened by better governance must not be permitted to undermine effective local officials, such as committed district police chiefs and district governors.

Buying Love versus Sustainable Development

Economic development efforts by the international community, including alternative livelihood efforts, have been plagued by vacillation between two competing understandings of the purpose of economic development projects. Is the purpose to buy off the population and wean it away from insurgents or to produce long-term sustainable development?

The buy-off concept has included quick-impact projects carried out by the U.S. military with money from the Commander's Emergency Response Program (CERP) or through the PRTs[32] as well as so-called "economic stabilization projects," also known as the District Delivery Program or District Stabilization Framework, carried out by USAID. The latter were designed as short-term cash-for-work programs lasting weeks or, at most, months. Their goals have been to keep Afghan males employed so economic necessities do not drive them to join the Taliban and to secure the allegiance of the population, which ideally will provide intelligence on the insurgents. Under this concept, U.S. economic development efforts have prioritized the most violent areas. Accordingly, the vast majority of the $250 million 2010 USAID budget for Afghanistan went to only two provinces, Kandahar and Helmand.[33] In Helmand's Nawa district, for example, USAID spent upward of $30 million within 9 months in what some dubbed "[the] carpet bombing of Nawa with cash."[34] With Nawa's 75,000 people, such aid amounts to $400 per person, while Afghanistan's per capita income is only $300 a year.

Although U.S. Government officials emphasize that these stabilization programs have generated tens of thousands of jobs in the south, many of the efforts have been short lived, such as canal cleaning, grain storage, road building, and small grants to buy seeds and fertilizers.[35] Characteristically, they collapse as soon as the money runs out, often in the span of several weeks. Nor has adequate consideration been given to the development of assured markets. Consequently, much of the produce cultivated under the USAID-contracted programs could rot before it was sold.

There is also little evidence that these programs have secured the allegiance of the population to either the Afghan government or to ISAF forces, or that they resulted in increased intelligence about the Taliban.[36] As many of these programs were budgeted to run only through October or December 2010, their closure sometimes antagonized the population by disappointing raised expectations. Another half billion dollars of U.S. aid was allocated for southern Afghanistan in 2011.

Because of the complexity and opacity of Afghanistan's political, economic, and contracting scene, many of these international programs have continued to flow to problematic, discriminatory, and corrupt power brokers, generating further resentment within the population and intensifying Afghanistan's rampant corruption and lack of accountability. At other times, they have spurred new tribal rivalries and community tensions.[37]

Nor have these programs yet addressed the structural deficiencies of the rural economy including the drivers of poppy cultivation. A microcredit system, for example, continues to be lacking throughout much of the country. In fact, many of the stabilization efforts, such as wheat distribution and grant programs, directly undermine some of the long-term imperatives for addressing the structural market deficiencies, such as the development of microcredit or the establishment of local seed banks and seed markets and rural enterprise and value-added chains. Shortcuts such as the Food Zone in Helmand and similar wheat distribution schemes elsewhere are symptomatic of the minimal short-term economic and security payoffs (but substantial medium-term costs) mode with which the international community has operated. The result has been persistent deep market deficiencies and compromised rule of law.[38]

There is a delicate three-way balance among long-term development, the need to generate support within the population and alleviate short-term economic deprivation, and state-building. A counternarcotics "alternative livelihoods" program provides a telling example. Aware of the deeply destabilizing effects of poppy suppression in the absence of alternative livelihoods and yet under pressure to reduce poppy cultivation, Helmand Governor Mohammad Gulab Mangal, widely acclaimed as a competent and committed governor, launched a wheat-seed distribution project during the 2008–2009 growing season. Farmers were handed free wheat seeds to discourage them from growing poppies. This program proved popular with the segments of the Helmand population who received the free wheat, and the program was emulated throughout Afghanistan and continued in 2010.

Poppy cultivation did decrease in Helmand in 2009, and many enthusiastically attributed it to the wheat distribution program rather than low opium prices. Yet there are good reasons to doubt the effectiveness of the program, at least concerning development and even governance. Because of land density issues, the lack of sustainability of the favorable wheat-to-opium price ratios under which the program took effect, and the limited ability of wheat cultivation to generate employment, wheat turned out to be a singularly inappropriate replacement crop.[39] Indeed, much of the wheat seed ended up being sold in markets rather than sown.

Due to the insecurity prevailing in Helmand at the time, the program was undertaken without any field assessment of what drives poppy cultivation in particular areas of the province and in Afghanistan more broadly.[40] Although this was a deficient process in which policy was developed without understanding the causes of the problem it was trying to address, the program was popular because most people welcome free handouts. It also became politically manipulated by local administrators and tribal elders who sought to strengthen their power. Although the program was deficient from a development perspective, it brought immediate political benefits to those who sponsored it including the political machinery of President Hamid Karzai, who was seeking reelection. Good governance was thus equated with immediate handouts and their political payoffs without regard for long-term economic development, best practices, and optimal decisionmaking processes.

At the same time, the wheat program and other economic stabilization programs often heightened expectations of free handouts from the central government and the international community without being economically viable and sustainable and without requiring commitments from the local community. Thus, many of the CERP and stabilization programs have encouraged the Afghans to expect payoffs for any activity consistent with the interests of the international community even if the activity is also in their own interest.

The Centers for Disease Control and Prevention, on the other hand, have required strong community participation and commitments to the development projects. Modeled similarly, the approach of the Dutch PRT in Uruzgan (at least until its forces withdrew in 2010) was particularly effective in limiting both the locals' expectations of free handouts and communal and intertribal tensions over the distribution of external assistance. The Dutch insisted that the entire community sanction any economic project and that the PRT only contribute the resources or technical knowledge that the community lacked. Thus the community had to identify and carry out all that it could execute in the project on its own, and the Dutch PRT and partner nongovernmental organizations (NGOs) would supply only the rest.[41]

Despite such examples, political pressures from the bottom up continue to reinforce ISAF's predilection for short-term, quick-impact projects. Sustainable development requires a lot of time, but the Afghan population has been highly impatient to see some minimal improvements and has often demanded handout programs without regard for long-term sustainability and desirability.[42] At least some Afghan government officials, however, have become dissatisfied with the short-term cash-for-work programs and are demanding that foreign aid be instead structured as capacity-building and long-term development projects.

Persistent though reduced insecurity even in high-profile focus areas such as Marja and Arghandab can threaten even the limited short-term "stabilization" programs. The Taliban has strongly intensified its campaign to assassinate government officials, international contractors, and NGO representatives and their Afghan counterparts who are cooperating with ISAF or the Kabul government; and both the implementers and the Afghan beneficiaries of these programs have been killed. This intimidation campaign has scared some Afghans away from participating in the programs and may result in local Afghan officials and internationals once more locking themselves up in their compounds as they did before the surge.[43] U.S. and ISAF officials emphasize that in cleared areas in the south, shops have reopened on the streets and bazaars seem livelier. Yet shopkeepers often say they are trying to make as much money as possible in a short window of opportunity because they expect security to deteriorate again.[44] Thus, even for these stabilization programs, security is a critical prerequisite.

Post-2014 Political and Poppy Dispensation

How sustainable and effective the current counternarcotics and anticorruption efforts will be overwhelmingly depends on the stability of the country and its political dispensation post-2014. Although major security improvements have been achieved in the south where the surge forces were concentrated, the east of the country has become at best a strategic stalemate. A major withdrawal of U.S. forces from the south to insert them into the troubled east can jeopardize all the fragile and costly improvements already made. Among other issues, any increased insecurity in the south will undermine counternarcotics efforts. Meanwhile, security in the north has been steadily deteriorating over the past 3 years with the Taliban insurgency, insecurity linked to official and unofficial militias[45] such as the Afghan Local Police, and accelerating crime. All three sources of insecurity reinforce conditions for the return of large-scale illicit economies, be it poppy cultivation as in the early 2000s, marijuana cultivation, or other forms of smuggling and extortion. How much and where security deteriorates following U.S. troop withdrawal will in turn depend on the effectiveness of the Afghan National Security Forces the United States has been trying to build up. Their strength and ability to withstand ethnic fragmentation have yet to be seen.

Overall, 2014 will bring about a triple shock. Not only will ISAF forces be substantially reduced, but U.S. funding will inevitably decline due to the drawdown of the military presence as well as economic conditions at home. For a country that is still overwhelmingly dependent on foreign aid and illegal economies for its revenues, the outcome will be a massive economic shrinkage. One likely result is intensification of Afghanistan's multiple illegal economies and a greater dependence on them for jobs. Although various initiatives are now under way to cushion the shock, such as efforts to build a new Silk Road through Afghanistan and exploit the country's vast mineral riches, there are no easy ways to generate revenues and employment over the next 3 years. Moreover, 2014 is also the year of another presidential election and hence will see major power infighting whether or not Karzai seeks to retain power. The fight over the remaining rents of the ending political dispensation and the need to consolidate support camps in anticipation of the shaky future—and hence to deliver spoils to them to assure their allegiance—will not be conducive to good governance. Even with the new U.S. emphasis on negotiations with the Taliban, the most likely scenario under current conditions is a civil war.[46] Even short of such a disastrous outcome, any deterioration of security and shrinking of the current legal economies in Afghanistan will undermine counternarcotics efforts.

Should a civil war be avoided and should counternarcotics efforts suddenly and miraculously destroy the poppy economy in Afghanistan—whether through fungicide-based eradication or speedily effective rural development—two critical "what-then" questions need to be built into policy analysis. First, what illicit economy would replace the existing opium economy in Afghanistan? And second, to what country would opium production shift?

What Replacement Illicit Economy Would Emerge?

Illicit economies rarely simply disappear. The presence of a large illicit economy not only satisfies the socioeconomic needs of the population, it also generates widespread smuggling knowhow, extensive criminal networks, and numerous powerful actors with vested interests in the preservation of the illicit economy. These actors include criminal and belligerent groups, corrupt government officials, and political power brokers. This infrastructure of crime can easily be transferred from one illicit economy to another. In Colombia, the drug trade built on several decades of smuggling cigarettes, household goods, marijuana, and emeralds. Many of the original smugglers emerged as prominent Colombian drug capos.[47] In Myanmar, the eradication of opium poppies since the late 1990s (crucially helped by overproduction in Afghanistan) gave rise to an extensive production of methamphetamines, rampant illicit logging, and a massive increase in the illegal trade in wildlife. In Thailand, where the most effective alternative livelihoods programs in the world to date eliminated poppy cultivation, heroin smuggling and meth production have been thriving. In Afghanistan itself, the illegal drug economy was built on decades of smuggling including extensive smuggling of legal goods and gems, illegal logging, and illegal trade in wildlife.

Apart from cannabis, which has already replaced opium cultivation in areas subject to bans, such as Balkh,[48] the least dangerous and potentially most easily suppressible illicit replacement economy would be just such an increase in the illicit trade of licit goods. This traffic exists as a result of the Afghan Transit Trade Agreement, under which goods can be imported into Pakistan duty-free for reexport into Afghanistan. Goods are then smuggled with profit back into Pakistan, and smugglers avoid Pakistani tariffs.[49] Although profits from the illicit traffic (at times $1 billion per year)[50] could rival those from drugs, if Pakistan and Afghanistan set their tariffs at the same level, this trade would disappear. Already today, as in the 1990s, smuggling generates extensive revenues for the Taliban and others.

A more ominous illicit replacement economy would be the production of synthetic drugs such as methamphetamines. Afghanistan would face stiff competition from Myanmar, Thailand, and Mexico, but the global market for synthetic drugs is rapidly growing, and Afghanistan could likely cut in on it. In that case, the country would still suffer from all the same political, economic, and social vices of illegal drug production, as is the case with opium cultivation. But the rural population would be left destitute since the production of synthetic drugs is much less labor-intensive than the cultivation of opium and hence could employ only a tiny fraction of the farmers and laborers of the opium economy. At the same time, large traffickers, corrupt government officials, and belligerent groups could easily maintain the level of income the opium economy affords them. Moreover, the production of synthetic drugs would be considerably harder to detect and disrupt. Rather than hoping that the overall criminal economy in Afghanistan can be eliminated, policymakers need to ask themselves what type of illicit economy is least detrimental to the objective of stabilizing Afghanistan and other key objectives. Correspondingly, efforts to disrupt the other illegal economies—those most disruptive of the key objectives—need to be maximized.

Where Would Poppy Cultivation Move?

The second what-then question of vital importance for the United States is what country opium cultivation would shift to. Given high world demand for illicit opiates, suppression of poppy cultivation in Afghanistan would not leave a highly lucrative market unsatiated but would simply move the industry elsewhere. Unlike coca, the opium poppy is a very adaptable plant that can be grown under a variety of climactic conditions. Theoretically, its cultivation could spread to many areas—Central Asia, back to the Golden Triangle of Southeast Asia, or West Africa.[51]

By far the worst scenario from the U.S. strategic perspective would be the shift of poppy cultivation to the Federally Administered Tribal Areas (FATA), Khyber-Pakhtunkwa, or even Punjab in Pakistan. For over 20 years, Pakistan has been a major heroin refining and smuggling hub in the region. It has an extensive *hawala* system that includes moving drug profits. Today, these territories also have extensive and well-organized Salafi insurgency and terrorist groups that seek to limit the reach of the Pakistani state and topple the government. A relocation of extensive poppy cultivation there would be highly detrimental to U.S. interests since it would contribute to a critical undermining of the state and fuel jihadi insurgency. Such a shift would not only increase profit possibilities for Pakistani belligerents, but also provide them with significant political capital by allowing them to become important local employers sponsoring a labor-intensive economy in areas with minimal employment opportunities.

Nor is Pakistan a newcomer to the drug trade. During the heyday of illicit poppy cultivation in Pakistan in the 1980s, the opium poppy was grown in FATA and the then–Northwest Frontier Province (now renamed Khyber-Pakhtunkwa). Opium poppy cultivation often involved entire tribes and represented the bulk of the local economy in these highly isolated (geographically, politically, and economically) places.[52] Pakistan was also the locus of heroin production and smuggling, with prominent and official actors such as Pakistan's military and the Inter-Services Intelligence directorate deeply involved in the heroin trade.[53]

U.S.-sponsored eradication during the 1980s generated violent protests and political costs too high even for the military dictatorship of General Zia ul Haq.[54] In the 1990s, strong emphasis was thus placed on generating legal economic alternatives to wean Pakistani tribes from drugs. Consisting mainly of small rural infrastructure projects and special economic opportunity zones (similar to those for textiles promoted by the current U.S. administration in Pakistan), the programs linked isolated areas better with the rest of Pakistan and increased local populations' identification with the Pakistani state.

In 2002, the United Nations Office on Drugs and Crime (UNODC) declared Pakistan cultivation-free. However, the dominant reason for the decline in opium poppy cultivation was not counternarcotics efforts, whether eradication or alternative development, but rather the wholesale shift of cultivation to Afghanistan during the 1990s.

Moreover, the positive political and economic effects of alternative development efforts in Pakistan frequently proved ephemeral and failed to generate sustainable employment. Many participants have continued to be consigned to subsistence agriculture, trucking, and smuggling and to migration to other parts of Pakistan such as Karachi, or to Dubai.[55]

This extensive drug-trade network, the history of poppy cultivation, and poor central-government control over the border regions with Afghanistan make Pakistan a likely candidate for vastly increased poppy cultivation if Afghan production is disrupted. Some opium cultivation has already emerged in Baluchistan, Khyber, Kohistan, and Kala Dhaka. Given the lack of systematic drug surveys in those and other areas of Pakistan, the extent of cultivation there is difficult to gauge, but some assessments report a resurgence of cultivation up to 2,000 hectares in recent years. It may well be more, given the lack of economic alternatives in the area, the history of opium poppy cultivation there, and the fact that the level of poppy cultivation in Kashmir on both sides of the Line of Control is estimated at 8,000 hectares.[56]

There is little evidence today that either the Afghan Taliban or the Pakistani Taliban (including Tehrik-i-Taliban-Pakistan and Tehrik-e-Nafaz-e-Sharia-Mohammadi) has systematically penetrated the slightly resurgent opium poppy cultivation in FATA and Khyber-Pakhtunkwa, even though they may have penetrated trafficking in drugs and precursor agents in Pakistan. Instead, it appears that the main sources of the Pakistani Taliban's income include: smuggling in legal goods, charging tolls and protection fees, taxation of all economic activity in the areas in which they operate—some being highly profitable, such as marble mining, theft and resale of North Atlantic Treaty Organization supplies heading to Afghanistan via Pakistan, illicit logging, and fundraising in Pakistan and the Middle East.[57] While profits from such a diverse portfolio can equal or even surpass profits from drugs, the main downside from the perspective of belligerent actors is that these economic activities are not labor-intensive. Consequently, unlike when belligerent groups sponsor the highly labor-intensive cultivation of opium poppies, the jihadi groups in Pakistan cannot present themselves as large-scale providers of employment to the local population.

If extensive poppy cultivation shifted to Pakistan, the consequences for U.S. national security would be extremely serious. FATA and parts of Khyber-Pakhtunkwa, as the jihadi takeover of Swat and Malakand in spring 2009 revealed, are already the hub for anti-American jihadists. Salafi insurgency and global terrorism networks have been leaking into and taking root in southern Punjab and go beyond the Lashkar-i-Taiba or Jaish-i-Mohammad presence.

Not only could al Qaeda and affiliated terrorist groups there profit financially from drug trafficking and money laundering, but ready access to cultivation (which these groups, unlike the Taliban, do not have as long as cultivation is centered in Afghanistan) would allow them to provide an economically superior livelihood to vastly undeveloped regions in Pakistan and thus obtain significant political capital within the population. Their calls to jihad against the Pakistani state would gain greater resonance with the tribal population. What these groups now can provide to the population are ideological succor and promises of martyrdom.

If production shifted to Pakistan, the sponsorship of cultivation would allow these groups to distribute significant real-time economic benefits to the population, a key source of legitimacy. Just as happened in Afghanistan in the 1980s, the jihadists would be able to outperform traditional tribal elites in providing for the population's needs. The sponsorship of relocated opium cultivation would allow the jihadists to offset the potential losses of support resulting from these attacks on the tribal elite.

Government efforts at eradication would generate protests and uprisings, cementing the bond between the jihadists and the population and weakening the already tenuous legitimacy of Islamabad. Weak central government presence there (military and otherwise) would compromise counternarcotics efforts, but eradication would greatly undermine even modest counterterrorism and stabilization efforts by the government. Given the existence of militancy in the likely poppy regions, forced eradication would greatly fuel militancy and generate far greater negative security externalities than it did in the 1980s and early 1990s when social protest had not congealed into a highly organized form, social networks were not premobilized, and pernicious political entrepreneurs were not at the ready to capitalize on social discontent.

Because of the continuing geographic, political, and social isolation of these areas, the lack of rule of law and the paucity of productive assets (both physical resources and human capital), generating employment opportunities there will be highly challenging under the best of circumstances.

A large-scale shift of opium poppy cultivation to Pakistan in the near and medium term would thus contribute to a further critical weakening of the state and undermine its control of and even reach to some of the most jihadi-susceptible areas in Pakistan. Such a large-scale shift of cultivation would also likely leak into Baluchistan, where heroin processing facilities and trafficking networks are already extensive. It would thus enable Baluchi nationalists to tap into the drug economy and strengthen the Baluchi insurgency in a multifaceted way, further threatening the territorial integrity of Pakistan and diverting the state's attention from the jihadi threat. Assisting the government of Pakistan today in both rural development in the critical regions and overall in enhancing the effectiveness of its interdiction and law enforcement capacity has the potential of reducing the security and political threats that could result from such a relocation.

Critically, in devising counternarcotics policy in any particular locale, policymakers also need to consider where the drug economy is likely to shift and whether that would be even more detrimental to their objectives than the current location of the illicit economy with all the problems it generates there.

Conclusions and Broader Policy Lessons

The Obama administration is finally in synch with the counterinsurgency effort. But persisting insecurity and often problematic operationalizing of the overarching counternarcotics strategy on the ground have limited its effectiveness in reducing Afghanistan's drug economy. Much of the strategy, such as rural development, ultimately depends on substantial and lasting improvements in security. Even then, substantial reduction in the size and significance of the drug economy will take several decades.

What lessons can be drawn from the case of Afghanistan for the involvement of international peacekeeping forces in tackling illicit economies? And how do efforts against illicit economies, such as the drug trade, interact with state-building efforts?

During conflict situations, peacekeeping and counterinsurgency forces should resist calls for their involvement in the eradication of illicit crops. Such involvement will antagonize the rural population, which is the center of gravity of counterinsurgency, and inhibit intelligence gathering. Although counternarcotics policy is not the sole determinant of the population's allegiance, and other factors, such as avoiding civilian casualties and providing consistent and robust security against the insurgents' reprisals, are also critical, counternarcotics policies are crucial since they strike at the economic survival of the population. Ultimately, the ability to feed one's family is as important from the perspective of survival as the threat of physical violence.

For success against illicit economies, it is essential to address corruption and poor governance. Indeed, strengthening good governance is a requisite for counterinsurgency and stabilization as well as economic reconstruction and counternarcotics efforts themselves. At the core of most of these efforts lies a process of building positive linkages between the population and the state.

Thus, for example, efforts against illicit economies such as the drug trade should be left only to specialized, constantly-vetted, and closely-monitored law enforcement units. To avoid the temptations of corruption, attacks on police by organized crime and militants, and alienation of the broader population from the police, regular police should be removed from counternarcotics efforts, including interdiction. Instead, broad police reform should be geared toward building trust so the population can see official law enforcement as a legitimate state representative delivering essential security, order, and justice.[58]

It is critical that policies against illicit economies be cognizant of the complex political dynamics that illicit economies generate. Embracing policies against illicit economies without paying close attention to their complex and multiple political effects can lead to counterproductive entanglement in local disputes, especially in places where the complexity and intensity of tribal, ethnic, and regional tensions and other local cleavages overlay the illicit economy. Illicit economies are deeply embedded in local social and political arrangements, especially where legal economies, official political arrangements, law enforcement capacity, and overall state presence are weak. Without recognizing this social embeddedness, policies to tackle illicit economies easily turn counterproductive and generate negative externalities and unanticipated second and third order effects that can negatively reverberate in other domains and within other networks.

Nonetheless, peacekeeping and counterinsurgency forces do have a large and, indeed, fundamental role in reducing illicit economies—delivering security. No matter what counternarcotics efforts will be undertaken, be they iron-fisted eradication or alternative livelihood, they will not be effective in reducing the illicit economy unless firm security throughout the entire territory has been established. The state needs to be strengthened and conflict must be ended before efforts against illicit economies, such as large-scale eradication of illicit crops, can be achieved.

International peacekeeping forces need to understand that the more the legal economy in the theater of intervention is destroyed, the more robust and deeply ensconced the illicit economy will be. Prominent military and political actors in the region—possible allies or proxies of the intervention forces—will also very likely be deeply involved in the illicit economy, and their power will be inextricably linked to their ability to use the illicit economy to provide for the population's elemental needs.[59] Conversely, however, the engagement of intervention forces with such actors will have profound effects on the shape of and power distribution within the illicit economy, and thus within the country itself.

Expanding the mission of international peacekeeping forces beyond the provision of security to direct efforts to reduce illicit economies and corruption demands that the international peacekeeping forces have a very detailed understanding of the intricacies of the local illicit networks and economy and their nexus to violent conflict, to the political and socioeconomic structures in the country, and to peace. Accordingly, enlargement of the traditional, more limited role of peacekeeping forces requires that the mission has a continual and robust information-gathering component that constantly monitors the effects of the policies against illicit economies on the political and economic distribution of power and on stability and development. It is therefore important to provide the peacekeeping forces with a robust analytical support component that would include economic, agricultural, and anthropology experts. But it is also important to recognize that the staying power of the international peacekeeping forces will always be inherently limited and that efforts to suppress illicit economies will only be sustainable if the population and its political representatives have the economic and political incentives to support such policies.

Regarding nonmilitary operations, such as economic reconstruction, it is vital that the international community scale down its expectations of how rapidly legal economies can replace illicit ones. Even when the basic economic infrastructure is present and intact, the growth of the legal economies may well coincide with the continuing flourishing of the illegal enterprises. But certainly in areas where the basic structural requirements for a legal economy are absent, as is true in the majority of the world's large-scale drug cultivation areas, efforts to boost alternative livelihoods are likely to take decades.

Moreover, a seeming success in suppressing an illicit economy in a particular region can easily lead to its transformation into a differently-organized illicit economy, which could be no less dangerous to the state and possibly the larger international community than the original economy. Nonetheless, efforts to boost licit livelihoods represent the only available source-country option to reduce illicit economies without resorting to substantial, lasting, and costly repression.

Counternarcotics efforts are indeed a key component of stabilization and reconstruction in Afghanistan and in any country where licit livelihoods have been decimated and an illicit narcotics economy thrives and intermingles with violent conflict. However, premature and inappropriate efforts against such an illicit economy, be it drugs or other commodities, greatly complicate counterterrorism, counterinsurgency, and stabilization objectives. Hence, they ultimately also jeopardize economic reconstruction and political consolidation.

Notes

[1] See, for example, Anne Barnard and Farah Stockman, "U.S. Weighs Role in Heroin War in Afghanistan," *Boston Globe*, October 20, 2004; and Michael E. O'Hanlon, "A Flawed Masterpiece," *Foreign Affairs* 81, no. 3 (May/June 2002), 47–63.

[2] Adam Pain, "Opium Trading Systems in Helmand and Ghor," Afghan Research and Evaluation Unit (AREU), Issues Paper Series (Kabul: AREU, January 2006), available at <www.areu.org.af/Uploads/EditionPdfs/601E-Opium%20Trading%20Systems-IP-print.pdf>.

[3] John F. Burns, "Afghan Warlords Squeeze Profits from the War on Drugs, Critics Say," *The New York Times*, May 5, 2002.

[4] Pain.

[5] Carlotta Gall, "Taliban Rebels Still Menacing Afghan South," *The New York Times*, March 2, 2006.

[6] Robert B. Charles, Assistant Secretary for International Narcotics and Law Enforcement Affairs, "Afghanistan: Are the British Counternarcotics Efforts Going Wobbly?" Testimony Before the House Committee on Government Reform Subcommittee on Criminal Justice, Drug Policy, and Human Resources, 108th Cong., 2nd sess., April 1, 2004, available at <www.useu.be/Article.asp?ID=90CBBA80-50B2-4939-8E76-FC1C3C041D2B>.

[7] United Nations Office on Drugs and Crime (UNODC), "The Opium Situation in Afghanistan as of 29 August 2005," (UNODC 2005), available at <www.unodc.org/pdf/afghanistan_2005/opium-afghanistan_2005-08-26.pdf>.

[8] David Mansfield, "Pariah or Poverty? The Opium Ban in the Province of Nangarhar in the 2004–05 Growing Season and Its Impact on Rural Livelihood Strategies," Project for Alternative Livelihoods (PAL), PAL Internal Document No. 11 (Jalalabad: German Agency for Technical Cooperation Project for Alternative Livelihoods in Eastern Afghanistan, June 2005).

[9] For details, see Vanda Felbab-Brown, *Shooting Up: Counterinsurgency and the War on Drugs* (Washington, DC: The Brookings Institution, 2010), 149–154.

[10] Author interviews in Uruzgan and Kandahar, Spring 2009.

[11] See David Mansfield and Adam Pain, "Evidence from the Field: Understanding Changing Levels of Opium Poppy Cultivation in Afghanistan," AREU Briefing Paper Series (Kabul: AREU, November 2007); and Barnett R. Rubin and Jake Sherman, *Counter-Narcotics to Stabilize Afghanistan: The False Promise of Crop Eradication* (New York: Center on International Cooperation, February 2008), available at <www.cic.nyu.edu/afghanistan/docs/counternarcoticsfinal.pdf>.

[12] UNODC, Afghanistan Opium Survey 2006, September 2006, available at <www.unodc.org/unodc/index.html>; UNODC, Afghanistan Opium Survey 2007.

[13] See David Mansfield, "The Ban on Opium Production across Nangarhar—A Risk Too Far," author's copy, Harvard University, March 8, 2010.

[14] For analysis of the social control structures, see Olivier Roy, *Islam and Resistance in Afghanistan* (Cambridge: Cambridge University Press, 1990).

[15] See, for example, Joel Hafvenstein, *Opium Season: A Year on the Afghan Frontier* (Guilford, CT: The Lyons Press, 2007); and Holly Barnes Higgins, "The Road to Helmand," *The Washington Post*, February 4, 2007.

[16] See, for example, Adam Pain, "'Let Them Eat Promises': Closing the Opium Poppy Fields in Balkh and Its Consequences," AREU Case Study Series (Kabul: AREU, December 2008).

[17] Provincial governors in Afghanistan can choose to engage in their own eradication efforts. During 2010, 2,316 hectares were eradicated in Afghanistan under this governor-led program; UNODC, Afghanistan Opium Survey 2010, 1.

[18] Night raids have been extremely controversial. Many Afghans believe the raids violate both key cultural norms as well as human rights. The North Atlantic Treaty Organization, on the other hand, has considered the raids one of its most important counterinsurgency tactics. The controversy escalated until, on January 2012, President Hamid Karzai delayed signing a strategic partnership agreement with the United States because of his refusal to tolerate further night raids. For a human rights perspective on the night raids, see Open Society Foundations and the Liaison Office, *The Cost of Kill/Capture: Impact of the Night Raid Surge on Afghan Civilians* (New York: Open Society Foundations, September 19, 2011), available at <www.opensocietyfoundations.org/>.

[19] For details, see David Mansfield, "Managing Concurrent and Repeated Risks: Explaining the Reductions in Opium Production in Central Helmand Between 2008 and 2011" (Kabul: AREU, August 2011), available at <www.areu.org.af/EditionDetails.aspx?EditionId=558&ParentId=7&ContentId=7&Lang=en-us>.

[20] On banking sector corruption and Western anti-money-laundering efforts, see, for example, Matthew Rosenberg, "Corruption Suspected in Airlift of Billions in Cash from Kabul," *The Wall Street Journal*, June 25, 2010; Greg Miller and Ernesto Londono, "U.S. Officials Say Karzai Aides Are Derailing Corruption Cases Involving Elite," *The Washington Post*, June 28, 2010.

[21] Miller and Londono.

[22] See, for example, interview with unnamed U.S. officials in Greg Jaffe, "U.S. to Temper Stance on Afghan Corruption," *The Washington Post*, September 4, 2010; Greg Miller and Joshua Partlow, "CIA Making Secret Payments to Members of Karzai Administration," *The Washington Post*, August 27, 2010; and Dexter Filkins and Mark Mazzetti, "Key Karzai Aide in Corruption Inquiry Is Linked to CIA," *The New York Times*, August 25, 2010.

[23] Author's interview with an International Security Assistance Force official, Kandahar, September 2010.

[24] As of January 2010, the United States alone had allocated more than $51 billion to Afghanistan, with more than half going to Afghan National Security Forces. See Special Inspector General for Afghanistan Reconstruction (SIGAR), "Quarterly Report to the United States Congress," July 30, 2010, available at <www.sigar.mil/pdf/ quarterlyreports/2010-07-30qr.pdf>, 43; and Kenneth Katzman, *Afghanistan: Post-Taliban Governance, Security, and U.S. Policy*, RL30588 (Washington, DC: Congressional Research Service, 2010), available at <www.fas.org/ sgp/crs/row/RL30588.pdf>, 77–78. In 2011, the plan was to raise the U.S. contribution to $71 billion. See, for example, Brett Blackledge, Richard Lardner, and Deb Reichmann, "After Years of Rebuilding, Most Afghans Lack Power," *Huffington Post*, July 19, 2010. The majority of the aid goes to building Afghan security forces.

[25] Matthew Rosenberg.

[26] See, for example, Dexter Filkins, "With U.S. Aid, Warlord Builds Afghan Empire," *The New York Times*, June 5, 2010.

[27] Following congressional inquiries into the corruption associated with the international contracts, the Shafafiyat anticorruption task force is also supposed to tackle this type of corruption. See Carlotta Gall, "Kandahar, a Battlefield Even Before U.S. Offensive," *The New York Times*, March 26, 2010.

[28] For how intense such "insecurity-pays" dynamics are in central Afghanistan, see International Crisis Group (ICG), *The Insurgency in Afghanistan's Heartland*, Asia Report No. 207, June 27, 2011, available at <www.crisisgroup. org/en/regions/asia/south-asia/afghanistan/207-the-insurgency-in-afghanistans-heartland.aspx>.

[29] Author's interviews in Kandahar, Uruzgan, and Kabul, Spring 2009 and Fall 2010.

[30] For details, see Dexter Filkins, "The Afghan Bank Heist," *The New Yorker*, February 14, 2011, available at <www.newyorker.com/reporting/2011/02/14/110214fa_fact_filkins>.

[31] Author's interviews in Helmand, Spring 2009. For a detailed study of the ethnic factionalization in the Afghan National Army, see ICG, *A Force in Fragments: Reconstituting the Afghan National Army*, Asia Report No. 190, May 12, 2010, available at <www.crisisgroup.org/en/regions/asia/south-asia/afghanistan/190-a-force-in-fragments-re-constituting-the-afghan-national-army.aspx>.

[32] Provincial Reconstruction Team (PRT) leaders have the authority to disburse up to $25,000 for individual projects and up to $100,000 per month.

[33] Rajiv Chandrasekaran, "In Afghan Region, U.S. Spreads the Cash to Fight the Taliban," *The Washington Post*, May 31, 2010; and Karen DeYoung, "Results of Kandahar Offensive May Affect Future U.S. Moves," *The Washington Post*, May 23, 2010.

[34] Ibid.

[35] Author's interviews with U.S. Agency for International Development (USAID) contractors and implementing contractor, nongovernmental organization (NGO) representatives, and Afghan government officials, maliks, and businessmen in Kandahar and Kabul, September 2010.

[36] See, for example, Andrew Wilder, "A 'Weapons System' Based on Wishful Thinking," *The Boston Globe*, September 16, 2009.

[37] Chandrasekaran; USAID also envisions spending about $140 million to help settle property disputes. That would be an immense and highly strategic accomplishment, yet it is a very difficult undertaking, as is building a land cadastre in Afghanistan.

[38] For details about the problematic and inadequate wheat distribution and similar programs, see David Mansfield, "Sustaining the Decline: Understanding the Nature of Change in Rural Livelihoods of Opium Growing Households in the 2008/09 Growing Season," Report for the Afghan Drugs Interdepartmental Unit of the UK Government, May 2009, available at <www.fco.gov.uk/resources/en/pdf/pdf21/drivers-report-0809>; Vanda Felbab-Brown, "The Obama Administration's New Counternarcotics Strategy in Afghanistan: Its Promises and Potential Pitfalls," Brookings Policy Brief Series, No. 171 (Washington, DC: The Brookings Institution, September 2009), available at <www.brookings.edu/papers/2009/09_afghanistan_felbabbrown.aspx>; and Joel Hafvenstein, "The Helmand Food Zone Fiasco," *Registan.net*, August 26, 2010, available at <www.registan.net/index. php/2010/08/26/helmand-food-zone-fiasco/>.

[39] For details, see Felbab-Brown, "Counternarcotics Policy"; Mansfield, "Sustaining the Decline"; and Christopher Ward and William Byrd, "Afghanistan's Opium Drug Economy," Working Paper Series, Report No. SASPR-5 (Washington, DC: The World Bank, December 2004).

[40] Author's interviews with counternarcotics officials in southern Afghanistan and Washington, DC, Spring 2009.

[41] Author's interviews with Dutch PRT members and Uruzgan government officials and NGOs, Uruzgan, Spring 2009.

[42] For the difficult choices in balancing a number of objectives and considerations in deciding between "quick-impact" but unsustainable projects or long-term development, see, for example, the complex decisionmaking regarding whether to bring power generators to Kandahar to satisfy the population, complicate the Taliban movements in the city, and reduce crime, even though the generators will be dependent on an outside fuel supply—or wait without electricity while the Kajaki Dam and power lines are completed. The latter approach is sustainable, but insecurity prevents the completion of the project. For details, see Rajiv Chandrasekaran, "U.S. Military, Diplomats at Odds Over How to Resolve Kandahar's Electricity Woes," *The Washington Post*, April 23, 2010; and Blackledge, Lardner, and Reichmann.

[43] Author's interviews with USAID contractors, NGOs, and their Afghan counterparts in Kandahar and Kabul, September 2010. See also Rajiv Chandrasekaran, "'Still a Long Way to Go' for U.S. Operation in Marja, Afghanistan," *The Washington Post*, June 10, 2010.

[44] Author's interviews with Afghan shopkeepers, September 2010.

[45] For a prominent critique of the militia program, which nonetheless does not sufficiently distinguish between the Afghan Local Police and the other militias, see Human Rights Watch, "'Just Don't Call It a Militia': Impunity, Militias, and the 'Afghan Local Police,'" (New York: Human Rights Watch, September 2011), available at <www.hrw.org/sites/default/files/reports/afghanistan0911webwcover.pdf>.

[46] For details on the various political and security trends as well as negotiations with the Taliban, see Vanda Felbab-Brown, "Afghanistan Ten Years after 9/11: Counterterrorism Accomplishments while a Civil War Is Lurking?" (Washington, DC: The Brookings Institution, September 6, 2011).

[47] Francisco E. Thoumi, *Illegal Drugs, Economy, and Society in the Andes* (Washington, DC: Woodrow Wilson Center Press, and Baltimore: The Johns Hopkins University Press, 2003).

[48] Cannabis production has also spread to areas of poppy cultivation since under current prices its profits can surpass opium poppies and since it is subject to far less law enforcement suppression than the opium poppy economy. For details, see UNODC, Afghanistan Cannabis Survey 2010, June 2010.

[49] For details on how such smuggling generates profits, see Felbab-Brown, *Shooting Up*, 122–124.

[50] Frederik Balfour, "Dark Days for a Black Market: Afghanistan and Pakistan Rely Heavily on Smuggling," *Business Week*, October 15, 2001, available at <http://businessweek.com/magazine/content/01_42/b3753016.htm>.

[51] For a discussion of these drug markets and their history of production and trade, see Vanda Felbab-Brown, "The Drug-Conflict Nexus in South Asia: Beyond Taliban Profits and Afghanistan," in *The Afghanistan-Pakistan Theater: Militant Islam, Security, and Stability*, ed. Daveed Gartenstein-Ross and Clifford May, 90–112 (Washington, DC: Foundation for Defense of Democracies, May 2010); and Vanda Felbab-Brown, "West African Drug Trade in the Context of Illicit Economies and Poor Governance" (Washington, DC: The Brookings Institution, October 14, 2010), available at <www.brookings.edu/speeches/2010/1014_africa_drug_trade_felbabbrown.aspx>.

[52] Amir Zada Asad and Robert Harris, *The Politics and Economics of Drug Production on the Pakistan-Afghanistan Border* (Burlington, VT: Ashgate, 2003); and Nigel J.R. Allan, "Opium Production in Afghanistan and Pakistan," in *Dangerous Harvest: Drug Plants and the Transformation of Indigenous Landscapes*, ed. Michael K. Steinberg, Joseph J. Hobbs, and Kent Mathewson, 133–152 (Oxford: Oxford University Press, 2004).

[53] Alfred W. McCoy, *The Politics of Heroin: CIA Complicity in the Global Drug Trade* (New York: Lawrence Hill Books, 2003, rev. ed.), 484–485; and Ikramul Haq, "Pak-Afghan Drug Trade in Historical Perspective," *Asian Survey* 36, no. 10 (October 1996), 945–963.

[54] Lawrence Lifschultz, "Inside the Kingdom of Heroin," *The Nation*, November 14, 1988.

[55] Author's interviews with former civilian and military officials in North-West Frontier Province, Fall 2008 and Spring 2009.

[56] Author's interviews with UNODC, Indian, and Pakistani officials, in New York, Kashmir, India, and Washington, DC (Spring, Summer, and Fall 2008).

[57] See, for example, Syed Irfan Ashraf, "Militancy and the Black Economy," *Dawn*, March 22, 2009; Sabrina Taversine, "Organized Crime in Pakistan Feeds Taliban," *The New York Times*, August 29, 2009; and Pir Zubair Shah and Jane Perlez, "Pakistan Marble Helps Taliban Stay in Business," *The New York Times*, July 14, 2008.

[58] For more details, see Vanda Felbab-Brown, "Conceptualizing Crime as Competition in State-Making and Designing an Effective Response," speech at Conference on Illicit Trafficking Activities in the Western Hemisphere, Center for Hemispheric Defense Studies and Office of National Drug Control Policy, available at <www.brookings.edu/speeches/2010/0521_illegal_economies_felbabbrown.aspx>.

[59] For a more extensive discussion of power brokers' and states' linkages to an illicit economy, including a postwar illicit economy, see Vanda Felbab-Brown, "Rules and Regulations in Ungoverned Spaces: Illicit Economies, Criminals and Belligerents," in *Ungoverned Spaces: Alternatives to State Authority in an Era of Softened Sovereignty*, ed. Harold Trinkunas and Anne Clunnan, 175–192 (Stanford, CA: Stanford University Press, 2010).

Part IV.
Fighting Back

Chapter 12
Fighting Networks with Networks

David M. Luna

> *There are not more than five musical notes, yet the combinations of these five give rise to more melodies than can ever be heard. There are not more than five primary colors (blue, yellow, red, white, and black), yet in combination they produce more hues than can ever be seen.... In battle, there are not more than two methods of attack—the direct and the indirect; yet these two in combination give rise to an endless series of maneuvers.*

> —Sun Tzu, *The Art of War*

Over the past decade, transnational organized crime has expanded in size, scope, and menace, destabilizing globalized economies and markets alike and creating insecurity in communities around the world. As criminal entrepreneurs and transnational illicit networks hijack the technological, financial, and communications advances of globalization for illicit gains, they continue to present new harms to the governance and security of all nations. The proliferation of these networks and the convergence of their illicit activities threaten not only the interdependent commercial, transportation, and transactional systems that facilitate free trade and the movement of people throughout the global economy, but are jeopardizing governance structures, economic development, security, and supply chain integrity.

Today's global threat environment is characterized by convergence: the merging and blending of an ever-expanding array of illicit actors and networks. In an interconnected world, the pipelines linking these threat actors and networks cut across borders, infiltrate and corrupt licit markets, penetrate fragile governments, and undercut the interests and security of our partners across the international community. The direct links among specific illicit actors are of increasing concern, as well as the growing illegal economy that supports and enables corrupt officials, criminals, terrorists, and insurgents.

The global illicit economy is becoming increasingly flush with cash derived from a wide spectrum of illicit activities: narcotics, kidnapping-for-ransom, arms trafficking, human smuggling and trafficking, the trade in stolen and counterfeit goods, bribery, and money laundering. According to some estimates, the illegal economy accounts for 8 to 15 percent of world gross domestic product. The wide availability of unregulated cash provides not only a safe haven and exploitable sanctuaries for illicit forces, but also illicit liquidity for corrupt officials, criminals, and terrorists to mingle, operate, and do business with, either with one another or on their own. And too often, these criminal and illicit actors and networks are staging operations without fear of reprisal from law enforcement.

Convergence defines the global economy today. The world we live in is one in which legal business transactions and legitimate commerce both facilitate and feed off the illegal economy, one in which illegal arms brokers and narcotics kingpins are acting in practice as the new chief executive officers and venture capitalists. From Wall Street to other financial centers across the globe, illicit networks are infiltrating and corrupting licit markets.

In this globalized, networked world, the uneven application of cross-border enforcement enables illicit networks to arbitrage differences in regulatory policies to extract maximum illegal profits. Policy levers activated in one country frequently have the undesired effect of driving crime from that country into another where regulation and enforcement are less stringent. The trade in illicit goods, people, arms, and services behaves much like a tide as it crosses oceans, crashing around obstacles, seeking cracks in the levees, converging with other flows, and flooding anywhere that is not protected.

From Sinaloa to Tirana to Caracas to Lagos, as transnational criminal organizations and "third-generation gangs"—gangs that have morphed from local groups of individual actors to cross-border, networked entities that toe the line between crime and war—increasingly forge alliances with corrupt government officials, undermine competition in key global markets, and diversify their illicit portfolios with ventures into legitimate commerce, they are unraveling the social fabric of the community of nations.

It is therefore critical that the international community work together in a coordinated manner to staunch this flow and dismantle the criminal opportunity structure at every node, pipeline, and channel across the global illicit landscape. By combining forces in response to the relentless convergence of illicit threat networks and reducing their ability to exploit market opportunities, we will have a much greater chance of success if we target their center of gravity in this landscape including their financial flows—the whole indeed can be greater than the sum of all parts. Collective action can be harnessed via an array of responses to disassemble today's formidable criminal and terrorist adversaries by disrupting their pipelines and ill-gotten financial assets. This chapter outlines some of the innovative approaches and partnerships the U.S. State Department is developing to combat today's illicit networks and converging threats globally.

Translating Threat Awareness into Threat Management

The United States has recently taken steps to make countering the convergence of illicit threats a national security priority. On July 25, 2011, the White House released the *Strategy to Combat Transnational Organized Crime: Addressing Converging Threats to National Security* (SCTOC), which aims to protect Americans and citizens of partner nations from violence and exploitation at the hands of transnational criminal networks.

Before an audience that included heads and senior officials from the U.S. Departments of Justice, Homeland Security, State, the Treasury, and other law enforcement, security, and intelligence agencies, at the White House ceremony to unveil the SCTOC, then–Under Secretary of State Williams Burns (now Deputy Secretary of State) articulated the adverse impact of transnational criminal threats on U.S. foreign policy:

> *Organized crime, in its many forms, is a threat to decent, hardworking people across the world. It empowers warlords, criminals, and corrupt officials. It erodes stability, security and good governance. It undermines legitimate economic activity and the rule of law. It undermines the integrity of vital governmental institutions meant to protect peace and security. It costs economies tax revenue and promotes a culture of impunity. It undercuts our fight against poverty and slows sustainable development.*

While the problem of transnational illicit networks is as ancient as the trade routes that many such networks still employ today, the United States and its partners recognize the importance of net-centric partnerships to confront converging threats and the lethal nexus of organized crime, corruption, and terrorism along global illicit pathways and financial hubs.

Of growing concern are illicit financial hubs and the complicity of today's banks and market-based facilitators and super fixers—such as lawyers, accountants, black market procurers of commodities and services, and cross-border transport movers. Illicit financial hubs and sanctuaries help to create the permissive environment that enables illicit funds to enter through vulnerable points in the system and be transferred very rapidly, often with little control or regulation, anywhere in the world. All it takes is a single illicit actor or bank to accept an unsavory client for illicit funds or goods to spread and disguise themselves across the globe.

These illicit channels are also allowing kleptocrats, criminals, and in some cases terrorists or their sympathizers to inject billions of dollars of illicit wealth into the stream of licit commerce and business, corrupting markets, financial institutions, officials, and communities. The international community can no longer turn a blind eye on the complicity of facilitators and super fixers who perhaps unwittingly construct illicit financial hubs and permissive sanctuaries by converting illicit financial flows into licit investments and legal liquid assets, or injecting illicit goods into legitimate commerce.

Breaking the corruptive power of transnational illicit networks will be a key area of focus for the United States under the new SCTOC, and the State Department will work with international partners to expose criminal activities hidden behind legitimate fronts, investigate actors and networks that run afoul of U.S. laws, protect strategic markets and the integrity of the global financial system, and employ all elements of national power and public-private partnerships to dismantle criminal organizations and cross-border networks.

For those nations that may lack the necessary means and capabilities to combat today's transnational criminal threats, the United States is committed to assisting these partners—especially those that have the will to fulfill their international law enforcement commitments—to develop stronger law enforcement and criminal justice institutions necessary for ensuring the rule of law. As the SCTOC underscores, the State Department will spearhead a renewed multilateral diplomacy, leveraging bilateral and global partnerships to elevate the importance of combating transnational organized crime as a key area of cooperation internationally.

Globalization and Horizontal Diversification of Illicit Networks

Since the end of the Cold War, the world has witnessed the expansion of transnational criminal organizations beyond their traditional boundaries. They are quick to identify new opportunities and spread into new geographic areas where national and international responses have yet to pose a credible threat to the survival of their operations. No region is immune. Today, the major organized crime groups have become even more global in reach, operating not only in the United States and Latin America, but also in West Africa, Southeast Europe, Asia, and Russia, integrating within and across networks in all regions of the world.

Central America and West Africa have become central areas of concern as safe havens for converging threats where trafficking in drugs, people, weapons, and other illicit commodities fuels instability and insecurity. Colombian drug cartels, which have long been viewed as a major security threat in the Western Hemisphere, are now expanding their activities beyond traditional areas of operation. Multi-ton shipments of Colombian cocaine now flow through West Africa as a transit point for moving the product to Europe and beyond. At the same time, West African drug traffickers have recently been spotted moving South American cocaine as far away as South Asia. West African drug mules have been caught at ports of entry from Florida to Thailand. Piracy in the waters off the Horn of Africa remains a serious challenge to humanitarian aid and commerce, and the problem is getting worse. Piracy presents a clear and present danger to the international maritime shipping industry, on which much of the global economy depends; to fragile political progress and development in Somalia; and to humanitarian assistance and trade in the region.

The security threats of today are more sophisticated and complex than in the past; converging threat networks are forming alliances where previously the parties acted alone. The illicit proceeds of global drug trafficking are among the funds used for arms trafficking to the Taliban in Afghanistan and the Revolutionary Armed Forces of Colombia, while drug traffickers themselves engage in countless other illicit activities—from antiquities smuggling to human trafficking—to launder money. Ideologically motivated terrorists such as al Qaeda in the Islamic Maghreb, Lebanese Hizballah, and al-Shabaab in the Horn of Africa are in some instances engaging with or benefiting from criminal organizations following a strict profit motive, and in other situations becoming criminal entrepreneurs in their own right, engaging in actions such as kidnapping for ransom.

Traders in illicit goods poison legitimate supply chains, markets, and communities with low-quality counterfeits, inherently illicit goods such as narcotics, and illegally procured goods such as blood diamonds that are traded on the black market. In some cases, organized criminal groups seize control of a legitimate supply chain or exploit the legitimate global financial infrastructure to launder the proceeds of crime. Today's enhanced communications and cyber technologies allow cyber criminals in one part of the world to cooperate with virtual accomplices across the globe in real time. Most worryingly, threats such as international arms and narcotics trafficking, transnational organized crime, and terrorism are no longer acting in isolation, but converging faster than governments can deploy responses to keep up with them.

Third-party facilitators grease the revolving door between converging threat networks. One type of criminal activity often depends on, stimulates, and feeds the capacity for a wide range of illicit conduct. Transnational criminal organizations may strike up a relationship with technical experts who occupy a gray zone between the underworld and the legitimate market, such as lawyers, business owners, bankers, scientists, and other specialists. Some of the best-known facilitators are full-time brokers. These facilitators are crucial to the ability of illicit networks to move goods, launder money, and acquire funding, falsified documents, weapons, and logistical support.

Other groups have entrenched themselves in so many diversified criminal activities that they can no longer be identified by a single specialty. Some drug cartels previously specializing in cocaine are now also involved in intellectual property rights theft, wildlife trafficking, and human smuggling. Human smugglers are involved in the production of fraudulent documents, corrupting public officials, money laundering, drug smuggling, and facilitating clandestine terrorist travel. Local mafias are involved in international financial fraud schemes. Somali pirates are believed to be involved in many other criminal activities including smuggling, arms trafficking, and human trafficking. The interwoven strands of such illicit transactions make it almost impossible to separate one from the other, like unzipping a double helix of DNA using a pair of scissors.

These networks are structured more fluidly than traditional hierarchical organized crime syndicates, and this makes them strategically more difficult to target. Network organizers are harder to identify than syndicate leaders, and the loose links among network elements can impede efforts to identify the full range of activities in which they are involved. Some transnational criminal groups, such as the Mara Salvatrucha (MS-13) gang, have been networked, loosely coordinated organizations since their inception, while traditional organized crime groups such as the Albanian Mafia, Chinese Triads, and Japanese Yakuza are increasingly restructuring their vertical organizations into horizontal networks.

The Convergence of International Crime and Terror Pipelines

The nexus between organized criminal activity and terrorist groups has long been debated and remains a threat requiring urgent attention. The Department of Justice reports that 31 of the 71 organizations on its fiscal year 2011 Consolidated Priority Organization Targets list, which includes the most significant international drug-trafficking organizations threatening the United States, are associated with terrorist groups. Criminal syndicates have long been known to support terrorist groups by facilitating their clandestine transborder movements, weapons smuggling, and forging of documents. Of particular concern would be the ability of some of these groups to acquire radioactive materials or chemical and biological weapons.

Despite differences in motivation and ideology, terrorists and insurgents increasingly turn to criminal networks to acquire financial and logistical support. These crime-terror pipelines run two ways. In one direction, terrorist organizations overcome differences in motivation and ideology as they turn to transnational illicit networks such as drug-trafficking organizations or facilitators for logistical support. In the other direction, terrorists and insurgent groups are increasingly evolving into criminal entrepreneurs in their own right, engaging in a wide range of illicit activities to finance their operations. Some criminal networks, such as Los Zetas in Mexico, are not only expanding their operations, but also diversifying their methods to include terrorist-like tactics and campaigns of violence.

U.S. and Colombian law enforcement investigators have recently uncovered some links between South American cocaine traffickers and some individuals associated with the Lebanon-based terrorist organization Hizballah. According to news reports, this transnational ring launders hundreds of millions of drug dollars each year, funneling illicit funds through free trade zones in Panama and Hong Kong in order to "clean" the money and using other methods of laundering in other parts of the world including Africa and Canada. The South American ring has allegedly paid Hizballah a percentage of its profits.

Owing to pressure from the United States and committed international partners to implement a no-tolerance policy against the state sponsorship of terror since September 11, 2001, al Qaeda and other terrorist groups have resorted to illicit activities to finance their operations including drug dealing (prior to the Madrid bombings of 2004), credit card theft, and insurance scams. It does not take much: the BBC, *Washington Post*, and other investigative news sources have reported that the cost of executing a terrorist attack has fallen dramatically over the past decade, from Bali to Amman to Madrid to London.

Further media reports suggest that despite their ideological aversion to common criminal activity, terrorist organizations operating in the Middle East and North Africa do engage in crime when it is necessary to achieve their objectives, particularly arms and narcotics trafficking, smuggling goods and commodities, migrant smuggling, trafficking in persons, extortion, kidnapping, intellectual property theft, counterfeiting, fraud, credit theft, armed robbery, and money laundering, which is nearly identical to terrorist financing. It has also been reported in the press that drug kingpin Dawood Ibrahim, head of the D-Company hybrid organization, has helped finance the terror activities of Lashkar-e-Taiba, which has launched terrorist attacks in Mumbai and other cities across South Asia.

The illegal economy that has grown globally is making it increasingly difficult to shut off the spigot used to finance terrorism, at least through traditional means. As terrorist groups mimic the tactics of organized crime, international counterterrorism efforts need to incorporate law enforcement tools as part of a global strategic response. Along these lines, preventing the criminal facilitation of terrorist activities was identified as a priority in the 2011 *Strategy to Combat Transnational Organized Crime*, and preventing the collaboration between criminal and terrorist networks and depriving them of their critical resources and infrastructure, such as funding, logistical support, safe havens, and the procurement of illicit materiel, were highlighted as major objectives.

Smart Power Diplomacy: Fighting Networks with Networks

The illicit networks and converging threat networks outlined above necessitate strong responses and partnerships. Fighting transnational crime and dismantling illicit networks is not something that any one government or agency can do alone. As underscored by President

Obama, "This strategy is organized around a single, unifying principle: To build, balance, and integrate the tools of American power to combat transnational organized crime and related threats to our national security—and urge our partners to do the same."

The State Department is working across the U.S. Government to improve internal cooperation and build domestic security and law enforcement networks to fight illicit networks. In support of the SCTOC, the State Department is working to "build international consensus, multilateral cooperation, and public-private partnerships to defeat transnational organized crime." This involves a broad range of bilateral, regional, and global programs and initiatives that help to strengthen international cooperation with other committed partners.

The State Department's Bureau of International Narcotics and Law Enforcement Affairs (INL), which is responsible for international counternarcotics and countercrime issues, leads diplomatic efforts to raise awareness of the destabilizing impact of transnational organized criminal activities to the community of nations and to strengthen global efforts to combat these threats. Enhanced coordination enables the international community to dismantle criminal networks and combat the threats they pose not only through law enforcement efforts, but also by building up governance capacity, supporting committed reformers, and strengthening the ability of citizens—including journalists—to monitor public functions and hold leaders accountable for providing safety, effective public services, and efficient use of public resources.

CARSI, WACSI, CACI

At the regional level, the State Department is developing innovative partnerships with governments through platforms such as the Central America Regional Security Initiative (CARSI), the West Africa Cooperative Security Initiative (WACSI), and the Central Asia Counter-narcotics Initiative (CACI) to coordinate investigations, support prosecutions, and build collective capacity to identify, disrupt, and dismantle transnational organized crime groups.

In places like Central America, traffickers and criminal gangs now facilitate the flow of up to 95 percent of all cocaine reaching the United States. CARSI will help to improve citizen safety by reducing the ability of criminal organizations to destabilize governments, threaten public safety, and spread illicit drugs, guns, and other transnational threats to Central America, its neighbors, and the United States. In partnership with the Central American Integration System (SICA), the United States and other donor nations and multilateral organizations participating in the Group of Friends of Central America are working more closely than ever before to ensure that citizen security assistance is well coordinated and aligned with regional priorities set forth in SICA's Central America Security Strategy.

West Africa faces a growing danger from transnational criminal organizations, particularly narcotics traffickers, whose activities threaten the collective security and regional stability interests of the United States, its African partners, and the international community. Illicit markets and those who profit from them weaken public institutions, foster corruption, and foment violence. To address this threat, U.S. Government agencies collaborated to create WACSI. In consultation with African and international partners, the United States will seek opportunities to complement and enable regional and national initiatives that seek to achieve similar objectives. WACSI will implement a multilayered approach that will target certain nations to strengthen anchor country capacities, invest in future anticrime partnerships, and build on regional and interregional networks to effectively support the efforts of the United States' African partners to address their assorted challenges.

Central Asian states on the Afghan border also face a significant threat from illicit narcotic drugs transiting from Afghanistan. Violent extremist groups from Afghanistan and Pakistan threaten stability in the region, with drug trafficking providing a significant source of their funding. The U.S. Government is partnering with Central Asian states directly to counter these threats. The CACI will improve the ability of Central Asian countries to disrupt drug trafficking networks originating in Afghanistan and dismantle related criminal organizations through effective investigation, prosecution, and conviction of mid- to high-level traffickers. CACI will also focus on regional cooperation and help establish counternarcotics task forces that will serve as an impetus for further reform, facilitate increased information-sharing, and form a foundation for further institutional capacity-building.

Transpacific Networks on Dismantling Transnational Illicit Networks

The State Department is also developing more dynamic interregional partnerships that strengthen cross-border cooperation including with Australia, New Zealand, and other committed jurisdictions. For example, in November 2009, the State Department sponsored, with U.S. Immigration and Customs Enforcement, the Trans-Pacific Symposium on Dismantling Transnational Illicit Networks, in Honolulu. This event brought together participants from approximately 25 economies from Asia, Latin America, and the Pacific Islands, as well as senior representatives from international organizations such as the Asia-Pacific Economic Cooperation (APEC), the Organization of American States (OAS), the Asia-Pacific Group on Money-Laundering, the International Criminal Police Organization commonly known as Interpol, and the United Nations Office on Drugs and Crime (UNODC). The United States was represented by senior officials from the Departments of Justice, Treasury, Homeland Security, State, Defense, and their law enforcement components including the Joint Interagency Task Force–West and U.S. Pacific Command.

Out of the Trans-Pacific Symposium emerged many ideas on ways to better equip participating countries against illicit threats across the region. These include devising more strategic frameworks and information-sharing arrangements, leveraging international cooperation across regional and global law enforcement networks, strengthening capacity-building in law enforcement, coordinating joint investigations to target "money conduits" and corruption nodes, and securing borders.

In November 2010, the United States and New Zealand cohosted a follow-up workshop to the 2009 Trans-Pacific Symposium, in Christchurch, which brought together partners to discuss further interagency cooperation on combating transnational illicit networks including pressing challenges such as narcotics trafficking, counterfeit medicines, trafficking in persons, illegal logging, money laundering, and the corruption that facilitates illicit trade. Transpacific participants have now formed a Trans-Pacific Network, which convenes regularly to work across borders on the transnational criminal threats across the Pacific and to develop joint and coordinated strategies and share case studies and effective practices to help achieve better interagency coordination at the national level.

In October, the Trans-Pacific Network met for a third time, in Phuket, Thailand. More than 125 law enforcement and other government officials from 30 Asia-Pacific economies and representatives of regional and international organizations participated. The Phuket workshop was cohosted by the governments of Thailand and the United States, in partnership with The Colombo Plan and other international partners. Law enforcement officials from across the transpacific region underscored the growing sophistication and increasing joint ventures among illicit organizations from regions such as Asia, West Africa, Latin America, the Middle East, and Eurasia.

It was noted that while regional law enforcement agencies had made significant narcotics seizures, arrests, and confiscations and recovery of bulk cash related to drug trafficking in their cross-border operations, participants were growing increasingly concerned about the expansion and influence of Latin American cartels, West African gangs, and Iranian-based criminal organizations in the Asia-Pacific region. In the other direction, discussion also focused on the growing threat posed by Chinese organized crime syndicates across the Americas. Another concern raised was the high purity of drugs such as Afghan heroin, compared to the impurities and toxic cutting agents present in Latin American crack cocaine, causing a public health crisis that aggravates regionalized drug epidemics.

The Trans-Pacific workshop in Phuket also placed a priority on environmental crimes, specifically illicit logging and associated trade. The objective of this session was to provide participants with a greater understanding and ability to practically respond to the broad array of illegal logging and related enforcement issues in the region. Discussion focused on ways to establish an interregional network to enhance cross-sector regional cooperation to combat criminal networks illicitly trading in illegally harvested or stolen timber. Partners emphasized the need to reinforce commitments to strengthen forest law enforcement sectors and criminal justice communities both within and across national borders. This included the need to take concrete measures to combat illicit logging and associated trade in three areas: prevention, detection, and suppression.

Falsified or fake medicines, medical products, and other dangerous counterfeits and defective and tainted products imperil the safety of our citizens and shake market confidence. The illicit trade of these counterfeit products is another profitable area for transnational criminals, especially given its high reward/low risk calculus. Another of the streams at the Trans-Pacific workshop in Phuket focused on how dangerous counterfeits continue to enter regional and global supply chains and markets, with harmful impacts on communities, healthcare institutions, and businesses. Participants have carried the momentum of this discussion into other fora such as APEC, where the Anti-Corruption and Transparency Experts Working Group has begun to focus on counterfeit pharmaceuticals as a major threat to the health and safety of APEC communities and markets.

Transpacific partners agreed, in essence, to shut down illicit markets, put criminal entrepreneurs out of business, and continue to cooperate across borders to dismantle transnational criminal threats and illicit networks. In addition to showcasing case studies and sharing best practices, partners agreed to further leverage intelligence- and information-sharing arrangements; promote mutual legal assistance that enables evidence-sharing to assist in carrying out investigations and prosecutions; expand capacity-building efforts at the interregional, subregional, and bilateral levels that help to combat cross-border crime and corruption; and synchronize regional transnational crime units with fusion centers and other intelligence-based regional policing efforts. The United States is working with Australia, New Zealand, Thailand, and other partners to coordinate a fourth meeting of the Trans-Pacific initiative, in 2013.

Trans-Atlantic Symposium on Dismantling Transnational Illicit Networks

The United States hosted a similar event in May 2011—the Trans-Atlantic Symposium—with the European Union (EU) to generate partnerships to address threat convergence across the trans-Atlantic region. Held in Lisbon, Portugal, the EU and U.S. joint initiative brought together 300 senior law enforcement and judicial officials from 65 countries including representatives from the United States, the European Union and its member states, Latin America, Canada, the Caribbean, and West Africa. Senior representatives of international and regional organizations included officials from the G8, UNODC, Economic Community of West African States, OAS, the World Customs Organization, Interpol, European Police Office (Europol), and the Intergovernmental Action Group Against Money Laundering in West Africa.

The Symposium launched an interregional dialogue among senior European, U.S., West African, and Latin American law enforcement and judicial officials on ways to strengthen international cooperation to combat transnational criminal threats and illicit networks that span the Atlantic. It focused on cross-border crimes and illicit routes including drugs, arms, human smuggling/trafficking, money laundering and illicit finance, corruption, and maritime crimes.

The co-organizers were the EU's European External Action Service working in close collaboration with the European Commission and the State Department's Bureau of International Narcotics and Law Enforcement.

The theme of the Trans-Atlantic Symposium was "Fighting Networks with Networks," building on the theme of the Trans-Pacific Symposium. Participants agreed on a new cooperative platform to counter transnational criminal threats that includes:

+ launching an informal transatlantic network of regional networks of law enforcement and judicial practitioners that could be interconnected to facilitate intelligence- and information-sharing arrangements and mutual legal assistance to assist in carrying out cross-border investigations and prosecution

+ developing a clearer picture of transnational crime trends across the Atlantic including the presence and activities of illicit networks and their illicit financial flows so government agencies can develop more effective strategies to address transnational threats and deploy the most effective investigative techniques to disrupt them

+ combating high-level corruption and denying safe haven including visas, to illicit actors

+ examining the establishment of sound mechanisms to trace the flow of criminally derived funds and to prevent illicit networks from benefitting from the proceeds of crime

+ strengthening collective capacities to counter transregional threats including synergies between multilateral and regional efforts as well as operational and information centers, with emphasis on West Africa, Latin America, and the Caribbean

+ improving our understanding of emerging threats, especially identifying gaps and needs, and promoting public awareness of the real costs to governments and their citizens of the insecurity, the loss of revenue, trade, and investment, and the risks to health and safety caused by the activities of illicit networks

+ promoting the implementation and best use of the available international instruments including the United Nations Conventions regarding the Control of Drugs and Corruption (UNCAC) and Transnational Organized Crime (UNTOC), as well as the Financial Action Task Force principles on combating money laundering and terrorist financing

+ leveraging partnerships with key Latin American states to help build capacity in West Africa

+ examining ways in which international scientific collaboration could promote research, data collection, and analysis of the scale and scope of transnational organized crime and the evaluation of law enforcement practices to ensure their effectiveness.

The network of law enforcement and judicial officials launched at the Trans-Atlantic Symposium has continued to meet and coordinate on the most pressing security threats. The United States, United Kingdom, and other committed jurisdictions have been especially active at countering the convergence of illicit threats as part of a shared global security agenda. In June 2012, the United States hosted the Trans-Atlantic Dialogue to Combat Crime-Terror Pipelines, in Washington, DC, to strengthen international cooperation to identify and disrupt the collusion of criminal and terrorist networks across the Atlantic threat theater. The Dialogue examined ways to explore smart law enforcement tools, strategic capabilities and sanctions, and levers to shine the light on an unholy alliance of illicit actors and their money pipelines, maximize information sharing, and facilitate the development of concrete policy recommendations. Among the Dialogue's final recommendations for further discussion are:

+ build a transatlantic network: combine joint national capabilities to develop an anticipatory approach to crime-terror interaction, and coordinate actions to mitigate the current threats posed by adaptive actors and hybrid networks

+ paint the crime-terror panorama: reevaluate the traditional separation of terrorism and organized crime as distinct threats

- drain the illicit economy: strengthen nonkinetic methods, especially financial tools and criminal justice responses, to target corrupt actors and illicit pathways and follow the money to disrupt and dismantle pipelines, target their facilitators, and eliminate their financial resources

- elevate transnational crime as a national and international security threat: governments should treat transnational organized crime as a national and international security priority, taking into account the relative long-term impacts of organized crime and terrorism based on a common understanding of shared threats; possible actions may include sanctioning terrorist groups as transnational criminal organizations where there is intelligence and evidence that specific terrorist groups are engaged in criminal enterprise

- target the facilitators: examine legislative powers and other means to combat corruption; some participants argued in favor of treating facilitators of crime and terrorism—from individuals such as professional arms brokers to corporate entities such as banks engaged in money laundering or facilitating terrorism financing—as criminal actors in their own right

- expose safe havens and sanctuaries: coordinate efforts to identify and uproot safe havens and exploitable sanctuaries that enable criminals, terrorists, and other illicit actors and networks to corrupt governments, access illegal markets, and stage operations without fear of reprisal from law enforcement; expose and prevent conditions for nesting illicit forces with criminalized states

- invest in evidence-based research and analysis: develop evidence-based research to target and dismantle crime-terror pipelines at key nodes and along major pathways including mapping, data analysis, network analysis, forecasting, and other analytic tools in both unclassified and classified formats

- strengthen international consensus: build a common understanding of crime-terror pipelines across borders and promote bilateral and multilateral partnerships to carry out operations and investigations and share information and intelligence

- safeguard private sector investment: form public-private partnerships to enable the public and private sectors to exchange relevant information on organized crime groups in certain markets that threaten the integrity of supply chains or the ability of companies to make sustainable, responsible investments

- promote international development: employ all tools of power and persuasion by integrating policies and programs aimed at improving governance, economic development, and foreign direct investment with policies targeting crime and terror networks.

Combating Wildlife Trafficking and Dismantling Transnational Illicit Networks

The discussion on environmental crimes that took center stage at the Trans-Pacific Symposium in Phuket has continued, focusing on specific threats on a regional and transregional basis. In April 2012, the United States partnered with the government of Gabon to host a Sub-Regional Workshop for Central Africa on Wildlife Trafficking and Dismantling Transnational Illicit Networks, in Libreville, Gabon, to address growing concerns about the threats posed to communities, ecosystems, institutions, and markets by poaching and trafficking in protected and endangered wildlife.

Today, 30 percent of the species hunted in the Congo Basin are threatened. Heavily armed poachers operating in central Africa—who have attacked law enforcement and military personnel—have become a threat to the national security of Central African countries. Furthermore, poaching and wildlife trafficking are largely intertwined with other criminal activities of transnational illicit networks that contribute to the insecurity and instability of economies globally and hinder sustainable development strategies including efforts to preserve national resources and the promotion of ecotourism as a source of revenue for governments and communities.

Approximately 150 Central African government officials, law enforcement personnel, and members of nongovernmental and international conservation organizations from Central African and Asian countries including China worked together during the 3-day workshop in Libreville to share ideas and best practices for antipoaching. The workshop facilitated the exchange of information and shared best practices to foster and develop innovative responses to stem the poaching and cross-border trafficking of endangered and protected wildlife by involving agencies throughout governments. In addition, participants discussed ways to protect biodiversity through leveraging partnerships with other countries and nongovernmental and international organizations from other regions to dismantle illicit networks. At the end of the workshop, participants committed to establishing a wildlife enforcement network to intensify and coordinate antipoaching efforts in Central Africa. The Gabon workshop provided key momentum within the international community on combating wildlife trafficking and related corruption and cross-border crime.

Other Global Partnerships and Multilateral Cooperation

By building cooperative platforms and networks incrementally, the United States continues to promote smart power diplomacy to generate greater collective action, joint cases, and strategic approaches with international partners to combat transnational criminal threats. Washington is also working multilaterally and bilaterally with committed international partners to combat the growing wave of crime and other illicit threats and to employ diplomatic tools and technical assistance to disrupt and dismantle transnational threat networks, including helping to strengthen law enforcement, judicial, legal, and security institutions.

G-8. In recent years, the Group of Eight (G-8) has played a leading international role in articulating the harms and destabilizing factors posed by transnational threats—terrorism, organized crime, nuclear proliferation, and corruption—and how they contributed to domestic and international destabilization and challenges to global security, required greater synergies and ideas for the G-8 and international community to enhance the security toolbox to preempt and counter transnational threats, and encouraged stronger joint intergovernmental coordination to combat the assets of the criminal and terrorist organizations. From G-8 summits in L'Aquila (Italy) in 2009, Huntsville, Ontario (Canada) 2010, Deauville (France) 2011, and Camp David, Maryland (United States) 2012, G-8 leaders have continued to reaffirm their commitment to combating transnational organized crime and the UNTOC.

OAS Summit of the Americas. At the Fifth and Sixth Summits of the Americas, held in Port-of-Spain, capital of the Republic of Trinidad and Tobago, in 2009, and in Cartagena, Colombia, in 2012, President Barack Obama and other hemispheric leaders agreed to fight all forms of transnational organized crime, drug and arms trafficking, trafficking in persons and migrant smuggling, money laundering, corruption, terrorism, kidnapping, gang violence, and technology related crimes including cyber crime. The United States is also working with OAS member states to implement a Hemispheric Plan of Action to Combat Transnational Organized Crime, which was approved by the Permanent Council of the OAS in 2006 and includes a series of actions to enhance coordination, increase the capacity of law enforcement to implement mechanisms to combat organized crime, and promote information sharing, among others. In Cartagena, the heads of state and government acknowledged how transnational criminal organizations were increasingly expanding beyond drug trafficking to other criminal activities, with increased levels of violence. They expressed concern that national responses were not sufficient to combat cross-border threats and agreed to join forces and leverage capabilities against transnational organized crime.

North Atlantic Treaty Organization (NATO). The North Atlantic Council has long expressed its willingness to cooperate on combating "global threats, such as terrorism, the proliferation of weapons of mass destruction, their means of delivery and cyber attacks" and to work together on "other challenges such as energy security, climate change, as well as instability emanating from fragile and failed states," which have a negative impact not only on NATO member states but the international community as a whole.

APEC/ASEAN. The United States is working with other APEC and Association of Southeast Asian Nations (ASEAN) economies to combat corruption and illicit trade, dismantle transnational illicit networks, and protect economies from abuse of the U.S. financial system by leveraging financial intelligence and law enforcement cooperation related to corrupt payments and illicit financial flows. Economies across the Asia-Pacific region continue to focus on ways to combat terrorism, prevent the proliferation of weapons of mass destruction, and strengthen international capacities to combat piracy and armed robbery at sea. The United States is working with other economies in the region to strengthen interregional capabilities and to develop transpacific cooperation to combat transnational terrorism, crime, illicit finance, and corruption and to dismantle threat networks including through the Trans-Pacific Network. The United States and Cambodia will cohost an APEC-ASEAN Pathfinder Project in 2013 to combat corruption and illicit trade.

Other International Organizations. The United States continues to advance leadership internationally and to strengthen global partnerships to combat transnational threats globally including at the United Nations, World Bank, Financial Action Task Force, Organisation for Economic Co-operation and Development (OECD), Interpol, Contact Group on Piracy off the Coast of Somalia, and numerous regional intergovernmental fora around the world.

Leveraging International Conventions

As transnational organized crime globalizes, the United States and other jurisdictions have made great strides in globalizing the fight against it and also the corruption that facilitates it. UNTOC (the UN Convention against Transnational Organized Crime) and its Protocols for the first time require criminalization of a broad range of activities by crime groups, while UNCAC (the UN Convention against Corruption), requires the criminalization of bribery and other corrupt conduct and taking preventive measures. These two international legal instruments complement each other to establish a broad, virtually global framework for inter-governmental legal cooperation.

These instruments have already been used successfully by many governments to facilitate extraditions, mutual legal assistance, the transfer of sentenced persons, joint investigations, and special investigative techniques to combat transnational organized crime, but their reach must be extended. By ratifying and, more importantly, fully implementing both UNTOC and UN-CAC, jurisdictions not only demonstrate their commitment to staying ahead of the criminals, but arm themselves with the assistance and information other signatory states can provide.

Ratifying the global treaty is not sufficient, however. The State Department encourages and helps states strengthen domestic laws to deny criminals, their associates, and family members safe haven and access to illicit assets including use of visa authority to deny visas for entry. Out of UNTOC and UNCAC have come laws prohibiting organized criminal actors and kleptocrats from enjoying the fruits of their illicit profits, including owning real estate, establishing businesses, or investing abroad. To dismantle their networks, the State Department aims to mobilize all available legal means to disrupt their logistics, supply chains, and army of professionals—money launderers, lawyers, accountants, and others—who help move money and bribe corrupt public officials.

Bilateral Cooperation: High-Level Meetings and Technical Assistance

President Obama and Secretary of State Hillary Clinton continue to have frank discussions and elevate the importance of combating transnational threats with their counterparts, including recently with Afghanistan, Australia, Brazil (through the Open Government Partnership), China, Colombia, India, Iraq, Mexico, Nigeria, Pakistan, Russia, the United Kingdom, and others, with the goal of strengthening bilateral and international cooperation in areas such as terrorism, organized crime, corruption, narcotrafficking, and threat convergence areas.

To counter transnational threats, INL has employed an integrated approach of U.S. assistance programs, from traditional prevention, law enforcement, and counternarcotics programs, to anticorruption, judicial reform, antigang, community policing, and corrections efforts. In addition to a whole-of-government approach, INL coordinates effectively with others in the U.S. Government who work with communities, civil society, and the private sector, recognizing that security solutions require a whole-of-society approach, and also ensuring that partner nations have the ability to protect their citizens and deal with crime and violence so these issues are handled effectively by law enforcement and other means and do not become global security threats.

The following represent examples of bilateral partnerships the State Department has helped facilitate.

Mexico. News reports arrive daily from Mexico with shocking stories of killings and violence, but the Mexican government, with assistance from the United States through the Mérida Initiative, has achieved significant results. Approximately $1.6 billion in contributions for Mérida since its inception have helped Mexico, together with its U.S. partner agencies, to continue turning the tables on the cartels. Through bilateral law enforcement cooperation, 47 high-value targets have been arrested or removed in Mexico since December 2009, including 25 of the country's top 37 most wanted criminals. This aggressive and coordinated approach to dismantle and disrupt the drug cartels has included an institutional focus on all elements of the justice sector and civil society. Thanks to the Mérida Initiative, the government is empowered to transform the security forces and strengthen government institutions in order to confront trafficking organizations and associated crime and maintain public trust and citizen security.

Colombia. In Colombia, the United States is working to reduce the amount of cocaine entering the United States by supporting Colombian-led counternarcotics programs that are closely coordinated with alternative development. As follow-on to Plan Colombia, the United States has continued its partnership with the government of Colombia to consolidate the gains made over the past decade. Out of this cooperation has arisen the Colombia Strategic Development Initiative (CSDI), which provides for civilian institution-building, rule of law, and alternative development programs, as well as security and counternarcotics efforts in those areas where poverty, violence, and illicit cultivation or drug trafficking persist and have historically converged. The State Department is supporting these endeavors with significantly reduced resource levels. Further resources will be needed to sustain and consolidate past gains.

U.S. efforts in Colombia pay dividends regionally as well. With the capacity that the government of Colombia has developed over the years, Colombia is now contributing its hard-learned lessons to address similar security concerns elsewhere in the region. Colombia today is no longer just a recipient of security assistance, but an active exporter. Since 2009, the Colombian National Police, the closest U.S. partner in promoting citizen security throughout the region, has trained nearly 10,000 police from across Latin America in areas such as criminal investigation skills, personnel protection, and anti-kidnapping, among other critical law enforcement disciplines. Colombia's participation in improving security and reducing instability throughout the hemisphere by providing needed training is an enormous return on the U.S. investment in that country and is precisely the type of smart power diplomacy and approach to security promoted by Secretary Clinton.

Afghanistan. Despite progress by the Afghan government in reducing opium poppy cultivation by 22 percent to 123,000 hectares through the provision of alternative sources of income and increased public information about the perils of poppy, the drug trade continues to undermine economic reconstruction and weaken the rule of law, and poses a threat to regional stability. INL continues to support programs that will expand interdiction and drug demand reduction activities into additional provinces, and strengthen the political will of the central and provincial governments to increase support for licit agriculture. The United States will continue to disrupt and dismantle the narcotics-insurgent-corruption nexus by enhancing the capacity and sustainability of specialized investigative and interdiction units of the Counter-Narcotics

Police of Afghanistan to collect intelligence, target drug traffickers, and disrupt processing operations and trafficking networks. INL is also working to tackle the culture of impunity and expand access to the state justice sector by increasing gender justice capacity, reducing corruption, and building public demand for rule of law and individual legal rights.

Indonesia. Indonesia is a strategic partner of the United States with growing regional and global influence. It is also home to the terrorist group Jemaah Islamiyah and has been the target of several deadly terrorist attacks over the past decade. Since the start of INL assistance in 2000, the government of Indonesia has embraced institutional reform of its law enforcement organizations and criminal justice system. The continued development of an effective civilian police force and support of prosecutorial and judicial reform will ensure that Jakarta remains a key partner with Washington in combating transnational crime and terrorism. For example, INL continues to work closely and successfully with the national police. The first police units that responded to the July 2009 attacks on the Marriot and Ritz Carlton hotels in Jakarta were trained through INL programs. The unit that ultimately brought down the mastermind behind those bombings, Noordin Top, was also trained by and worked closely with the United States for many years. Noordin had ties to Jemaah Islamiyah as well as to al Qaeda.

International Law Enforcement Academies. INL reinforces these bilateral programs and others through a worldwide network of International Law Enforcement Academies (ILEAs), which provide training for local and regional law enforcement officials to combat international drug trafficking, criminality, and terrorism. Serving four regions—Europe, Africa, South America, and Asia—the ILEAs help protect American citizens and business by improving law enforcement at the source, buttressing democratic governance through the rule of law, enhancing the functioning of free markets through improved legislation and law enforcement, and increasing social, political, and economic stability.

Converging Public-Private Partnerships to Fight Converging Threats

It is vital that the international law enforcement community begin to think creatively in examining threat linkages. Even as the capability of the United States and its partners to disrupt illicit networks develops, criminals are constantly adapting their practices to avoid detection. The problem is too large for any one government to solve. It requires a net-centric approach at the bilateral, subregional, regional, and global levels based on information-sharing and coordination to break the financial strength of criminal and terrorist networks, disrupt illicit trafficking networks, defeat transnational criminal and terrorist organizations, fight government corruption, strengthen the rule of law, and bolster judicial and security systems.

Effective public-private partnerships are the key to achieving a whole-of-society response to crime. The United States is reaching out directly to the international business community and the general public to drive home the fact that nations and individuals share a common enemy in transnational organized crime and have a stake in addressing it.

Strategic public-private partnerships are key for one obvious reason: the overwhelming majority of industries and businesses are privately owned. Far more often than not, it is the private sector that first encounters intrusions, theft, extortion, and other forms of organized criminal disturbance. Over 90 percent of the world's information technology infrastructure rests in private hands. Private sector actors are the first responders in dealing with transnational organized crime. As such, they have timely information at their disposal on how organized crime is attempting to hijack the global economy from the ground up.

Governments can play a role in encouraging industry groups to develop and share best practices and standards for preventing organized criminal infiltration. This is a growing problem around the world. Certain key industries are increasingly facing unfair competition from organized crime networks seeking to burrow their way into the legitimate economy. Organized crime must not be allowed to capitalize on the gains of globalization and undermine the benefits of rule-based, competitive trading systems.

Governments must help businesses perform due diligence in ensuring that criminally influenced businesses are kept out of the global supply chain. Businesses must adopt internal security measures to make themselves harder targets for global crime. The growing infiltration of transnational crime into licit industry and commerce is a threat to the expansion of open markets and global trading regimes that underpin the modern global economy. Bringing the rule of law to the international flow of goods and services is essential in order to have a long-term impact against transnational crime.

Public-Private Partnerships to Combat Illicit Trade and Illegal Economy

The illegal economy poses an existential threat when it begins to create criminalized markets and captured states, which launches a downward, entropic spiral toward greater insecurity and instability. In countries that have been corrupted by criminal networks, market- and state-building become less attainable, economic growth is stunted, efforts toward development and poverty eradication are stifled, and foreign direct investment is deterred.

The United States is supporting the OECD, the World Economic Forum (WEF), and other international partners to provide knowledge-based platforms for international public and private stakeholders to raise awareness of the threat posed by illicit trade and illegal economy to economic growth, development, and global security, and to share experience on practical approaches to the control of illicit activities as well as of the negative externalities of the illicit economy. Engaging the public and private sectors through innovative public-private partnerships will be particularly important for securing the integrity of the global supply chains and for ensuring long-term sustainable licit commerce and productive markets.

The steep rise in mobility of goods, people, capital, and information that has accompanied globalization is largely comprised of lawful and beneficial exchanges, but an increasing share is illicit. Criminal entrepreneurs and illicit networks sometimes use or exploit legitimate businesses and legitimate global supply chains to carry out financial frauds, industrial espionage, money laundering, and other illicit activities. Hundreds of billions of dollars of revenue from these activities flow through the global economy every year, distorting local economies, diminishing legitimate business revenues, deteriorating social conditions, and fueling conflicts.

Public-private partnerships ensure the support and cooperation of the businesses and other market actors whose legitimate revenues are threatened by illicit trade and the illegal economy. These partnerships can also help create legitimate markets to replace and shut down illegal markets. Nurturing a network of government and business leaders who support innovation in the vibrant sectors of tomorrow—such as computer and high engineering industries and biomedical, green energy, and other emergent technologies—and harnessing the ideas, talents, and human potential of communities, can help governments and entrepreneurs alike create the right governance conditions across sectors for new markets and investment frontiers to thrive globally.

Public-private partnerships such as those advanced by the OECD and WEF will help articulate the harms and risks posed by transnational illicit networks and provide the mutual expertise that is critical to understanding the interaction among illicit and legitimate flows of goods, people, capital, and information. By identifying and assessing key risks and harms, both public and private sectors can better focus on what market vulnerabilities they need to mitigate with the goal of driving criminal entrepreneurs and illicit networks out of business and incentivizing willing compliance with laws and regulations to promote and grow legitimate global trading networks.

Conclusion

> *Secretary Clinton has often spoken of the need to build what she calls a "global architecture of cooperation" to solve the problems that no one country can solve alone. Certainly, this is true of the challenge before us. Transnational organized crime is a threat that endangers communities across the world including our own. The State Department remains determined, working closely with all of our interagency partners, to translate common interest into common action that makes us all safer.*
>
> —Under Secretary of State William J. Burns, July 20, 2011

Through multilateral fora and bilateral partnership, the United States is joining forces with global partners to take the fight directly to today's transnational threats, dismantle their networks, unravel the illicit financial nodes that sustain a web of criminality and corruption, develop strong law enforcement and multidisciplinary threat mitigation capabilities, and enhance cooperation with businesses through public-private partnerships.

The State Department remains committed to working with other governments and other committed international partners to meet the cross-border security challenges from transnational criminal and terrorist organizations facing all nations. As President Obama underscored in the July 2011 *Strategy to Combat Transnational Organized Crime*, the United States is keen to forge new partnerships based on shared responsibility. The threats and challenges facing the world today are real and complex. As illicit networks converge, the solution for governments is to converge along with them, synchronize efforts across borders, and develop networks to fight and defeat a common threat, protect our borders and homeland, and anchor sustainable partnerships vital for shared prosperity and security across economies and markets.

Author's Note

In honor and memory of Ambassador Chris Stevens and all patriots, civilian and military, who give their lives for our country to courageously advance our democratic values and ideals including to end tyranny and combat terrorism and violent crime, promote freedom and the rule of law, and safeguard security so that people can build a better world that nurtures sustainable peace, progress, and human dignity. Their sacrifice is never forgotten and strengthens our resolve.

Chapter 13

The Department of Defense's Role in Combating Transnational Organized Crime

William F. Wechsler and Gary Barnabo

Transnational organized crime (TOC) is altering the face of national security and changing the character of the battlespace. Loose networks of criminals operate across international boundaries with impunity, and their actions are more violent, deadly, and destabilizing than in any previous time. Once viewed primarily through the lens of law enforcement as drug cartels, transnational criminal organizations now engage in a web of overlapping illegal activities, from cybercrime to trafficking in persons, money laundering, and distribution of materials for weapons of mass destruction. As the power and influence of these organizations have grown, their ability to undermine, corrode, and destabilize governments has increased. The links forged among criminal groups, terrorist movements, and insurgencies have resulted in a new type of adversary: hybrid, ever-evolving networks that are criminal, violent, and, at times, politically motivated and blend into the landscape of a globalization-dominated world. Hybrid networks adapt their structures and activities faster than countries can combat their threats. These adversaries have become the new normal, compelling governments toward integrated, innovative approaches to countering the growing dangers posed by transnational organized crime.

The Character of Transnational Organized Crime

For decades, drug trafficking was the dominant lens through which the United States viewed TOC. In the 1970s and 1980s, the flow of illicit narcotics into the United States was deemed a major risk to the health and safety of Americans, and the government expended massive resources to curtail both the supply of and the demand for illegal drugs. Supply side reduction strategies emphasized degrading the capabilities of Western Hemisphere drug-trafficking organizations—highly capable, violent, centralized, and hierarchical organizations often led by charismatic kingpins. Pablo Escobar and the Medellín Cartel, for example, were emblematic of the type of threat facing the United States.

The drug-trafficking organization–centered paradigm for understanding the threat from TOC was already anachronistic by the 1990s as the character of transnational criminal groups began changing in three critical ways. First, drug-trafficking organizations diversified their activities. Taking advantage of the hallmarks of globalization such as open borders, rapid increases in the volume and speed of global trade, and the dissemination of technology tools, criminal organizations that once dealt almost exclusively in narcotics began trafficking in small arms and light weapons, people, counterfeit goods, and money, while continuing to ship vast quantities of drugs along an expanding set of transportation routes. These groups recognized the additional profits and operational flexibility that a broader range of trafficking activities could provide.

Second, transnational criminal organizations harnessed new methods of doing business. Criminal groups began penetrating the licit global marketplace, subverting and distorting legitimate markets and economic activity. These groups became adroit at harnessing information technology tools, seizing the opportunities presented by the accelerating velocity of information flows, the proliferation of online money transfer, and the general anonymity of virtual exchange to increase the scale and scope of their activities while spreading or reducing the risk of detection. Cyberspace is now a central arena for TOC: it enables vital elements of criminal transactions, such as communications and financial exchange, at low risk. Moreover, the increased virtual nature of some transnational criminal organizations has created deep overlaps and synergies between legal and illegal activities: the same networks that facilitate legitimate international business transactions are used for crime. Consequently, it is increasingly difficult for governments to identify, target, and degrade the activities of transnational criminal groups.

Third, the structure of transnational criminal organizations changed. Reflective of globalization's trends, centralized criminal groups have, in many cases, yielded to dispersed networks of criminal agents. They have become loose, amorphous, highly adaptable networks that are decentralized and flat, and emphasize ad hoc partnerships defined by the opportunity of the moment rather than top-down, command-and-control hierarchies. These characteristics have rendered transnational criminal organizations more flexible and agile than the government agencies chartered with degrading or defeating them. In the context of the Department of Defense (DOD), criminal networks cut across geographic combatant commands, which are responsible for the employment of military capabilities in distinct regions. Criminals who operate across multiple combatant commands present challenges for DOD in synchronizing its efforts to understand and address their activities. These realities demand new approaches that are networked, whole-of-government, and adaptable.

Together, these shifts have generated a more threatening set of challenges to national and international security. Transnational criminal organizations co-opt government officials and elites, corrode the legitimacy of government institutions, and increasingly have the propensity and ability to use national infrastructures to further their activities. In some cases, these groups threaten and challenge the state's control of its territory. Guinea-Bissau in Africa is on the verge of becoming a narco-state. Transnistria, a region of Moldova controlled by a separatist regime, has been deeply penetrated by organized crime. And closer to home, certain governments in the Western Hemisphere are increasingly challenged to preserve citizen security from the insidious effects of criminal organizations, such as corruption and large-scale violence.

Criminal networks are increasingly merging, blending, and cooperating with other nefarious and violent nonstate actors including terrorist organizations and insurgent movements. Crime is used to finance terrorism and is often how insurgencies raise revenue to pay individual cadres. Terrorists and insurgents can tap into the global illicit marketplace to underwrite their activities and acquire weapons and other supplies vital to their operations. Criminals, in their search for profits, will turn almost anywhere for revenue. The net effect is a crime-terror-insurgency nexus.

Examples of this nexus abound. As of November 2011, the Drug Enforcement Administration (DEA) had linked 19 of the 49 organizations on the State Department's list of Foreign Terrorist Organizations to the drug trade.[1] Similarly, in 2010, 29 of the 63 organizations on the Justice Department's Consolidated Priority Organization Targets list, which includes key drug trafficking and criminal organizations, had associations with terrorism.[2] In January 2012, Director of National Intelligence James Clapper concluded that "Terrorists and insurgents will increasingly turn to crime and criminal networks for funding and logistics, in part because of U.S. and Western success in attacking other sources of their funding. The criminal connections and activities of Hezbollah and al-Qaida in the Islamic Maghreb illustrate this trend."[3] Global crime networks coalesce with members of radical political movements in ungoverned border regions in Central and South America; similar blending has occurred in the Balkans.

The crime-terror-insurgent nexus is changing the global security landscape and, more importantly for DOD, is also altering the character of the battlespace. In Afghanistan, the Taliban and a variety of criminal networks often operate in lockstep; illicit activities are an important funding stream for the Taliban and its offshoots. Moreover, drug incomes corrupt politicians, alienating populations who then become vulnerable to the messages of the Taliban and its ilk. Crime and corruption thus perpetuate the Afghan insurgency and threaten coalition forces' ability to consolidate security gains and establish conditions for long-term stability. Of direct concern to DOD are the criminal networks that provide the materials and financial and logistical support for improvised explosive devices, which are among the most significant threats to U.S. and allied troops in Afghanistan.

More generally, hybrid criminal-terror-insurgent threats are forcing DOD to adapt its approaches to conflict and adjust the type and level of resources it deploys around the world. War against nebulous, networked threats that operate at the seams of war, instability, and peace; blur the illicit and licit; and simultaneously display the characteristics of criminals, terrorists, and insurgents requires DOD to think innovatively about how it plans, organizes, and operates.

A New Impetus for Combating Transnational Organized Crime

Largely because of the changing character of transnational criminal organizations, the national security threats posed by TOC have expanded and become more dangerous. This finding is at the core of the Obama administration's July 2011 *Strategy to Combat Transnational Organized Crime*, which concludes that "criminal networks are not only expanding their operations, but they are also diversifying their activities. The result is a convergence of threats that have evolved to become more complex, volatile, and destabilizing."[4] In a clear articulation of the U.S. Government's position, the strategy declares that transnational organized crime poses "a significant threat to national security."[5] It directs the government to "build, balance, and integrate the tools of American power to combat transnational organized crime and related threats to national security—and to urge our foreign partners to do the same."[6]

The strategy is a call to action: it compels the U.S. Government, including DOD, to reframe and refocus its efforts in combating TOC. Importantly, the strategy addresses organized crime and drug trafficking as increasingly intertwined threats, laying the foundation for a broad approach to global crime that rightfully retains a core focus on counternarcotics but widens the aperture by recognizing the diversification of global criminal activity and its relationship with other irregular threats.

Transnational criminal organizations are the essence of a networked threat. Not only are criminal entities themselves networked, but they also rely on support networks—transportation, financial, human capital, logistics, and communications—to enable their activities. It follows that U.S. Government efforts to combat organized crime should emphasize counternetwork approaches. Disrupting networks, particularly those associated with financial flows—vital enablers of any criminal activity—through multifaceted targeting of key network nodes, be they individuals, institutions, or even specific trafficking routes, can have a broad positive impact on degrading criminal organizations. Employing these counternetwork approaches, however, requires the government to assume the characteristics of a network, integrating the capabilities of diverse interagency stakeholders and adjusting approaches as rapidly as the adversary changes tactics. Only when our policy responses and actions in the field are as adaptive as the threat will we see true progress against transnational organized crime.

DOD's Supporting and Evolving Role

DOD supports law enforcement, other U.S. agencies, and foreign partners in combating transnational organized crime. The U.S. Government and its partners address transnational criminal threats primarily through law enforcement channels, capacity-building efforts, diplomacy with key partner states, intelligence and analytical efforts, and the use of innovative policy tools such as counterthreat finance capabilities. Because it is in a supporting role, DOD rarely leads these efforts, but it offers unique capabilities and resources that strengthen overall U.S. Government approaches against criminal networks. As the threat has morphed, however, there have been important instances of law enforcement agencies supporting DOD operations. The Afghanistan Threat Finance Cell, for example, is run by law enforcement entities and provides direct support to military operations in Afghanistan. Support that cuts both ways—from DOD to law enforcement and from law enforcement to the military—is a strategically important evolution in how the U.S. Government organizes to combat TOC.

The Department of Defense can provide intelligence and equipment to support law enforcement efforts against transnational criminal organizations. DOD's focus, however, has typically been on counternarcotics missions—historically in the Western Hemisphere and over the past decade in Afghanistan and the surrounding region. As described above, the nexus among criminals, terrorists, and insurgents is increasingly visible as our adversaries forge hybrid organizational and operational models, suggesting that counterterrorism initiatives in the future should incorporate a greater emphasis on combating the nexus with TOC.

Moreover, the traditional emphasis on counterdrug missions will also continue to shift due to the blurring of illicit narcotics production and trafficking and other forms of transnational crime. Criminal networks that traffic in drugs more often than not also deal in small arms, light weapons, precursor chemicals, people, or bulk cash. This is certainly the case of organized crime groups such as Los Zetas, the Sinaloa Cartel, and Mara Salvatrucha (MS-13), which operate across the Western Hemisphere and have tentacles reaching into Europe and Asia, and the illicit networks that crisscross Afghanistan, Pakistan, Iran, and Central Asia. Not all transnational criminal organizations undertake the full spectrum of illegal activity, but it is increasingly rare to see major crime groups confine their business strictly to one type of criminality.

Spectrum of Support

The evolution of many drug-trafficking organizations into diversified criminal groups has driven a shift in DOD's mission focus. The Department now provides support to law enforcement, other U.S. Government agencies, and foreign partners across a spectrum of operations and activities: military-to-military assistance, capacity-building and training to partner states, intelligence support to law enforcement, support to the development of institutions that convene interagency stakeholders in whole-of-government approaches to combating the national security threats TOC present, and, as a last resort, direct military action against terrorist or insurgent groups that also engage in crime. DOD is not often the most visible government agency combating transnational organized crime, and this is entirely proper for a Department in a supporting role. But the lack of attention it receives for its efforts against TOC does not diminish the importance of its contributions.

The extent and success of DOD support in the fight against TOC are best understood through examples at various points on the spectrum.

Military-to-Military Support

In the 1980s and 1990s, Colombia faced a full-blown challenge to the viability of the state. Several armed criminal-insurgent groups waged an active campaign against the government, engaging in major terrorist attacks, kidnapping, extortion, and all phases of the drug trade. In the late 1990s, there was widespread belief that the Revolutionary Armed Forces of Colombia (FARC) would seize Bogota. For 20 years, Colombian drug-trafficking organizations exported cocaine to the United States in staggering quantities, and it appeared almost impossible to make a dent in the drug trafficking and organized crime emanating from that country.

Over time, however, the political will among Colombian leaders, supported by a sustained U.S. Government interagency effort, resulted in a remarkable impact against the FARC and related criminal-insurgent activity in Colombia. DOD was a vital enabler of Colombia's success, helping its security forces overcome critical capability gaps. The Department's sustained counternarcotics and security assistance delivered military training, tactical and operational support, capacity-building on intelligence sharing, and information operations, equipment, and human rights training. The Department's engagements assisted the Colombian military in turning the tide against the FARC and other violent criminal groups that sought to capture the state. Colombia is now an exporter of security in the region, providing military training at the request of regional and international organizations and collaborating with both neighboring states and other countries as far afield as Turkey to share its lessons learned and provide capacity-building assistance for combating narcotics and TOC, as well as citizen security, peacekeeping operations, disaster response, and the strengthening of defense institutions. Its special operations forces are now considered to be among the best of their kind in many circles. The lessons learned from Colombia's successes are regularly applied to U.S. Government efforts in support of Mexico, Afghanistan, and other partners. DOD was just one U.S. Government actor among many that provided support to Colombia, but its military-to-military collaboration was a vital part of perhaps the first whole-of-government approach to combating TOC.

Building Partner Capacity

DOD's capacity-building efforts with foreign partners are widespread. Capacity-building and training of security forces, which serve a variety of national security objectives, are often areas in which DOD has a comparative advantage vis-à-vis other departments and agencies. As transnational criminal organizations expand their presence in regions such as West Africa and Southeast Asia, the Department's ability to support an ever-greater array of partner nations is more and more valuable. DOD's role in training, equipping, and in some cases operating alongside partner countries' security forces is the bedrock of building these states' capacity to combat TOC over the long term. DOD also partners with U.S. law enforcement agencies to support the improvement of policing capacity in key partner states. Given resource constraints, support for military-to-military and law enforcement capacity-building is in many cases a particularly effective use of limited resources. It amounts to a long-term investment and provides partner countries with the skills and tools that allow them, rather than the United States, to assume the burden of combating TOC in and around their territory.

The impact of DOD initiatives is clear in numerous cases. In Indonesia, a sprawling archipelago across which multiple criminal organizations and terrorist groups operate, Joint Interagency Task Force–West (JIATF-W) activities have increased the capacity of Indonesian security forces to fight a broad range of transnational threats including criminal organizations trafficking in narcotics and precursor chemicals and terrorist and insurgent movements seeking safe space from which to operate. JIATF-W support to Indonesia's National Narcotics Board has resulted in the construction of facilities at a counternarcotics academy used by various elements of the Indonesian security forces. In Liberia, a strategic location in a region that has experienced growth in trafficking and transnational crime, DOD support has focused on capacity-building in two principal areas: maritime security (primarily through rebuilding the Liberian Coast Guard) and support to the newly formed Transnational Crime Unit. United States Africa Command (USAFRICOM) supported the renovation of a Liberian Coast Guard base of operations, which included construction of a boat ramp/launch facility, floating pier, and security fence. In addition, USAFRICOM has sponsored attendance by Liberian Coast Guard personnel at various U.S. Coast Guard training courses that support Liberia's maritime security operations.

Military Intelligence Support to Law Enforcement

Network mapping has become widely recognized as a key enabler in combating transnational organized crime. Criminal organizations use multiple, overlapping transportation, logistics, and financial networks to do business. Financial networks are particularly vital since money is central to all criminal activities. Consequently, DOD has expanded its military intelligence support to law enforcement on counterthreat finance initiatives that seek to identify, penetrate, and exploit the critical nodes in the complex financial systems that enable global illicit networks. DOD recognizes that countering the financial flows that facilitate trafficking groups, criminal organizations, terrorists, and insurgents will produce cascading positive effects across the range of networks these nefarious actors use.

A particular strength of DOD is integrating various forms of all-source intelligence to facilitate a network-based approach to targeting. This is the approach used in the Afghan Threat Finance Cell, through which DOD supports the DEA, Federal Bureau of Investigation, Department of the Treasury, and other U.S. Government partners to identify insurgent financiers, disrupt front companies, develop actionable financial intelligence, freeze and seize illicit funds, and build criminal cases. These actions are vital to degrading and diminishing the power of criminal-terrorist networks by enabling a judicial endgame—the prosecution of adversaries, which often removes them from the battlefield for far longer than typical military-led raids. The Afghan Threat Finance Cell is also designed to provide direct support to the government of Afghanistan's efforts to reclaim tainted money moving around the country, which further undercuts the power of insurgent and terrorist networks. Importantly, these types of approaches can be replicated and DOD's experiences in Afghanistan have equipped it with new tools, techniques, and institutional skill sets that add to the overall set of capabilities the Department offers in support of law enforcement.

Direct Military Action with a Counter-Network Focus

In Afghanistan, DOD is taking direct action against a hybrid insurgent-criminal adversary in partnership with law enforcement agencies and the government of Afghanistan. Over the past 10 years, there has been an evolution in how the Department thinks about and approaches its missions there. DOD—and the interagency—have shifted from the paradigm that crime oriented around narcotics production was an indelible feature of the national landscape but disconnected from operations against al Qaeda and the Taliban, to a recognition that corruption, narcotics production and trafficking, criminal organizations, and insurgency and terror are mutually enabling and reinforcing and cannot be addressed in isolation. Ultimately, organized crime and corruption threaten the viability of the Afghan state.

Consequently, DOD actions in Afghanistan emphasize counternetwork approaches: attacking, degrading, and defeating the criminal networks that sustain the insurgency and undermine the legitimacy and capacity of the Afghan government. Counterinsurgency, counter–transnational organization crime missions, counterthreat finance activities, and capacity-building efforts are part of an integrated U.S. Government effort led, in many cases, by DOD. The Department's role in the operation of the Afghan Threat Finance Cell is a critical contribution to whole-of-government approaches to combating hybrid criminal-insurgent enemies in Afghanistan. The Interagency Operations Coordination Center, a joint U.S.–United Kingdom center that provides law enforcement targeting support and operational coordination for counternarcotics and counter–illicit networks operations in Afghanistan, is a broader example of successful whole-of-government initiatives to counter organized crime that also include the participation of key U.S. allies.

The Department of Defense has also adapted its policies in Afghanistan to address the convergence of crime and insurgency more effectively. In 2008, DOD changed its policy to enable U.S. forces to provide direct support to counternarcotics missions conducted by the DEA and its Afghan counterparts. This shift in policy has led to a number of successful operations, such as *Khafa Khardan*, during which U.S. and coalition forces operating under the International Security Assistance Force enabled 94 counterdrug operations by Afghan counterdrug law enforcement units and their mentors in a 30-day period.

DOD possesses significant convening power outside of Afghanistan. Through organizations such as the Joint Interagency Task Force–South (JIATF-S) and JIATF-W, DOD has brought key U.S. agencies together in pursuit of common objectives. These hubs of collaboration facilitate fast and flexible responses to the range of threats posed by transnational organized crime. They are prime examples of the government assuming the characteristics of a network—essentially creating a network among myriad U.S. Government agencies in one location—to combat a networked threat. While its traditional emphasis has been on counternarcotics, particularly the interdiction of illegal drugs, JIATF-S is a resounding success. In 2009, it accounted for more than 40 percent of global cocaine interdiction, and in the past two decades it has deprived criminal organizations of nearly $200 billion in profits.[7] Its success stems in large part from the long-term, sustained, and institutionalized integration of effort among law enforcement, DOD, and intelligence. Indeed, it may be that DOD's role in establishing the institutional architecture that enables agencies to work together as part of a unified, whole-of-government approach is actually the most significant contribution it has made in the fight against transnational organized crime and related national security threats.

Entities such as the JIATFs are not easy to create, and success does not stem from standing up an organization, calling it "joint," and staffing it from across the interagency. Impact and sustainability require a set of conditions, beginning with a clear overlap in vital interests so that stakeholders see the value of enduring participation, and a shared commitment to unified action around the issues affecting those interests. Next, it is critical to develop and implement strong cross-organizational structures that bring order and discipline to execution of critical tasks. Finally, the organization must be adaptive, adjusting its approaches based on evolutions, transformations, and shocks to the "issue set" it is chartered to address. DOD leadership has fostered these conditions in the JIATFs, turning them into highly effective interagency hubs that are models for replication.

New Challenges and Paradigms of Support

Five years ago, the Mexican government directed its military to increase its traditional role of supporting law enforcement while Mexico undertook reforms to strengthen police, the judicial system, and related institutions. These measures were largely necessitated by Mexican transnational criminal organizations' diversification beyond drug trafficking into kidnapping, extortion, smuggling persons, and other crimes. As the Mexican government disrupted transnational criminal organizations, these groups increasingly fought for market share, escalating violence against one another, targeting Mexican security forces using military-like weapons, tactics, and techniques, and endangering the civilian population in the process. Mexican criminal organizations also expanded their operations into Central America and have reached into the United States, penetrating cities such as Chicago, Detroit, and Atlanta, far from the border. The diversification of criminal activity and the increase in violence forced the Mexican military to assume even more prominent roles, while simultaneously transforming themselves to become more capable of protecting the public from hybrid, irregular threats.

The Way Forward

As the spectrum demonstrates, DOD tailors the type of support it provides to law enforcement, other U.S. Government agencies, and foreign partners. Given the ever-evolving character of transnational organized crime, this adaptability allows DOD to have significant strategic impact in a range of circumstances. There are risks, however, of overextension and redundancy. Providing support across a wide spectrum of efforts demands that DOD carefully prioritize the specific activities it undertakes—a prescient requirement as resources shrink.

DOD has taken important steps toward setting clear priorities for its actions against transnational organized crime. The Department's Counternarcotics and Global Threats Strategy, derived from national strategic and military guidance, is a key departure point for how DOD combats transnational organized crime. But this strategy is not exhaustive. There are manifestations of organized crime such as cybercrime, financial crimes, and state-sponsored illicit activities that DOD has not yet fully integrated into its strategy and policy development. Doing so will be critical to ensuring that the resources expended against criminal organizations have maximum strategic impact.

The Department must also link its efforts against transnational organized crime with other national security priorities. The strategies and policies created and implemented to combat the threats posed by criminal networks should be connected to other focal points for DOD and the U.S. Government—particularly cybersecurity, counterterrorism, counterproliferation, building partner capacity, and strengthening governance. Holistic approaches that recognize transnational organized crime not as a stovepiped problem but as a nefarious feature of the global security environment that touches the whole world and impacts a multitude of vital U.S. interests are the types most likely to succeed over the long term.

DOD understands that the most important prerequisite to success against transnational organized crime is organizational flexibility and adaptability. As the character of transnational criminal organizations continues to change and DOD and its U.S. Government partners hone their understanding of the types of national security threats these groups pose, we must be able to shift our strategies, priorities, policies, operations, and tactics for combating them. To fail in our mission because DOD is unable to organize itself properly is unacceptable, yet it remains a risk if we do not assume the characteristics of flexible, flat organizations.

Notes

[1] Statement of Derek S. Maltz, Special Agent in Charge of the Special Operations Division, Drug Enforcement Administration, before the U.S. House of Representatives Committee on Terrorism, Non-Proliferation, and Crime, November 17, 2011.

[2] *Strategy to Combat Transnational Organized Crime: Addressing Converging Threats to National Security* (Washington, DC: The White House, July 2011), 6.

[3] James Clapper, Unclassified Statement for the Record on the Worldwide Threat Assessment of the U.S. Intelligence Community for the Senate Select Committee on Intelligence, January 31, 2012.

[4] *Strategy to Combat Transnational Organized Crime*, cover letter by President Barack Obama.

[5] Ibid., 1.

[6] Ibid.

[7] Evan Munsing and Christopher J. Lamb, *Joint Interagency Task Force–South: The Best Known, Least Understood Interagency Success*, Strategic Perspectives No. 5, Center for Strategic Research, Institute for National Strategic Studies (Washington, DC: National Defense University Press, June 2011), 3.

Chapter 14

Collaborating to Combat Illicit Networks Through Interagency and International Efforts

Celina B. Realuyo

Introduction

This chapter examines collaborative efforts among U.S. agencies, partner nations, and multilateral organizations to address the hybrid threats posed by illicit networks through the diplomatic, development, and defense lenses of national security in the Western Hemisphere. The resourcefulness, adaptability, impunity, and ability of illicit networks to circumvent countermeasures make them a formidable foe for governments. Since illicit actors have expanded their activities throughout the global commons, in the air, land, sea, and cyberspace domains, nations must devise comprehensive and multidimensional strategies and policies to combat the transnational threats posed by these illicit networks.

The initiatives discussed here focus on programs in the Western Hemisphere to tackle the converging threat of transnational organized crime (TOC) and other illicit networks, undertaken by the United Nations Office on Drugs and Crime (UNODC), Organization of American States (OAS), Central American Regional Security Initiative (CARSI), and U.S. Southern Command's Joint Interagency Task Force–South. Each of these endeavors represents models for collaborations that require political will, institutional capacity, mechanisms, resources, and measures of effectiveness for success. These instructive examples of interagency and international collaboration in the Americas underscore the continued need to foster communities of interest to comprehend illicit networks and devise more effective strategies to counter them as they further evolve and threaten the national security of the United States and its allies.

Based on analysis of these examples, the following recommendations are offered for consideration by policymakers:

+ Promote collaborative models for security and development that include the following critical elements: political will, institutions, mechanisms to assess threats and deliver countermeasures, resources, and measures of effectiveness to ensure success against illicit networks.

+ Ensure that international conventions, agreements, and strategies are accompanied with robust action plans and are adequately resourced in order to restrict illicit actors' operations and enablers.

+ Encourage greater donor coordination so that all these security and development programs complement rather than duplicate each other. Given the level of political will demonstrated across these organizations, there should be an interest in allocating resources and building capacity in the specific geographic areas most vulnerable to transnational organized crime.

The Nature of the Threat of Illicit Networks

Throughout history, governments have been responsible for ensuring and promoting security, prosperity, and governance. The unprecedented pace of change in a globalized world has challenged these basic missions of the nation-state. From demographic pressures to an increasingly interconnected global economy, and from the race for resources to information overload, governments are struggling to cope with what Harvard University scholar Joseph Nye calls "the diffusion of power" in the 21ˢᵗ century. Governments are faced with a broad spectrum of national security threats that emanate from nonstate actors as well as traditional nation-states. Illicit networks that include transnational crime organizations, drug traffickers, gangs, and terrorist groups are among these nonstate actors. While illicit activities have been with us since ancient times, what is novel today is the pervasive, prolific, and converging nature of illicit networks around the world. These networks threaten the rule of law, government institutions, the economy, and society. In the words of Supreme Allied Commander Europe, Admiral James Stavridis, "Just as legitimate governments and businesses have embraced the advances of globalization, so too have illicit traffickers harnessed the benefits of globalization to press forward their illicit activities."[1] Illicit activities such as drug, arms, contraband, and human trafficking are nothing new; however, their velocity, scale, and associated violence as a result of globalization have made these transnational crimes national security concerns.

Illicit Networks Capitalize on Global Supply Chains

Illicit networks seek to navigate, infiltrate, and/or dominate global supply chains to further their activities and enhance their power. They actually thrive in open societies with the free flow of goods, people, and capital. Just like licit businesses, illicit networks are matching the supply and demand for goods, services, capital, and information for their clients. Illicit actors utilize and even seek to control or co-opt supply chains around the world to facilitate the movement of "bad people and bad things" such as drugs, guns, and counterfeit goods.

Regardless of industry or geography, there are four critical elements of any global supply chain whose integrity must be preserved and protected at all costs:

1. *Materiel:* What is being moved through the supply chain? People, goods, commodities, services, data? Where are those materials coming from and destined for?

2. *Manpower:* Who controls and staffs the supply chain? Who are the key enablers of that supply chain? Who is in control of these mechanisms or modes of conveyance?

3. *Money:* Who is funding the supply chain? What business model is being used to generate revenue? Where is the financing originating from and directed to?

4. *Mechanisms:* What modes of conveyance are used by the supply chain? Are people, goods, and services moving by land, air, sea, or cyberspace? How is the supply chain organized?[2]

Figure 1. The New Global Security Enviroment National Security Threats from Illicit Networks

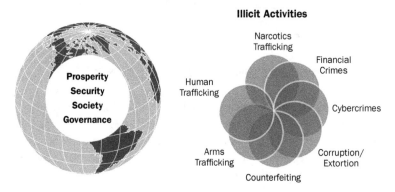

While each of these four critical elements of global supply chains has its unique features, the security of each is vital to safeguarding supply chains. Who is securing these elements? Global competition has led the private sector to identify and adopt the most efficient means of matching supply and demand for goods, services, and information, and to incorporate risk management mechanisms;[3] similarly, illicit networks have adopted these best practices. Unfortunately, international safety standards and government control of global supply chains still lag far behind.

Illicit actors are well aware of these weaknesses and exploit them to further their interests; they capitalize on gaps in governance, regulations, and oversight to bolster their enterprises. Illicit networks easily adapt to their changing operating environment. They often operate as service providers for each other. For transnational criminal organizations, maximizing profits is paramount. Their activities distort global markets and pricing, undermine consumer confidence, and even endanger consumers around the world, as in the case of counterfeit drugs. As a result of globalization, the sheer volume and velocity of international trade make it virtually impossible to control and secure global supply chains. Nevertheless, governments must keep abreast of the ways illicit actors are exploiting the increasingly borderless world and identify the vulnerabilities of global supply chains. Accordingly, governments must develop measures to combat illicit networks by leveraging and safeguarding the four critical elements of supply chains: materiel, manpower, money, and mechanisms. New strategies to combat transnational organized crime and illicit networks, including those discussed below, are increasingly addressing these challenges.

Figure 2. Global Supply Chain Management Four Critical Elements

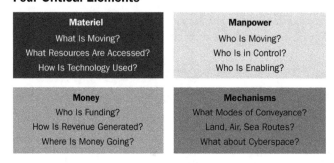

Strategies to Combat Transnational Organized Crime and Illicit Networks

Governments at the local, national, regional, and multilateral levels must devise holistic strategies to address the complex, adaptive, and converging transnational security threats from illicit networks. Several efforts are underway to promote interagency and international collaboration to address transnational organized crime and illicit networks and promote security in the Western Hemisphere. These regional efforts will be examined through the lenses of national strategy, diplomacy, development, and defense and security.

The May 2010 *National Security Strategy* aptly acknowledges and describes the convergence of threats posed by illicit networks:

> *Transnational criminal threats and illicit trafficking networks continue to expand dramatically in size, scope, and influence—posing significant national security challenges for the United States and our partner countries. These threats cross borders and continents and undermine the stability of nations, subverting government institutions through corruption and harming citizens worldwide. Transnational criminal organizations have accumulated unprecedented wealth and power through trafficking and other illicit activities, penetrating legitimate financial systems and destabilizing commercial markets. They extend their reach by forming alliances with government officials and some state security services. The crime-terror nexus is a serious concern as terrorists use criminal networks for logistical support and funding. Increasingly, these networks*

are involved in cyber crime, which cost consumers billions of dollars annually, while undermining global confidence in the international financial system.[4]

Combating transnational criminal and trafficking networks requires a multidimensional strategy that safeguards citizens, breaks the financial strength of criminal and terrorist networks, disrupts illicit trafficking networks, defeats transnational criminal organizations, fights government corruption, strengthens the rule of law, bolsters judicial systems, and improves transparency. While these are major challenges, the United States will be able to devise and execute a collective strategy with other nations facing the same threats.[5]

Recognizing transnational organized crime as a national security threat, the United States issued its most recent *Strategy to Combat Transnational Organized Crime*, in July 2011, which seeks to build, balance, and integrate the tools of American power to combat transnational organized crime and related threats to U.S. national security. The strategy posits that transnational organized crime threatens U.S. interests by taking advantage of failed states or contested spaces; forging alliances with corrupt foreign government officials and some foreign intelligence services; destabilizing political, financial, and security institutions in fragile states; undermining competition in world strategic markets; using cyber technologies and other methods to perpetrate sophisticated frauds; creating the potential for the transfer of weapons of mass destruction (WMD) to terrorists; and expanding narcotrafficking and human and weapons smuggling networks. Terrorists and insurgents increasingly are turning to criminal networks to generate funding and acquire logistical support. TOC also threatens the interconnected trading, transportation, and transactional systems that move people and commerce throughout the global economy and across our borders.[6]

The strategy's key policy objectives are to:

1. protect Americans and our partners from the harm, violence, and exploitation of transnational criminal networks

2. help partner countries strengthen governance and transparency, break the corruptive power of transnational criminal networks, and sever state-crime alliances

3. break the economic power of transnational criminal networks and protect strategic markets and the U.S. financial system from TOC penetration and abuse

4. defeat transnational criminal networks that pose the greatest threat to national security by targeting their infrastructures, depriving them of their enabling means, and preventing the criminal facilitation of terrorist activities

5. build international consensus, multilateral cooperation, and public-private partnerships to defeat transnational organized crime.

Key actions of the strategy include:

+ reduce the demand for illicit drugs in the United States, thereby denying funding to illicit trafficking organizations

+ continue to attack drug trafficking and distribution networks and their enabling means within the United States to reduce the availability of illicit drugs

+ sever the illicit flow across U.S. borders of people, weapons, currency, and other illicit finance through investigations and prosecutions of key TOC leadership, as well as through targeting enabling means and infrastructure of TOC networks

+ identify and take action against corporate and governmental corruption within the United States

+ work with Congress to secure ratification of the Inter-American Convention against the Illicit Manufacturing of and Trafficking in Firearms, Ammunition, Explosives and Other Related Materials

+ seek accession to the Protocol against the Illicit Manufacturing and Trafficking in Firearms, Their Parts and Components and Ammunition, supplementing the United Nations Convention against Transnational Organized Crime.

Other priority actions seek to: enhance intelligence and information sharing; protect the financial system and strategic markets against transnational organized crime; strengthen interdiction, investigations, and prosecutions; disrupt drug trafficking and its facilitation of other transnational threats; and build international capacity, cooperation, and partnerships.[7]

This new strategy elevates transnational organized crime to a national security threat and focuses attention on the destructive effects of TOC on U.S. interests at home and abroad. In the past, organized crime had been considered and handled as a local law enforcement issue. The new strategy seeks to better organize and coordinate interagency efforts within the U.S. Government to address transnational organized crime and promote international cooperation against this global threat.

Prior to the publication of this strategy, U.S. agencies such as the Departments of Homeland Security, Justice, State, and Treasury as well as the law enforcement and intelligence communities had been actively countering, pursuing, and prosecuting transnational criminal organizations for years. From border controls to banking regulations, roles and responsibilities to combat illicit activities have been carved out for these agencies in their respective areas of expertise. Those who investigate and research transnational organized crime and illicit networks, both inside and outside of government, expected bolder language and guidance on how this new strategy would synchronize and coordinate U.S. efforts to combat TOC across the government. However, as of this writing, the U.S. Government has not issued an implementation plan for this new strategy. The strategy does not create a new government agency, establish a White House czar position, allocate additional funding, or empower a lead agency.

Perhaps the greatest challenge to the implementation of the 2011 *Strategy to Combat Transnational Organized Crime* is not the adversary of illicit actors, but rather the fiscal state of the United States. Some, including former Chairman of the Joint Chiefs of Staff Admiral Mike Mullen, consider the national debt as the greatest threat to U.S. national security.[8] In this age of fiscal austerity, it is unclear how these agencies charged with combating TOC can match the resources and ingenuity of these illicit actors and "do even more with less." With a limited and if not diminishing budget, the U.S. interagency community will need to seek more innovative and efficient ways to confront and combat transnational organized crime.

On the Diplomatic Front: United Nations Initiatives

Palermo Convention and Protocols. The United Nations (UN) Convention against Transnational Organized Crime, known as the Palermo Convention, is the main international instrument to counter organized crime and promote coordinated international cooperation. It was adopted in December 2000 in Palermo, Italy, and was complemented by three protocols that target trafficking in persons, especially women and children, smuggling of migrants, and the illicit manufacturing and trafficking of firearms.[9] The Palermo Convention establishes transnational organized crime as a threat to all UN member nations and sets forth common definitions and language to characterize activities that constitute this phenomenon.

United Nations Office on Drugs and Crime. As the guardian of the Palermo Convention, the United Nations Office on Drugs and Crime (UNODC) helps countries create the domestic legal framework to investigate criminal offences related to organized crime and prosecute offenders, and adopt new frameworks for extradition, mutual legal assistance, and international law enforcement cooperation.[10] UNODC is a global leader in the fight against illicit drugs and international crime. Established in 1997 through a merger between the United Nations Drug Control Program and the Center for International Crime Prevention, UNODC is headquartered in Vienna, with 54 field offices covering more than 150 countries around the world. It relies on voluntary contributions, mainly from governments, for 90 percent of its budget.[11] UNODC is one of the world's leading sources of reliable data, analysis, and forensic science services related to illicit drugs and crime. Its publications include the annual *World Drug Report*, *U.N. Global Report on Trafficking in Persons, 2009*, the 2010 report *Corruption in Afghanistan: Bribery as Reported by Victims*, and the 2011 *Global Afghan Opium Trade: Threat Assessment*. UNODC also provides practical tools and resources for policymakers, legislators, and criminal justice professionals.[12]

The UNODC work program is supported by three pillars. First, *field-based technical cooperation projects* enhance the capacity of member states to counteract illicit drugs, crime, and terrorism. Second, *research and analytical work* increases knowledge and understanding of drugs and crime issues and expands the evidence base for policy and operational decisions. And third, *normative work* assists states in the ratification and implementation of the relevant international treaties, the development of domestic legislation on drugs, crime, and terrorism, and the provision of secretariat and substantive services to the treaty-based and governing bodies.[13]

UNODC offers specialized assistance and expertise in the following areas:

+ Organized crime and trafficking: UNODC helps governments react to the instability and insecurity caused by crimes such as the smuggling of illicit drugs, weapons, natural resources, counterfeit goods, and human beings between countries and continents. It also addresses emerging forms of crime such as cybercrime, trafficking in cultural artifacts, and environmental crime.

+ Corruption: Corruption is a major impediment to economic and social development. UNODC partners with the public and private sectors, as well as civil society, to loosen the grip that corrupt individuals have on government, national borders, and trading channels. In recent years, the office has stepped up its efforts to help states recover assets stolen by corrupt officials.

+ Crime prevention and criminal justice reform: UNODC promotes the use of training manuals and the adoption of codes of conduct and standards and norms that aim to guarantee that the accused, the guilty, and the victims can all rely on a criminal justice system that is fair and grounded on human rights values. A strong rule of law will also instill confidence among citizens in the effectiveness of the courts and the humanness of the prisons.

+ Drug abuse prevention and health: Through educational campaigns and by basing its approach on scientific findings, UNODC tries to convince young people not to use illicit drugs, drug-dependent people to seek treatment, and governments to see drug use as a health problem, not a crime.

+ Terrorism prevention: UNODC is moving toward a more programmatic approach that involves developing long-term, customized assistance to entities involved in investigating and adjudicating cases linked to terrorism.[14]

UNODC's consolidated budget for drugs and crime for the biennium 2010–2011 amounted to $468.3 million. Voluntary contributions are budgeted for $425.7 million, of which 64 percent is for the drugs program and 36 percent for the crime program. The United States was the second largest country contributor after Colombia, contributing $34.3 million to UNODC funding in 2010.[15]

The UNODC 2010 Annual Report "showed how health, security, and justice are the antidotes to drugs, crime and terrorism," according to UNODC Executive Director Antonio Maria Costa. It featured the organization's work on promoting drug treatment and alternative development, improving criminal justice, strengthening integrity, and reducing vulnerability to crime.[16] To combat transnational organized crime, UNODC developed the *Serious Organized Crime Threat Assessment Handbook* to give policymakers tools for assessing immediate threats, the direction of current trends, and likely future challenges so that they can implement effective strategies to combat organized crime. In April 2010, Brazil hosted the Twelfth United Nations Congress on Crime Prevention and Criminal Justice to consider comprehensive strategies for addressing global challenges to crime prevention and criminal justice systems, in particular those related to the treatment of prisoners, juvenile justice, the prevention of urban crime, smuggling of migrants, trafficking in persons, money laundering, terrorism, and cybercrime.

In its anticorruption efforts, state parties to the UN Convention against Corruption met in Doha, Qatar, in November 2009, and agreed to establish a mechanism to review implementation of the Convention against Corruption. Under the new mechanism, all state parties will be reviewed every 5 years on the fulfillment of their obligations under the convention. On the basis of self-assessments and peer review, the mechanism will help identify gaps in national anticorruption laws and practices. The convention's new monitoring mechanism represented a major breakthrough in the global campaign against corruption. To address human trafficking, UNODC developed, in cooperation with the Global Initiative to Fight Human Trafficking, a model law to guide UN member states in preparing national laws against trafficking in persons.[17]

In 2009, UNODC launched an international drug treatment and care initiative in partnership with the World Health Organization (WHO) that is a landmark in the development of a comprehensive, integrated, health-based approach to drug policy. The UNODC-WHO Joint Programme on Drug Dependence Treatment and Care aims to provide humane and accessible care to greater numbers of people with drug dependence and drug-related diseases (particularly HIV/AIDS) in low- and middle-income countries, resulting in their rehabilitation and reintegration into society.[18]

As an example of UNODC's more recent initiatives, UNODC Executive Director Yury Fedotov inaugurated the Center of Excellence for Crime Statistics on Governance, Victims of Crime, Public Security and Justice on October 1, 2011, in Mexico City. The center will develop field surveys, share knowledge in the area of crime statistics, and organize an annual international conference on statistics. Fedotov stated:

> As criminal gangs and drug trafficking rings become more sophisticated in their quest to avoid detection and escape justice, we support the Mexican Government in strengthening and sharing data collection methods relating to governance, victims of crime, public security and justice so that it—and other countries—can better respond to these threats.

The Executive Director also noted:

> We must all remember that organized crime has become transnational and borderless; it is no longer a problem of just one country. Therefore, this center of excellence will help to assist in the government's response to crime in Mexico, and its findings will also be crucial to regional and international authorities and organizations fighting transnational organized crime.[19]

UNODC's broad spectrum of activities, like the Center of Excellence for Crime Statistics in Mexico City, demonstrates constructive bilateral and multilateral engagements directed against transnational organized crime and illicit networks that should continue as these threats mutate and converge in the new global security environment. Since many of the UNODC programs are dedicated to strengthening institutions to fight drugs and crime around the world, and their results are usually not immediate, it is difficult to evaluate their effectiveness. Moreover, as most programs are executed in partnership with host governments and other multilateral or nongovernmental organizations, determining the return on UNODC investments is challenging. While the mission of UNODC is noble, it would be helpful if the organization could provide more information about its programming and institute measures of effectiveness to better demonstrate the impact against transnational organized crime, and determine better which programs actually worked.

On the Diplomatic Front: Organization of American States Initiatives

The Organization of American States (OAS) has been combating the phenomenon of illicit networks in the Americas at the multilateral level through three vehicles: the Hemispheric Plan of Action against Transnational Organized Crime, the Inter-American Drug Abuse Control Commission (CICAD), and the Inter-American Committee against Terrorism (CICTE). The OAS early on recognized that the threat of illicit activities was transnational in nature and required dynamic political, social, economic, and security initiatives to respond to illicit networks.

OAS Hemispheric Plan of Action against Transnational Organized Crime

The OAS Hemispheric Plan of Action against Transnational Organized Crime was adopted in October 2006 to promote the application by OAS member states of the United Nations Convention against Transnational Organized Crime (Palermo Convention) and the Protocols thereto: the Protocol to Prevent, Suppress and Punish Trafficking in Persons, Especially Women and Children; the Protocol against the Smuggling of Migrants by Land, Sea and Air; and the Protocol against the Illicit Manufacturing and Trafficking in Firearms, Their Parts and Components and Ammunition. Its general objectives are to:

1. prevent and combat transnational organized crime, in full observance of human rights, using as a frame of reference the Palermo Convention and the three additional protocols

2. enhance cooperation in the areas of prevention, investigation, prosecution of, and judicial decisions related to, transnational organized crime

3. encourage coordination among OAS bodies responsible for issues related to combating transnational organized crime and cooperation among those bodies with UNODC

4. strengthen national, subregional, and regional capacities and capabilities to deal with transnational organized crime.

The Hemispheric Plan of Action put forth specific actions in the following areas: national strategies against transnational organized crime, legal instruments, law enforcement matters, training, information-sharing, and international cooperation and assistance.[20]

Despite initial enthusiasm about the Plan of Action, the OAS Technical Group on Transnational Crime has only met on three occasions: Mexico City, in July 2007, Washington,

DC, in October 2009, and Trinidad and Tobago, in November 2011.[21] At the November 2011 meeting, OAS Secretary for Multidimensional Security Adam Blackwell said that international crime is "everyone's problem, it occurs in developed and developing countries where poverty, inequality, social and political exclusion and governance challenges are both causes and consequences of the problem." Those at the meeting debated and offered some recommendations on police management in the hemisphere, a fundamental issue in improving security for citizens in the Americas.[22] OAS Secretary General José Miguel Insulza explained that "police forces are tasked with the great responsibility of being the primary representatives of the state in guaranteeing compliance with the rule of law, within a community in a large or small community," for which he stressed that "the importance of fostering institutional capacity development to ensure effective response by the police in a democratic framework" cannot be overestimated.[23]

While the OAS Hemispheric Plan of Action against Transnational Organized Crime is well intentioned, it is not clear what has actually been accomplished, as data regarding its programming are limited. Many of these OAS meetings issue communiqués that sound promising, but actions speak louder than words. To further complicate matters, OAS member states are hard-pressed to fund and staff new initiatives in support of the Plan of Action given the global economic recession and resource constraints.

Inter-American Drug Abuse Control Commission

To address the challenging issues of narcotics trafficking and drug abuse in the Western Hemisphere, the OAS established an agency known as CICAD, the Inter-American Drug Abuse Control Commission, in 1986. It focuses on strengthening the human and institutional capabilities and channeling the collective efforts of its member states to reduce the production, trafficking, and use of illegal drugs.[24] CICAD received almost $3 million of its $7.8 million total 2010 budget from the United States to counter the trafficking and abuse of illegal drugs, including methamphetamine.[25] Its specific activities seek to:

+ prevent and treat substance abuse

+ reduce the supply and availability of illicit drugs

+ strengthen national drug control institutions and machinery

+ improve money-laundering control laws and practice

+ develop alternate sources of income for growers of coca, poppy, and marijuana

+ assist member governments in improving their data gathering and analysis on all aspects of the drug issue

+ help member states and the hemisphere as a whole measure their progress over time in addressing the drug problem.[26]

For more than 10 years, the OAS/CICAD Multilateral Evaluation Mechanism (MEM) has strengthened regional and subregional collaboration on all levels including drug awareness and treatment approaches, information data collection, sharing and harmonizing counter-narcotics and crime legislative models, extradition practices, and other control measures. The MEM deploys the expertise of independent peer reviewers from all OAS countries, which has resulted in hundreds of recommendations that individual countries and CICAD are taking concrete and effective measures to implement.[27]

Carrying out the MEM recommendations is an essential part of the toolkit that allows OAS countries to work as a cohesive force against the dual threats of drugs and crime. The MEM focuses on institution-building, demand reduction, supply reduction, control measures, and international cooperation. Notably, the MEM evaluation country reports, published in 2010, prompted the independent hemispheric experts who draft the reports to make over a third of their recommendations in the area of narcotics control measures. These recommendations for action include establishing and/or refining laws and regulations to control weapons, ammunition, and related material to stem the growing violence posed by illegal drugs and crime.[28]

On May 3, 2010, a new Hemispheric Drug Strategy was adopted by CICAD in Washington, DC, which updates the Anti-Drug Strategy in the Hemisphere, originally approved by the OAS General Assembly in 1997. Its 52 articles cover five aspects: Institutional Strengthening, Demand Reduction, Supply Reduction, Control Measures, and International Cooperation.[29] In May 2011, the OAS adopted the Hemispheric Plan of Action on Drugs, 2011–2015, to assist member nations in implementing the Hemispheric Drug Strategy[30] on the demand and supply sides of the problem.

Among its various programs, CICAD continued to carry out its activities in anti-money laundering and combating the financing of terrorism (AML/CFT) throughout Latin America and the Caribbean in 2010. CICAD's AML/CFT training programs seek to enhance the knowledge and capabilities of judges, prosecutors, public defenders, law enforcement agents, and financial intelligence unit analysts. CICAD's Anti-Money Laundering Section organized 17 seminars and workshops in 14 countries in 2010, training nearly 700 judges, prosecutors, law enforcement officers, financial intelligence unit analysts, and forfeited assets administration officers, among other participants. It partnered with UNODC, the Stolen Assets Recovery initiative of the World Bank, the World Bank Institute, Spain's ministry of interior, and the Bureau of International Narcotics and Law Enforcement Affairs of the Department of State, as well as the OAS's Inter-American Committee Against Terrorism (CICTE) and the governments of CICAD member states.

CICAD also coordinated with the UNODC Legal Assistance Program for Latin America and the Caribbean, Interpol, and the South American Financial Action Task Force (GAFISUD) in setting up the Asset Recovery Network of GAFISUD as a vehicle for exchanging information about the identification and recovery of assets, products, or instruments of transnational illicit activities. This initiative is based on the guidelines of CARIN (Camden Assets Recovery Inter-Agency Network) in Europe.[31]

In November 2011, CICAD celebrated its 50th session and the 25th anniversary of the commission, in Buenos Aires, where OAS Secretary Adam Blackwell said, "Ties between drugs, crime, violence and weapons are evident and demand greater coordination among the countries of the region to be able to face them with a comprehensive and multilateral approach." Secretary Blackwell added that "drug demand reduction through education" would be one of CICAD's working pillars. During this meeting, CICAD agreed to expand some of its most innovative programs to other regions, such as Central America and the Caribbean. The project for the establishment of Drug Treatment Courts was highlighted. These courts, which seek to improve the quality of treatment to drug-dependent offenders, received strong support from the member states, several of which requested that OAS implement pilot programs in their territories. The countries also highlighted the use of scientific evidence produced in the Inter-American Drug Observatory as a pillar for the establishment of public policies in the field, and they requested that the CICAD Executive Secretariat continue working toward strengthening national information systems.[32]

Inter-American Committee against Terrorism

The OAS also has a separate entity charged with the counterterrorism mission in the Americas. In 2002, CICTE (the Inter-American Committee against Terrorism) established an Executive Secretariat within the OAS General Secretariat after the terrorist attacks of September 11, 2001. Its mission is to promote and develop cooperation among member states to prevent, combat, and eliminate terrorism in accordance with the principles of the OAS Charter and the Inter-American Convention against Terrorism. The CICTE Secretariat has developed a full range of technical assistance and capacity-building programs to help OAS member states to prevent, combat, and eliminate terrorism. CICTE oversees 10 programs divided into six broad program areas: border controls, financial controls, critical infrastructure protection, legislative assistance and consultations, crisis management exercises, and policy development and coordination. The latter program is devoted to promoting international cooperation and coordination with other international, regional, and subregional bodies, as well as the private sector.[33]

More specifically, CICTE enhances the exchange of terror-related information among member states, including the establishment of an inter-American database on terrorism issues; assists member states in drafting appropriate counterterrorism legislation; compiles bilateral, subregional, regional, and multilateral counterterrorism treaties and agreements signed by member states; promotes universal adherence to international counterterrorism conventions; enhances border cooperation and travel documentation security measures; and develops activities for training and crisis management.

These OAS multilateral security programs to counter the threat of illicit networks (including drug-trafficking organizations and terrorist groups) illustrate the potential but also the inherent challenges of multilateral cooperation. Since many of the strategies and policies to address illicit phenomena including drug, arms, and human trafficking are national, the OAS is hard-pressed to dictate solution sets to sovereign nation-states. Several countries in the region have limited resources to confront drug-trafficking organizations that are often better trained and equipped than government forces. The situation is further complicated by the problematic history of those Latin American countries that have transitioned from military dictatorships to democratic governments responsible for safeguarding the rule of law. There remain deep-seated suspicions of the role of defense and security forces, and trepidation concerning abuses of power and corruption by these forces if they are employed against illicit networks. Therefore, the OAS serves as a forum and catalyst for member states to better understand the threat of transnational organized crime and illicit networks and to share resources and best practices to foster regional security, while incorporating measures of effectiveness to evaluate the real impact of OAS programming.

On the Development Front

Central America has become the unfortunate victim of the aggressive counternarcotics campaigns of the governments of Colombia and Mexico. As illicit actors have perceived more risks in operating in Colombia and Mexico, they have stepped up their activities in Central America. Increasing flows of narcotics through Central America have been contributing to rising levels of violence and the corruption of government officials, which are weakening citizens' support for democratic governance and the rule of law. Violence is particularly intense in the "northern triangle" countries of El Salvador, Guatemala, and Honduras, which have some of the highest homicide rates in the world. Citizens of nearly every Central American nation now rank public insecurity as the top problem facing their countries. Given the transnational character of criminal organizations and their abilities to exploit ungoverned spaces, some analysts assert that insecurity in Central America poses a threat to the United States.[34]

The Central American Integration System (SICA) is a regional organization with a Secretariat in El Salvador, composed of the governments of Belize, Costa Rica, El Salvador, Guatemala, Honduras, and Panama. Its Security Commission was created in 1995 to develop and carry out regional security efforts to combat illicit networks among other transnational threats. On April 14, 2011, Secretary General of SICA Juan Daniel Aleman presented the new regional security strategy to the Group of Friends in Support of the Central American Security Strategy, which included ministers and vice ministers of defense, security, and foreign affairs including Guatemalan Defense Minister Abraham Valenzuela, U.S. Assistant Secretary of State for Western Hemisphere Affairs Arturo Valenzuela, Guatemalan Deputy Minister of the Presidency Mauricio Boraschi, Canadian Assistant Deputy Minister for the Americas Jon Allen, and Nicaraguan National Police First Commissioner Aminta Granera.

The four pillars of the regional security strategy are Prevention, Combating Crime, Rehabilitation and Reintegration, and Institutional Strengthening. This strategy is based on the premise that security in Central America is a shared responsibility, and one of the key principles is that it implies an interconnected chain between local, national, and regional actions from a perspective of resource allocations and should take advantage of synergies, according to Aleman.[35] SICA's strategy will require close cooperation between Central American countries and the international community to achieve success in the fight against violence and is yet another example of multilateral partnerships to combat the convergence of threats from illicit networks.

The United States recognizes that its efforts in Colombia and Mexico have provided incentives for criminal groups to move into Central America and other areas where they can exploit institutional weaknesses to continue their operations. In response, the Obama administration has made ensuring the safety and security of all citizens one of the four overarching priorities of U.S. policy in Latin America and has sought to develop collaborative partnerships with countries throughout the hemisphere. These partnerships have taken the form of bilateral security cooperation with countries such as Colombia and Mexico, as well as regional programs such as the Caribbean Basin Security Initiative and the Central America Regional Security Initiative (CARSI).[36]

Central America Regional Security Initiative

On the diplomatic and development front, the U.S.-sponsored CARSI seeks to promote citizen security and socioeconomic development as a direct response to the growing threat of illicit networks. CARSI was originally created in 2007 as part of the Mexico-focused counterdrug and anticrime assistance package known as the Mérida Initiative.[37] CARSI not only provides equipment, training, and technical assistance to support immediate law enforcement and interdiction operations, but also seeks to strengthen the capacities of governmental institutions to address security challenges and the underlying economic and social conditions that contribute to them.[38]

The five goals of CARSI are to:

1. create safe streets for the citizens in the region

2. disrupt the movement of criminals and contraband within and between the nations of Central America

3. support the development of strong, capable, and accountable Central American governments

4. reestablish effective state presence and security in communities at risk

5. foster enhanced levels of security and rule of law coordination and cooperation between the nations of the region.[39]

CARSI's collaborative efforts to combat illicit networks in Central America include programming in the narcotics interdiction, law enforcement support, institutional capacity-building, and prevention arenas. In addition to traditional "train and equip" programs, U.S. assistance provided through CARSI also supports specialized law enforcement units that are vetted by, and work with, U.S. personnel to investigate and disrupt the operations of transnational gangs and trafficking networks. Led by the Federal Bureau of Investigation (FBI), Transnational Anti-Gang (TAG) units, which were first created in El Salvador in 2007, are now expanding to Guatemala and Honduras with CARSI support. Similarly, Drug Enforcement Administration, Immigration and Customs Enforcement, and the State Department's INL also have vetted unit programs throughout Central America. Among other activities, they conduct complex investigations into money laundering, bulk cash smuggling, and the trafficking of narcotics, firearms, and persons.[40]

In Central America through the CARSI program, the International Law Enforcement Training Academy in El Salvador trained approximately 450 law enforcement officers from the seven CARSI countries in 2010. In just 3 months of 2010, the TAG unit in El Salvador handled 141 investigative leads and disseminated information to domestic and international law enforcement agencies. Customs and Border Protection, working with CARSI national border forces, conducted assessments of more than 30 land, sea, and air entry points throughout the region and has provided training using nonintrusive inspection equipment provided by the State Department.[41]

State Department/INL and the United States Agency for International Development (USAID) community-policing programs are designed to build institutional capacity and local confidence in police forces by converting them into more community-based, service-oriented organizations. One such program, the *Villa Nueva* model precinct in Guatemala, is being replicated with CARSI funding as a result of its success in establishing popular trust and reducing violence. To improve the investigative capacity of Central American nations, CARSI has supported assessments of forensic laboratories, the implementation of the Bureau of Alcohol, Tobacco, Firearms, and Explosives' eTrace system to track firearms, and the expansion of the FBI's Central America Fingerprint Exchange, which assists partner nations in developing fingerprint and biometric capabilities. CARSI also seeks to reduce impunity by improving the efficiency and effectiveness of Central American judicial systems.[42]

In 2010, USAID continued its efforts in crime and violence prevention, working with local and national governments, civil society, and community leaders to build comprehensive prevention approaches and provide opportunities for youth who are at risk of becoming involved in the narcotics trade and substance abuse. Through its management of CARSI funds in the Economic and Social Development Fund for Central America, USAID supports prevention programs by providing educational, recreational, and vocational opportunities for at-risk youth. In El Salvador, for example, USAID's Community-Based Crime and Violence Prevention Project works in 12 municipalities to strengthen the capacities of local governments, civil organizations, community leaders, and youth to address the problems of crime and violence.[43] These comprehensive political, judicial, economic, and social programs under CARSI are intended to reinforce the rule of law, economic development, and democratic institutions. They are aimed at addressing the complex threats posed by illicit networks by working with different segments of the society.

Perhaps one of the greatest challenges for this important mission in Central America is U.S. funding in an age of fiscal austerity and the rate at which the programs are being implemented. Since fiscal year (FY) 2008, the United States has provided Central America with $361.5 million through Mérida/CARSI, and the Obama administration has requested an additional $100 million for CARSI in FY2012.[44] The Department of State has not released any information since March 2011, at which point 88 percent of the funds appropriated between FY2008 and FY2010 had been obligated and 19 percent had been expended.[45]

At the June 2011 SICA conference, Secretary of State Hillary Clinton announced that U.S. funding for the Central America Citizen Security Partnership would exceed $290 million in FY2011.[46] The $290 million pledge includes the $101.5 million being provided through CARSI as well as all other bilateral and regional U.S. assistance being provided to support security efforts in the region in FY2011. Many experts on Central America are concerned that this will be insufficient to help Central American nations to realize rule of law, prosperity, and democracy in the face of the formidable illicit networks operating in the region. Not only do

Central American countries lack financial resources; they also require human resources and capacity to implement these security and development programs.

On the Defense and Security Front

In the Western Hemisphere, narcotics trafficking and its associated violence represent the gravest threat from illicit networks and pose a formidable challenge for government, defense, and security forces. For decades, Washington has invested significant resources in both the demand and supply side of counternarcotics efforts known as the "war on drugs." The United States and its partner nations have made significant strides in interdicting illicit drug trafficking in the Americas, but the international drug trade continues to flourish. In June 2011, the Global Commission on Drug Policy, whose members include former UN Secretary General Kofi Annan, three former Latin American presidents (of Brazil, Mexico, and Colombia), former Federal Reserve chairman Paul Volcker, former U.S. Secretary of State George Schultz, Virgin Group Founder Richard Branson, and Greek Prime Minister George Papandreou declared that the global war on drugs has failed after 40 years.[47]

USSOUTHCOM's Counter Illicit Trafficking Mandate

The United States Southern Command (USSOUTHCOM), located in Miami, Florida, is one of nine joint unified combatant commands in the Department of Defense. It is responsible for providing contingency planning, operations, and security cooperation for Central and South America, the Caribbean (except U.S. commonwealths, territories, and possessions), and Cuba, as well as force protection of U.S. military resources at these locations. It is also responsible for ensuring the defense of the Panama Canal and canal area. According to the "2020 Command Strategy: Partnership for the Americas," issued in July 2010, USSOUTHCOM's strategic objectives are to defend the United States and its interests, foster regional security, and be an enduring partner of choice in support of a peaceful and prosperous region. Its areas of focus are Counter Illicit Trafficking (CIT), Humanitarian Assistance/Disaster Relief, and Peacekeeping Operations.

The primary mission of USSOUTHCOM's CIT efforts is to support the interdiction of drug trafficking. USSOUTHCOM collaborates with other agencies and nations to support interdiction of transnational criminal organizations through detection and monitoring, information-sharing, and partner nation capacity-building. All efforts are focused to achieve U.S. National Drug Control Strategy interdiction goals.

The Department of Defense is the lead Federal agency in efforts to detect and monitor aerial and maritime transit of illegal drugs toward the United States. Joint Interagency Task Force–South (JIATF-South) is the national task force that serves as the catalyst for integrated and synchronized interagency counter-illicit trafficking operations. It is responsible for the detection and monitoring of suspect air and maritime drug activity in the Caribbean Sea, Gulf of Mexico, and the eastern Pacific. JIATF-South also collects, processes, and disseminates counter-drug information for interagency and partner nation operations. Using information gathered by JIATF-South–coordinated operations, U.S. law enforcement agencies and partner nations take the lead in interdicting drug runners. U.S. military interdiction involvement, if any, is in support of those law enforcement agencies. Typically, U.S. military personnel are involved in supporting an interdiction during maritime operations in international waters, where U.S. Navy ships and helicopters patrol and intercept suspected traffickers. The actual interdictions—boarding, search, seizures, and arrests—are led and conducted by embarked U.S. Coast Guard Law Enforcement Detachments or partner nation drug law enforcement agencies.

The U.S. military commits a variety of forces in the region to support detection and monitoring efforts:

+ Maritime: Normally, U.S. Navy, U.S. Coast Guard, and partner nation (British, French, Dutch, Canadian, and Colombian) ships patrol the waters in the Caribbean Sea, Gulf of Mexico, and the eastern Pacific on a year-round basis. Embarked on U.S. and at times allied nation naval vessels are Coast Guard Law Enforcement Detachments who take the lead during operations to board suspected vessels, seize illegal drugs, and apprehend suspects.

+ Air: JIATF-South utilizes U.S. military, interagency, and partner nation aircraft that are strategically located throughout the region and at two Cooperative Security Locations in Comalapa, El Salvador, and in Curacao and Aruba, formerly part of the Netherlands Antilles. These aircraft, in cooperation with partner nations and U.S. agencies, fly persistent missions to monitor areas with a history of illicit trafficking. The U.S. aircraft offer unique surveillance capabilities that complement the counter–illicit trafficking efforts of U.S. and partner nation law enforcement agencies.

+ Other: USSOUTHCOM also provides support to partner nations through training, information-sharing, and technological and resource assistance.

JIATF-South: Structure and Operations

Focused on countering illicit trafficking, JIATF-South exemplifies perhaps the best model of interagency and international efforts to combat the convergence of illicit networks, with a proven 20-year track record. JIATF-South is a combined military-civilian task force charged with the mission to combat illicit trafficking in Latin America and the Caribbean. While JIATF-South reports to USSOUTHCOM, it pursues its four-pronged mission of detection, monitoring, interdiction, and apprehension with international partners.[48] Over the past 20 years, JIATF-South has arrested some 4,600 traffickers, captured nearly 1,100 vessels, and deprived drug cartels of an estimated $190 billion in profits.[49] Former USSOUTHCOM Commander and current Supreme Allied Commander Europe Admiral James Stavridis considers JIATF-South "a national treasure" and the "crown jewel of Southcom"[50] that demonstrates a whole-of-government approach to converging transnational threats such as illicit networks.

JIATF-South is a joint, interagency, international, combined, and allied team comprised of professionals from all four Services of the U.S. military, nine U.S. agencies (including intelligence and law enforcement), and 13 partner nations including Argentina, Brazil, Chile, Colombia, Dominican Republic, Ecuador, El Salvador, France, Mexico, the Netherlands, Peru, Spain, and the United Kingdom.[51] JIATF-South's area of responsibility covers nearly 42 million square miles. Using information from law enforcement agencies, JIATF-South detects and monitors suspect aircraft and maritime vessels in the Caribbean Basin and eastern Pacific and then provides this information to international and interagency partners who have the authority to interdict illicit shipments and arrest members of transnational criminal organizations. In 2010, JIATF-South and international and interagency partners were directly responsible for interdicting 142 metric tons of cocaine and 3,419 pounds of marijuana and making 309 arrests, denying transnational criminal organizations an estimated $2.8 billion in revenue.[52]

Figure 3. JIATF–South's Integrated Team

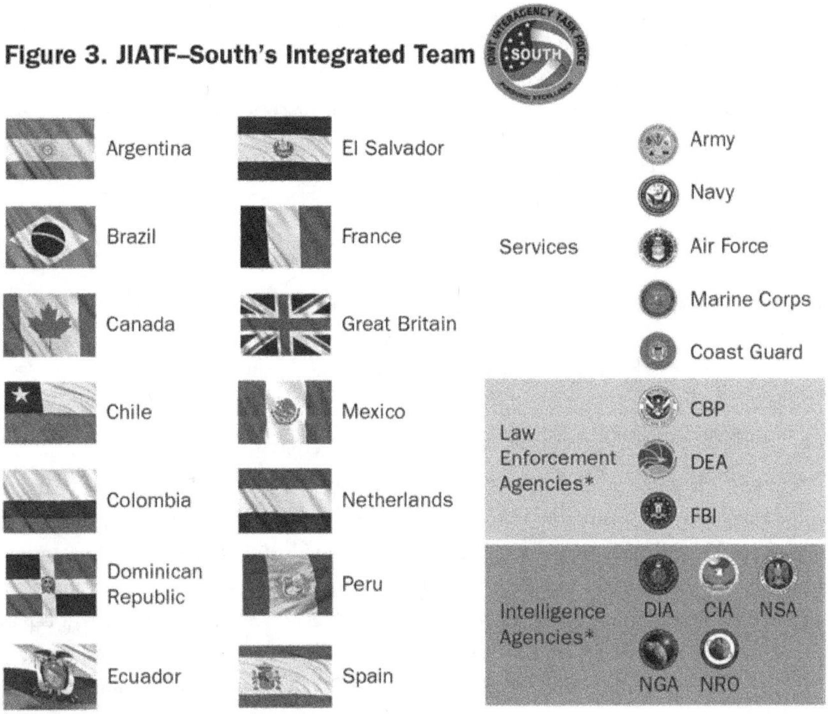

Argentina	El Salvador	Army
Brazil	France	Navy
Canada	Great Britain	Services — Air Force
Chile	Mexico	Marine Corps
Colombia	Netherlands	Coast Guard
Dominican Republic	Peru	Law Enforcement Agencies* — CBP, DEA, FBI
Ecuador	Spain	Intelligence Agencies* — DIA, CIA, NSA, NGA, NRO

*Not all participating law enforcement and intelligence organizations are represented here.

According to the 2011 publication *Joint Interagency Task Force–South: The Best Known, Least Understood Interagency Success,* the accomplishments of JIATF-South over the past two decades can be attributed to the following lessons learned:

+ *Get a mandate from higher authority.* The mission and team had sufficient legitimacy with a clear mandate from senior civilian and military authorities.

+ *Tailor a holistic solution set to a discrete problem.* The mission was discrete and clearly identified (stop drug trafficking from entering the United States) with measurable outcomes (that is, number of arrests, interdictions, vessels boarded, and drug confiscations).

+ *Know your partners.* A better understanding of partners, their interests, capabilities, and limitations fostered true collaboration and cooperation and unity of effort.

+ Get resources.

+ Build networks.

The report also noted several mistakes to avoid:

+ don't command the presence of interagency personnel on your team

+ don't segregate interagency staff in separate buildings

+ don't disrespect smaller partners; they can make big contributions

+ don't demand binding agreements on cooperation (at least initially)

+ don't ignore any partner's need to feel they make a contribution

+ don't make binding decisions without substantial vetting and support

+ don't forget to build a culture of trust and empowerment

+ don't take credit for collaborative success.[53]

JIATF-South is considered one of the best models of interagency and international cooperation to counter transnational threats. It is a team that blends experience, professionalism, and knowledge that is greater than the sum of its individual parts, according to Admiral Stavridis.[54] JIATF-South has executed successful interdiction operations that illustrate how a culture of collaboration that incorporates military, intelligence, and law enforcement capabilities can combat the convergence of threats posed by illicit networks. However, it is not apparent how easily this model with over 20 years of experience can be replicated in terms of human and financial resources as well as collaborative culture.

Fostering Collaborative Models to Combat Illicit Networks

Illicit networks that include criminals, terrorists, and facilitators have brokered strategic alliances to promote their interests, threatening the rule of law, global supply chains, and free and fair markets around the world. To counter the convergence of these threats, governments need to develop interagency and international strategies that leverage the diplomatic, development, intelligence, military, and law enforcement instruments of national power. To this end, collaborative models for security and development require the following critical elements: political will, institutions, mechanisms to assess threats and deliver countermeasures, resources, and measures of effectiveness to ensure success against illicit networks.

All of the international and interagency initiatives examined in this chapter demonstrate the political will to combat illicit networks. In 2000, the United Nations formally recognized

Figure 4. To Combat Illicit Networks
Foster Collaboration and Communities of Interest at the National, Regional, and International Levels

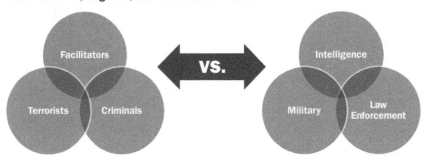

Figure 5. Critical Elements for Effective Collaborative Models

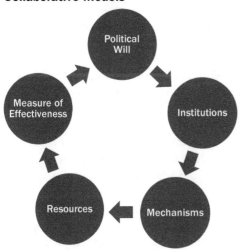

the threat from illicit networks with the adoption of the UN Convention against Transnational Organized Crime and has engaged in many activities to combat drug trafficking and organized crime through the UN Office on Drugs and Crime. On the U.S. front, the 2011 *Strategy to Combat Transnational Organized Crime* represents a significant development in acknowledging the national security threat and advocating interagency collaboration and international cooperation to counter global crime. However, an implementation plan with specific undertakings for the U.S. interagency and measures of effectiveness has yet to be published as of this writing. Publicly realizing the threat from illicit networks to the peace and prosperity of nation-states and global markets is not enough. These international conventions, agreements, and strategies must be accompanied with action plans and be adequately resourced in order to restrict illicit actors' operating environment and enablers.

Table 1. Critical Elements for Effective Collaborative Models in the Western Hemisphere

Collaborative Model	Political Will	Institutions	Mechanisms	Resources	Measures of Success
U.S. Counter–Transnational Organized Crime	Yes	Yes	Yes	Yes	Some
United Nations Office on Drugs and Crime	Yes	Yes	Yes	Yes	Unclear
Organization of American States	Yes	Yes	Some	Some	Unclear
Central America Regional Security Initiative	Yes	Yes	Limited	Insufficient	Unclear
Joint Interagency Task Force–South	Yes	Yes	Yes	Yes	Yes

In the Western Hemisphere, diplomatic, development, and defense initiatives to combat illicit networks are underway through the UN, OAS, CARSI, and USSOUTHCOM's JIATF-South as described above. These collaborative models focus on the operational and preventive countermeasures to confront illicit actors and promote security and development. Current efforts under the auspices of multilateral institutions such as UNODC and OAS are noteworthy and laudable, but they lack measures of effectiveness that make it difficult to determine their real countercrime impact and their ability to establish rule of law and socioeconomic opportunities. Meanwhile, CARSI will require more financial and human resources to carry out law enforcement and socioeconomic development programs to combat transnational criminal organizations. In contrast, JIATF-South has enjoyed a 20-year track record of progress against illicit trafficking. It has all the critical elements of a collaborative model: commitment, institutions, mechanisms, resources, and metrics, as well as a strong interagency and international culture that has developed over its lifetime. All of these initiatives to combat illicit networks in the Americas are ongoing but appear to be executed in parallel. In this resource-constrained environment, greater donor coordination would ensure that all these security and development programs complement rather than duplicate each other. Given the level of political will demonstrated across these organizations, there should be an interest in allocating resources and building capacity in the specific geographic areas most vulnerable to transnational organized crime.

Globalization has benefited the private, public, and civic sectors of society over the past few decades. We have witnessed extraordinary rates of economic growth with significant advances in international trade, capital markets, science, and technology. Meanwhile, illicit networks have leveraged the processes and new opportunities arising from globalization and capitalized on weak institutions and gaps in governance around the world to expand their enterprises. In the Western Hemisphere, illicit networks have been increasingly engaged in drug, arms, counterfeiting, and human-trafficking activities that have been accompanied by unprecedented levels of violence. Consequently, the threat of transnational organized crime and illicit networks has become a national security concern to the United States and its allies.

Since illicit networks threaten citizen safety and economic security, governments must engage the public, private, and civic sectors to diminish and defeat the power and influence of transnational organized crime and illicit actors. As John Aquilla, at the Naval Postgraduate School, explains, "It takes a network to defeat a network." In that vein, governments around the world must build networks that promote interagency and international collaboration to combat the convergence of illicit networks and the national security threats they pose to the international community. To ensure success, the initiatives undertaken to combat illicit networks must be supported by political commitment, institutions, mechanisms, financial and human resources, and measures of effectiveness to promote security and development.

Disclaimer

The views expressed in this chapter are those of the author and do not reflect the official policy or position of the Center for Hemispheric Defense Studies, the National Defense University, the Department of Defense, or the U.S. Government.

Notes

[1] James G. Stavridis, *Partnership for the Americas: Western Hemisphere Strategy and U.S. Southern Command* (Washington, DC: National Defense University Press, 2010), 84.

[2] Celina B. Realuyo, "Securing Global Supply Chains in an Age of Uncertainty: Focus on Food Safety," *George Mason University Center for Critical Infrastructure and Homeland Security Critical Infrastructure Report* 10, no. 1 (July 2011), available at <http://cip.gmu.edu/archive/CIPHS_TheCIPReport_July2011_SupplyChain.pdf>.

[3] Ibid.

[4] *National Security Strategy* (Washington, DC: The White House, May 2010), 49, available at <www.whitehouse. gov/sites/default/files/rss_viewer/national_security_strategy.pdf>.

[5] Ibid.

[6] *Strategy to Combat Transnational Organized Crime: Addressing Coverging Threats to National Security* (Washington, DC: The White House, July 2011), available at <www.whitehouse.gov/sites/default/files/Strategy_to_Combat_Transnational_Organized_Crime_July_2011.pdf>.

[7] Ibid.

[8] Michael J. Caden, "National Debt Poses Security Threat, Mullen Says," American Forces Press Service, August 28, 2010, available at <www.defense.gov/news/newsarticle.aspx?id=60621>.

[9] United Nations Convention against Transnational Organized Crime, December 2000.

[10] United Nations Office on Drugs and Crime (UNODC), "Promoting Health, Security, and Justice: Cutting the Threads of Drugs, Crime, and Terrorism," 2010 Annual Report, 19, available at <www.unodc.org/documents/frontpage/UNODC_Annual_Report_2010_LowRes.pdf>.

[11] Ibid.

[12] Ibid., 55.

[13] UNODC, "About UNODC," available at <www.unodc.org/unodc/en/about-unodc/index.html?ref=menutop>.

[14] Ibid.

[15] UNODC, "Funds and Partners," available at <www.unodc.org/unodc/en/donors/index.html>.

[16] UNODC, "UNODC 2010 Annual Report Released," April 8, 2010, available at <www.unodc.org/unodc/en/frontpage/2010/April/unodc-2010-annual-report-released.html>.

[17] UNODC, "Promoting Health, Security, and Justice," 22.

[18] Ibid.

[19] UNODC, "UNODC Head Inaugurates Centre of Excellence in Statistics in Mexico," October 1, 2011, available at <www.unodc.org/unodc/en/frontpage/2011/October/unodc-head-inaugurates-statistics-centre-of-excellence-in-mexico.html?ref=fs4>.

[20] Organization of American States (OAS), "Hemispheric Plan of Action against Transnational Organized Crime," Res. 908, October 25, 2006, available at <www.oas.org/consejo/resolutions/res908.asp>.

[21] OAS Committee on Hemispheric Security, "Transnational Organized Crime as a Multidimensional Threat: Promoting Full Implementation of the Hemispheric Plan of Action against Transnational Organized Crime," concept paper roundtable discussion, April 18–19, 2005, available at <www.oas.org/csh/english/TOC.asp>.

[22] "OAS meets to analyze progress against organized crime," *Infosur Hoy*, November 17, 2011, available at <http://infosurhoy.com/cocoon/saii/xhtml/en_GB/newsbriefs/saii/newsbriefs/2011/11/17/newsbrief-05>.

[23] "OAS analyses hemispheric progress against transnational organized crime," *Caribbean News Now*, November 18, 2011, available at <www.caribbeannewsnow.com/headline-OAS-analyses-hemispheric-progress-against-transnational-organized-crime-8793.html>.

[24] Inter-American Drug Abuse Control Commission (CICAD), "Mission Statement," available at <www.cicad.oas.org/Main/Template.asp?File=/Main/AboutCICAD/about_eng.asp>.

[25] OAS/CICAD, "2010 Annual Report of the Inter-American Drug Abuse Control Commission to the 41st Regular Session of the General Assembly of the OAS," available at <www.cicad.oas.org/AnnualReports/2010/1861-AnualReport_ENG-rev1.pdf>.

[26] Ibid.

[27] U.S. Department of State, Bureau for International Narcotics and Law Enforcement Affairs, "International Narcotics Control Strategy Report 2011," vol. 1, March 2011, 15, available at <www.state.gov/documents/organization/156575.pdf>.

[28] Ibid., 15.

[29] James F. Mack, "New Hemispheric Drug Strategy," OAS/CICAD, June 2010, available at <www.cicad.oas.org/en/basicdocuments/Hemispheric%20Drug%20Strategy100603.pdf>.

[30] OAS/CICAD, "Hemispheric Plan of Action on Drugs, 2011–2015," May 4–6, 2011, available at <www.cicad.oas.org/apps/Document.aspx?Id=1091>.

[31] U.S. Department of State, Bureau for International Narcotics and Law Enforcement Affairs, "International Narcotics Control Strategy Report 2011," vol. 2, March 2011, 29, available at <http://photos.state.gov/libraries/mauritius/320773/PDF_documents/incsr_2011.pdf>.

[32] OAS/CICAD, "Next Multilateral Evaluation Mechanism Round gets framework for evaluation," available at <www.cicad.oas.org/Main/Template.asp?File=/main/news/update_eng.asp>.

[33] OAS Inter-American Committee Against Terrorism (CICTE), "History," available at <http://cicte.oas.org/rev/en/about/History.asp>.

[34] Peter J. Meyer and Clare Ribando Seelke, "Central America Regional Security Initiative: Background and Policy Issues for Congress," *Congressional Research Service*, R41731, March 30, 2011, available at <www.fas.org/sgp/crs/row/R41731.pdf>.

[35] "SICA presents Central American citizen security strategy to the international community," *Inter-American Development Bank News*, April 14, 2011, available at <www.iadb.org/mobile/news/detail.cfm?lang=en&id=9345>.

[36] Testimony of Arturo A. Valenzuela, Assistant Secretary of State for Western Hemisphere Affairs, U.S. Department of State, before the Senate Subcommittee on the Western Hemisphere, Peace Corps, and Global Narcotics Affairs, February 17, 2011.

[37] Meyer and Seelke, "Central America Regional Security Initiative," March 30, 2011.

[38] *U.S. National Drug Control Strategy* (Washington, DC: The White House, 2011), 76, available at <www.whitehouse.gov/sites/default/files/ondcp/ndcs2011.pdf>.

[39] U.S. Department of State, Bureau of Public Affairs, "The Central America Regional Security Initiative: A Shared Partnership Fact Sheet," August, 5, 2010, available at <www.state.gov/r/pa/scp/fs/2010/145747.htm>.

[40] Meyer and Seelke, 25.

[41] "U.S. National Drug Control Strategy," 2011, 78.

[42] Meyer and Seelke, 25.

[43] Ibid.

[44] Ibid.

[45] Ibid., 28.

[46] Hillary Clinton, "Remarks at the Central American Security Conference (SICA)," U.S. Department of State, June 22, 2011.

[47] "Global war on drugs 'has failed' say former leaders," BBC, June 2, 2011, available at <www.bbc.co.uk/news/world-us-canada-13624303>.

[48] Dialogo staff, "The Tip of the Spear: JIATF-South in Action," 2010, available at <www.dialogo-americas.com/en_GB/articles/rmisa/features/viewpoint/2010/10/01/feature-02>.

[49] Stavridis, 68.

[50] Ibid., 69.

[51] Ibid., 68.

[52] Douglas M. Fraser, Commander, U.S. Southern Command, "Posture Statement before the 112th Congress House Armed Services Committee," March 30, 2011, available at <www.southcom.mil/newsroom/Documents/SOUTHCOM%202011%20Posture%20Statement.pdf>.

[53] Evan Munsing and Christopher J. Lamb, *Joint Interagency Task Force–South: The Best Known, Least Understood Interagency Success*, INSS Strategic Perspectives No. 5 (Washington, DC: National Defense University Press, 2011), 83–85, available at <www.ndu.edu/press/lib/pdf/strategic-perspectives/Strategic-Perspectives-5.pdf>.

[54] Stavridis, 68.

About the Contributors

Michael Miklaucic is the director of research, information, and publications in the Center for Complex Operations (CCO), Institute for National Strategic Studies, at the National Defense University. He is also the editor of *PRISM*, the journal of CCO. Prior to this assignment, Mr. Miklaucic served in various positions at the U.S. Agency for International Development (US-AID) and Department of State, including USAID representative on the Civilian Response Corps Inter-Agency Task Force; senior program officer in the USAID Office of Democracy and Governance; and rule-of-law specialist in the Center for Democracy and Governance. During 2002–2003, he served as Department of State deputy for war crimes issues. He received his education from the University of California, London School of Economics, and School for Advanced International Studies. Mr. Miklaucic is an adjunct professor of U.S. foreign policy at American University and adjunct professor of conflict and development at George Mason University.

Jacqueline Brewer is an analyst in the Center for Complex Operations, Institute for National Strategic Studies, at the National Defense University. She is responsible for a wide range of activities including research and project development targeted at preparing U.S. Government and multinational partners for interagency operations. She previously worked in the private sector supporting the State Department's Bureau of Near East Affairs/Iraq and the U.S. National Security Council on Iraq rule-of-law issues. She has also worked with various Department of Defense educational organizations to coordinate and design curriculum for operational training programs. Ms. Brewer received a B.A. in international studies and an M.A. in international affairs, both from American University.

Gary Barnabo is a consultant with Booz Allen Hamilton. He advises the Office of the Deputy Assistant Secretary of Defense for Counternarcotics and Global Threats on policy and strategy issues related to transnational organized crime and illicit networks. Mr. Barnabo has supported clients in the Office of the Secretary of Defense, Department of Homeland Security, and the intelligence community by analyzing and designing policy approaches to complex national security challenges. He also serves as the president of Young Professionals in Foreign Policy, a Washington, DC–based nonpartisan, nonprofit organization that fosters the next generation of foreign policy leaders.

Duncan Deville leads the commercial Anti-Money Laundering/Counter-Terrorist Financing Compliance Program (AML/CTF) at Booz Allen Hamilton. Previously, he was a senior financial enforcement advisor with the Treasury Department's Office of International Affairs and mentored various units on compliance with AML/CTF standards. He served with the Department of Defense in Baghdad, leading rule-of-law efforts by training Iraqi judges and prosecutors until sovereignty was transferred. Prior to this, he was a visiting scholar at Harvard Law School and consultant to the RAND Corporation on criminal justice matters. Mr. Deville served as a prosecutor for over a dozen years, including as assistant U.S. attorney with the Organized Crime Strike Force in Los Angeles. He also held long-term posts in Moscow and Yerevan, Armenia, for the Justice Department. Mr. Deville is an Adjunct Professor of Law at Georgetown University, with graduate degrees from Oxford, Harvard, and the University of Denver, and an undergraduate degree from the University of Louisiana. His articles have appeared in the law reviews of Columbia University, The George Washington University, Harvard, Stanford, and University of Chicago, as well as in the *Boston Globe, Moscow Times, The New York Times, Washington Times*, and numerous business publications.

Douglas Farah is a senior fellow at the International Assessment and Strategy Center and adjunct fellow at the Center for Strategic and International Studies. In his 18 years as a foreign correspondent and investigative reporter for *The Washington Post*, he covered the drug war in the Andean region, the Medellín and Cali cartels, the Colombian and Mexican drug mafias, Russian organized crime in Latin America and the Caribbean, Mexican drug cartels in the United States, and drug-related banks in the Caribbean. He has also written on the civil wars in Sierra Leone and Liberia, the network of agents who profited from those conflicts and the diamonds-for-weapons trade, al Qaeda's ties to the networks, and the financial network of Osama bin Laden. Mr. Farah has a B.A. in Latin American studies and B.S. in journalism from the University of Kansas. As bureau chief in El Salvador for United Press International, he covered the civil war there and U.S.-backed Contra rebels in Honduras, winning numerous prizes for investigative work on death squads and drug trafficking. He wrote *Blood from Stones: The Secret Financial Network of Terror* (Broadway Books, 2004) and, with Stephen Braun, *Merchant of Death: Money, Guns, Planes and the Man Who Makes War Possible* (J. Wiley and Sons, 2007).

Vanda Felbab-Brown is an expert on international and internal conflict issues and their management, including counterinsurgency. Her research centers on the interaction between illicit economies and military conflict. Dr. Felbab-Brown is a fellow in foreign policy and a fellow in the 21st Century Defense Initiative at the Brookings Institution where she focuses on South Asia, Burma, the Andean region, Mexico, and Somalia. She is the author of *Shooting Up: Counterinsurgency and the War on Drugs* (Brookings Institution Press, December 2009), as well as numerous articles and reports on illicit economies, crime, insurgency, and state-building around the world. She has conducted field work in Afghanistan, Burma, Colombia, Mexico, Burma, Brazil, India, and Morocco. A regular commentator in U.S. and international media, Dr. Felbab-Brown frequently testifies before Congress. She received a Ph.D. in political science from the Massachusetts Institute of Technology and a B.A. from Harvard University.

Nils Gilman is with the consulting firm Monitor 360, where he focuses on economic development and national security. He has led projects on topics such as the security implications of climate change, diverse cultures of hackers, and the global narcotics trade. Prior to joining Monitor 360 in 2006, Dr. Gilman spent 6 years leading competitive intelligence and product marketing teams at enterprise software companies. He holds a B.A., M.A., and Ph.D. in intellectual history from the University of California at Berkeley. He is the author of *Mandarins of the Future* (The Johns Hopkins University Press, 2003) and coeditor of *Deviant Globalization* (The Continuum International Publishing Group, 2011), which explores how globalized black market economies are challenging traditional state authority. Dr. Gilman is also coeditor of *Humanity*, an international journal of human rights, humanitarianism, and development.

Jesse Goldhammer is a partner at the Monitor Group and the firm Monitor 360, where he helps public- and private-sector clients engage in strategic, analytic, organizational, and institutional transformation. Dr. Goldhammer has spent the past 20 years bringing together unique people, ideas, and approaches to devise lasting and effective solutions to vexing problems. These solutions include developing novel analytic approaches to understand and reframe client challenges, using human networks to leverage alternative and unorthodox perspectives, and designing training programs to propagate new strategies and tradecraft. Having come to Monitor through Global Business Network, Dr. Goldhammer is also an expert in scenario planning, has taught scenario-planning courses, and published "Four Futures for China Inc." in *Business 2.0*. He holds a B.A. in social science from the University of California at Berkeley, an M.A. in political science from New York University, and a Ph.D. in political science from Berkeley. An accomplished instructor and expert in modern political theory, Dr. Goldhammer is the author of *The Headless Republic* (Cornell University Press, 2005) and coeditor of *Deviant Globalization* (The Continuum International Publishing Group, 2011).

Patrick Radden Keefe is a staff writer at *The New Yorker* and fellow at the Century Foundation where his research and writing focus on transnational crime, international security, and foreign affairs. He is the author of *The Snakehead: An Epic Tale of the Chinatown Underworld and the American Dream* (Knopf-Doubleday, 2009) and *Chatter: Uncovering the Echelon Surveillance Network and the Secret World of Global Eavesdropping* (Random House, 2005), along with numerous book chapters on espionage, national security, and illicit networks. From 2010 to 2011, he served as a special advisor in the Office of the Secretary of Defense for Policy where he focused on crime, corruption, and rule-of-law issues. His articles and op-eds have appeared in *The New York Times*, *The New York Review of Books*, *Legal Affairs*, *Boston Globe*, *WIRED*, *World Policy Journal*, and others. A former Marshall scholar and Guggenheim fellow, Mr. Keefe received an M.A. in international relations from Cambridge University, M.A. in new media and information systems from the London School of Economics, and J.D. from Yale Law School.

Danielle Camner Lindholm is a technical director in the BAE Systems Intelligence and Security Sector. As an expert in the economic and commercial aspects of national security, Ms. Lindholm plays a strategic role in activities related to hybrid/transnational threats, including threat finance, illicit networks, and anti-money-laundering issues. She has also worked as country manager for

Middle East and North Africa regions for the U.S. Trade and Development Agency; special assistant to the Under Secretary of State for Economic, Business, and Agricultural Affairs; advisor for international economic policy at the National Economic Council; and senior policy advisor to a Congressman, responsible for financial services, national and homeland security, and foreign affairs. Before joining BAE Systems, Ms. Lindholm was vice president for policy at Business Executives for National Security, where she worked with Treasury, the Defense Department, Drug Enforcement Administration, combatant commands, and the private sector to develop solutions and share information on potential vulnerabilities in the financial services sector. Ms. Lindholm holds a B.A. from Tufts University and a J.D. from the University of Miami. She is a member of the District of Columbia and Florida Bars and cochairs the American Bar Association's International Anti–Money Laundering Committee.

David M. Luna is director of Anti-Crime Programs at the Department of State's Bureau of International Narcotics and Law Enforcement Affairs (INL). Mr. Luna helps coordinate diplomatic initiatives on national security that disrupt and dismantle transnational criminal networks. He leads INL teams to combat international organized crime, corruption/kleptocracy, money laundering, terrorist financing, intellectual property rights and cyber-crimes, and smuggling/trafficking crimes that impact U.S. homeland security. Over 12 years at the Department of State, Mr. Luna has chaired numerous U.S. interagency working groups, National Security Council policy subcommittees, and multilateral policy coordinating groups. His previous experience includes professional stints with Congress and U.S. executive agencies responsible for public integrity investigations, as well as legal, legislative, domestic policy, foreign affairs, national security, and international economic development issues. A graduate of the U.S. Army War College, he received a B.A. from the University of Pennsylvania and a J.D. from The Columbus School of Law at the Catholic University of America.

Moisés Naím is senior associate in the international economics program at the Carnegie Endowment for International Peace and chief international columnist for *El País*, Spain's largest newspaper. Dr. Naím has written extensively on international economics and global politics, economic development, and the unintended consequences of globalization including books such as *Illicit: How Smugglers, Traffickers, and Copycats Are Hijacking the Global Economy* (Doubleday, 2006). He was Venezuela's minister of Trade and Industry, director of Venezuela's Central Bank, and executive director of the World Bank. He has held positions as professor of business and economics, dean of Venezuela's main business school, the Instituto de Estudios Superiores de Administración, and guest lecturer in many U.S. and European universities. Dr. Naím was editor-in-chief of *Foreign Policy* magazine. He is the chairman of the board of both the Group of Fifty (G-50) and Population Action International and a member of the board of directors of the National Endowment for Democracy and the International Crisis Group. He holds an M.A. and Ph.D. from the Massachusetts Institute of Technology.

Justin Picard is chief scientist at Advanced Track & Trace in France. He has invented disruptive technologies that protect billions of products and documents from counterfeiting and tampering. Mr. Picard has written numerous papers and patents on document security, anticounterfeiting, artificial intelligence, and illicit trade, and is a frequent speaker on anticounterfeiting technologies. In 2009, Mr. Picard and his company were selected as a World Economic Forum Technology Pioneer. Since 2009, he has been a member of the World Economic Forum Global Agenda Council on Illicit Trade, and is also a member of the Evian Group. Mr. Picard manages the Web site illicittrade.org, a compilation of information on illicit trade and of the work of the Global Agenda Council on Illicit Trade.

Celina B. Realuyo is an assistant professor of national security affairs in the Center for Hemispheric Defense Studies at the National Defense University. Ms. Realuyo teaches courses on U.S. national security, globalization, counterterrorism, and transnational criminal organizations. She is an expert on geopolitical risks in the 21st century, global supply chains, international financial systems, terrorist financing, and money laundering. Ms. Realuyo served as the director of counterterrorism finance programs at the State Department's Office of the Coordinator for Counterterrorism. Prior to returning to Washington, DC, Ms. Realuyo was a private banker with Goldman Sachs in London. She began her career as a U.S. Foreign Service Officer with overseas assignments in Madrid, Panama, and the U.S. Mission to the North Atlantic Treaty Organization in Brussels, and served in the State Department's Operations Center, National Security Council's White House Situation Room, and as special assistant to the Secretary of State. She is a member of the Council on Foreign Relations, International Institute for Strategic Studies, Women in International Security, and Professional Risk Managers International Association. Ms. Realuyo holds an M.B.A. from the Harvard Business School, an M.A. in international relations from The Johns Hopkins University School of Advanced International Studies, and a B.S. in foreign service from Georgetown University.

Louise Shelley is a professor in the School of Public Policy at George Mason University. She founded and directs the Terrorism, Transnational Crime, and Corruption Center (TraCCC). She is an expert on the relationship among terrorism, organized crime, and corruption as well as human trafficking. Dr. Shelley's research has focused primarily on the former Soviet Union. She is the recipient of the Guggenheim, National Endowment for the Humanities, IREX, Kennan Institute, and Fulbright Fellowships and received a MacArthur Grant to establish the Russian Organized Crime Study Centers. She is the author most recently of *Human Trafficking: A Global Perspective* (Cambridge University Press, 2010), as well as 10 other books. She has written numerous articles and book chapters on all aspects of transnational crime and corruption. Dr. Shelley has run large-scale programs in the former Soviet Union on organized crime and corruption for the last 15 years. She served on the Global Agenda Council on Illicit Trade of the World Economic Forum and was the first cochair of its Council on Organized Crime. Dr. Shelley holds an M.A. in criminology and a Ph.D. in sociology from the University of Pennsylvania.

Admiral James G. Stavridis, USN, assumed duties as commander of U.S. European Command and as Supreme Allied Commander Europe, in early summer 2009. He is a distinguished graduate of the U.S. Naval Academy. A surface warfare officer, Admiral Stavridis commanded the USS *Barry* (DDG 52) from 1993 to 1995. In 1998, he commanded Destroyer Squadron

21 and deployed to the Arabian Gulf, winning the Navy League's John Paul Jones Award for Inspirational Leadership. From 2002 to 2004, he commanded the USS *Enterprise* carrier strike group, conducting combat operations in the Arabian Gulf in support of Operations *Iraqi Freedom* and *Enduring Freedom*. He commanded U.S. Southern Command and has served as a strategic and long-range planner on the staffs of the Chief of Naval Operations and Chairman of the Joint Chiefs of Staff. He has served as the executive assistant to the Secretary of the Navy and the senior military assistant to the Secretary of Defense. Admiral Stavridis earned a Ph.D. and MALD from The Fletcher School of Law and Diplomacy at Tufts University, where he won the Gullion Prize as outstanding student in 1984. He is also a distinguished graduate of both the National and Naval war colleges.

John P. Sullivan is a career police officer. He currently serves as a lieutenant with the Los Angeles Sheriff's Department. He is also an adjunct researcher at the Vortex Foundation in Bogotá, Colombia; senior research fellow in the Center for Advanced Studies on Terrorism; and senior fellow with the *Small Wars Journal's El Centro*. He is coeditor of *Countering Terrorism and WMD: Creating a Global Counter-Terrorism Network* (Routledge, 2006), and *Global Biosecurity: Threats and Responses* (Routledge, 2010). He is coauthor of *Mexico's Criminal Insurgency: A Small Wars Journal–El Centro Anthology* (iUniverse, 2011). His current research focus is the impact of transnational organized crime on sovereignty in Mexico and other countries.

Steven Weber is a specialist in international relations with expertise in international and national security; the impact of technology on national systems of innovation, defense, and deterrence; and the political economy of knowledge-intensive industries. Trained in history and international development at Washington University and medicine and political science at Stanford, Mr. Weber joined the University of California at Berkeley faculty in 1989. As senior policy advisor with the Glover Park Group in Washington, DC, he consults with government agencies, multinational firms, and nongovernmental organizations on foreign policy, risk analysis, strategy, and forecasting. Mr. Weber's major publications include *The Success of Open Source* (Harvard University Press, April 2004), *Cooperation and Discord in U.S.-Soviet Arms Control* (Princeton University Press, 1991), and *Globalization and the European Political Economy* (Columbia University Press, 2001) as well as numerous articles and chapters in the areas of U.S. foreign policy, the political economy of trade and technology, politics of the post–Cold War world, and European integration. With colleague Bruce Jentleson of Duke University, Mr. Weber directs the New Era Foreign Policy Project. Their latest publication is *The End of Arrogance: America in the Global Competition of Ideas* (Harvard University Press, 2010).

William F. Wechsler is the deputy assistant Secretary of Defense for Special Operations and Combating Terrorism. Mr. Wechsler is the primary lead for Defense Department policies, plans, authorities, and resources related to special operations and irregular warfare, with special emphasis on counterterrorism, counterinsurgency, unconventional warfare, information operations, and sensitive special operations. Before taking this position, Mr. Wechsler served for over 3 years as deputy assistant Secretary of Defense for Counternarcotics and Global Threats. In that role, he focused on integrating law enforcement operations into our military campaign plans in Afghanistan, institutionalizing the Defense Department's counterthreat finance structures and doctrine, and substantially increasing budget support for capacity-building in Mexico. Mr. Wechsler also served as special advisor to the Secretary of the Treasury, on the staff of the National Security Council, and as special assistant to the Chairman of the Joint Chiefs of Staff. Before rejoining the Defense Department, he was managing director of Greenwich Associates, a management consultancy to the financial services industry, where he led the firm's global asset management practice.

Phil Williams is Wesley W. Posvar Professor and director of the Matthew B. Ridgway Center for International Security Studies at the University of Pittsburgh. Professor Williams has published extensively in the field of international security, including books on crisis management, the U.S. Senate, troops in Europe, and superpower détente. He has edited volumes on Russian organized crime, human trafficking, and combating transnational crime. From 2001 to 2002, he was a visiting scientist at the Community Emergency Response Teams program, where he worked on cybercrime and infrastructure protection. From 2007 to 2009, he was a visiting professor in the Strategic Studies Institute at the U.S. Army War College where he published two monographs: *From the New Middle Ages to a New Dark Age: The Decline of the State and U.S. Strategy* (2008), and *Criminals, Militias and Insurgents: Organized Crime in Iraq* (2009). In 2009, he coauthored (with James Cockayne) a study for the International Peace Institute titled *The Invisible Tide: Towards an International Strategy to Deal with Drug Trafficking Through West Africa*.